Springer Series on
Signals and Communication Technology

SIGNALS AND COMMUNICATION TECHNOLOGY

For further volumes:
http://www.springer.com/series/4748

John A. Richards

Remote Sensing with Imaging Radar

Springer

John A. Richards
ANU College of Engineering and Computer Science
The Australian National University
Canberra, ACT, 0200
Australia
john.richards@anu.edu.au

ISSN 1860-4862
ISBN 978-3-642-02019-3 e-ISBN 978-3-642-02020-9
DOI 10.1007/978-3-642-02020-9
Springer Heidelberg Dordrecht London New York

Library of Congress Control Number: 2009931106l

Cover design: WMXDesign GmbH, Heidelberg, Germany

Printed on acid-free paper

Springer is part of Springer Science+Business Media (www.springer.com)

Preface

This book is concerned with remote sensing based on the technology of imaging radar. It assumes no prior knowledge of radar on the part of the reader, commencing with a treatment of the essential concepts of microwave imaging and progressing through to the development of multipolarisation and interferometric radar, modes which underpin contemporary applications of the technology.

The use of radar for imaging the earth's surface and its resources is not recent. Aircraft-based microwave systems were operating in the 1960s, ahead of optical systems that image in the visible and infrared regions of the spectrum. Optical remote sensing was given a strong impetus with the launch of the first of the Landsat series of satellites in the mid 1970s. Although the Seasat satellite launched in the same era (1978) carried an imaging radar, it operated only for about 12 months and there were not nearly so many microwave systems as optical platforms in service during the 1980s. As a result, the remote sensing community globally tended to develop strongly around optical imaging until Shuttle missions in the early to mid 1980s and free-flying imaging radar satellites in the early to mid 1990s became available, along with several sophisticated aircraft platforms. Since then, and particularly with the unique capabilities and flexibility of imaging radar, there has been an enormous surge of interest in microwave imaging technology.

Unlike optical imaging, understanding the theoretical underpinnings of imaging radar can be challenging, particularly when new to the field. The technology is relatively complicated, and understanding the interaction of the incident microwave energy with the landscape to form an image has a degree of complexity well beyond that normally encountered in optical imaging. A comprehensive understanding of both aspects requires a background in electromagnetic wave propagation and vector calculus. Yet many remote sensing practitioners come from an earth sciences background within which it is unlikely they will have acquired that material. So that they can benefit from radar technology it is important that a treatment be available that is rigorous but avoids a heavy dependence on theoretical electromagnetism. That is the purpose of this book. It develops the technology of radar imaging, and an understanding of scattering concepts, in a manner suited to the background of most earth scientists, supported by appendices that summarise important mathematical concepts. That enables the treatment to move quickly to the practical aspects of imaging, since it does not require early chapters that focus on electromagnetic theory rather than radar itself.

In addition to being a resource book for the user this treatment is also intended to be used as a teaching text, at senior undergraduate or graduate level.

After providing a framework for the book in Chapter 1 including a commentary on the knowledge that is assumed on the part of the reader, Chapters 2 and 3 cover the fundamentals of radar and how images are formed. The material is set in the context of multipolarisation radar which is the hallmark of modern microwave imaging technology. Chapter 4 covers errors in radar data and how they can be corrected, while Chapter 5 is devoted to the landscape and how it responds to incident microwave energy; that chapter is central to using radar in remote sensing.

An important application of radar is interferometry, which allows the derivation of detailed topographic information about the landscape from a collection of radar images, and means by which landscape changes with time can be detected. This covered in Chapter 6.

An emerging technology in the remote sensing context is bistatic radar in which the source of radiant energy and the receiver are not necessarily collocated, as has been the case for most remote sensing imaging radars to date. An introduction to the technology of bistatic radar is the topic of Chapter 7.

Interpretation of imagery is a logical end point in most remote sensing studies. Chapter 8 covers the range of approaches to radar image interpretation in common use, including statistical and target decomposition methods for thematic mapping.

The book concludes with a brief coverage of passive microwave imaging in Chapter 9. There is sufficient natural microwave energy emanating from the landscape that it can be used to produce coarse resolution images that find particular value in soil moisture studies and sea surface assessment.

Appendices are included that introduce the reader to the concept of complex numbers, summarise essential results in matrices and vectors, demonstrate how images are formed from the recorded radar signal data, and provide other supplementary material.

The idea for this book arose during a graduate course on radar remote sensing taught at the University of California, Santa Barbara in 1985 during a sabbatical period spent with the late Professor David Simonett, one of the pioneers in remote sensing with radar. Following a lengthy intervening period in university administration, a further sabbatical year in 2008 gave time for the book to be written. I wish to record my appreciation to the Australian National University for this opportunity. I am also grateful to the Department of Engineering at the University of Cambridge, and Wolfson College Cambridge, for hosting me for a two month period in 2008 during which substantial progress was made possible.

Several colleagues provided imagery and other examples used in this text that are acknowledged at the appropriate locations. I am particularly grateful to Annie Richardson and Ben Holt at JPL for their help in locating good quality copy of early remote sensing images that have good didactic content, and to Ian Tapley for assisting with the provision of Australian AirSAR data. I wish also to record my appreciation to ITT Visual Information Solutions. They made available a copy of the excellent ENVI™ image processing package to assist in preparing some of the figures in this book.

As always the support of my wife Glenda is gratefully acknowledged, not only for the time together she had to forego during 2008, but for her constant and gentle encouragement, particularly when the task took on a magnitude that at times seemed a little insurmountable.

John Richards
The Australian National University
Canberra Australia
May 2009

TABLE OF CONTENTS

List of Symbols and Operators

symbol	meaning		units
a	rate of the transmitted chirped ranging pulse		rad $s^{-1}s^{-1}$
a_n	expansion coefficient for the nth degree Legendre polynomial		
A	anisotropy		
A_r	aperture of an antenna when receiving		m^2
b	rate of the Doppler induced azimuth chirp		rad $s^{-1}s^{-1}$
B	baseline		m
$B_\perp$	orthogonal baseline		m
B	strength of the earth's magnetic field		T (tesla)
$\mathcal{B}$	bandwidth		Hz
$\mathcal{B}_c$	chirp bandwidth		Hz
c	velocity of light (in a vacuum)	$299.792\text{x}10^6$	ms^{-1}
$\mathbf{C}$	covariance matrix		
d	flight line spacing in a tomographic SAR		m
$d_m(\mathbf{k})$	distance of target vector $\mathbf{k}$ from m^{th} class mean		
$\mathbf{e}$	electric field unit vector		
E	magnitude (phasor) of the electric field		Vm^{-1}
E_o	amplitude of the electric field		Vm^{-1}
$\mathbf{E}$	vector of electric field components		Vm^{-1}
$\mathbf{E}$	propagating electric field vector		Vm^{-1}
f	frequency		Hz
f_o	radar operating (or carrier) frequency		Hz
f_i	abundance coefficient in decompositional scattering models		
$f(h)$	vertical profile function in polarisation coherence tomography		
$\mathcal{F}$	Fourier transform operation		
$\mathcal{F}^{1}$	inverse Fourier transform operation		
g	distance within a pixel above a datum in SAR tomography		m
$g_m(\mathbf{k})$	discriminant function for class ω_m and target vector $\mathbf{k}$		
$\mathbf{g}_c$	coherency vector (of a wave or field)		
$\mathbf{g}_i$	eigenvector of the coherency matrix		
G_t	gain of an antenna in transmission		dimensionless
G_r	gain of an antenna when receiving		dimensionless
$\mathbf{G}$	unitary matrix of the eigenvalues of the coherency matrix		
h	Planck's constant	$6.62607\text{x}10^{-34}$	Js
h	topographic elevation		m

h_v	vegetation canopy depth		m
$\mathbf{h}$	magnetic field unit vector		
$\boldsymbol{\hat{h}}$	horizontal polarisation unit vector		
H	platform altitude		m
H	entropy		
H	magnitude (phasor) of the electric field		Am^{-1}
H_o	amplitude of the electric field		Am^{-1}
$\mathcal{H}$	Mueller matrix		
$\mathbf{H}$	propagating magnetic field vector		Am^{-1}
$\mathbf{J}$	wave coherency matrix		
k	Boltzmann's constant	$1.38065x10^{-23}$	JK^{-1}
k	wave number		rad m^{-1}
k_h	vertical (spatial) wave number (spatial phase constant)		rad m^{-1}
$\mathbf{K}$	Kennaugh matrix		
l_a	antenna length in the azimuth direction		m
l_v	antenna length in the vertical direction		m
$\boldsymbol{\ell}$	left circular polarisation unit vector		
l	surface correlation length in the small perturbation model		m
L	canopy loss (ratio)		dimensionless
L_a	synthetic aperture length		m
L_T	tomographic aperture		m
M	power density of radiant energy		Wm^{-2}
M_e	the solar constant	1.37	kWm^{-2}
M_λ	spectral power density, spectral radiant exitance		$Wm^{-2}\mu m^{-1}$
$\mathbf{M}$	Stokes scattering operator		
N	number of looks		
N_e	electron density (in the ionosphere)		m^{-3}
p	power density (average)		Wm^{-2}
p_p	peak power density		Wm^{-2}
p	co-polar ratio		
$p(\omega_m)$	prior probability that the class is ω_m		
$p(\omega\|\mathbf{k})$	posterior probability that the class is ω for the target vector $\mathbf{k}$		
$p(\mathbf{k}\|\omega)$	class conditional distribution function for the target vector $\mathbf{k}$		
$p(t)$	pulse of unit amplitude		
$\mathbf{p}^{ra}$	unit vector aligned to a receiving antenna		
$\mathbf{p}$	unit vector aligned to the polarisation of a wave		
P	power		W

P_r	received power	W
P_t	transmitted power	W
$P_n(x)$	Legendre polynomial in x, of degree n	
$\mathcal{P}$	degree of polarisation	
q	cross-polar ratio	
Q_e	extinction cross section in a volume scatterer	m^2
r	distance	m
r_a	azimuth resolution	m
r_g	ground range resolution	m
r_r	slant range resolution	m
$\boldsymbol{r}$	right circular polarisation unit vector	
R	slant range distance	m
R	power reflection coefficient	dimensionless
R_o	slant range at broadside	m
R_{oT}	transmitter slant range at broadside for bistatic radar	m
R_{oR}	receiver slant range at broadside for bistatic radar	m
R_{oRt}	receiver slant range at transmitter broadside for bistatic radar	m
s	exponential variate with unity mean	
$s(t)$	received signal as a function of time	
s	standard deviation of surface roughness	m
s	rms height variation of a surface	m
s_i	i^{th} Stokes parameter	
s_γ	exponential variate with mean γ	
$\mathbf{s}$	Stokes vector	
s	standard deviation of surface roughness	m
$\mathbf{s}_m$	modified Stokes vector	
S	swath width	m
SNR	signal to noise ratio	
S_{PQ}	scattering matrix element for P receive and Q transmit	dimensionless
$\mathbf{S}$	Poynting vector	Wm^{-2}
$\mathbf{S}$	scattering or Sinclair matrix	
t	time	s
t	trunk height	m
T_a	time duration of the synthetic aperture and coherent integration time	s
T	absolute temperature	K
T_o	physical temperature	K
T_B	radiometric brightness temperature	K

T_c	canopy brightness temperature	K
T_s	soil brightness temperature	K
T_H	radiometric brightness temperature for horizontal polarisation	K
T_V	radiometric brightness temperature for vertical polarisation	K
T_U	cross polarised radiometric brightness temperature	K
T_{V}	cross polarised radiometric brightness temperature	K
T	period (of a sinusoid)	s
T_a	azimuth chirp width (duration)	s
$\mathbf{T}$	coherency matrix	
$\mathbf{T}_6$	polarimetric interferometric coherency matrix (reciprocal media)	
$\mathbf{T}_8$	polarimetric interferometric coherency matrix (nonreciprocal)	
$\mathbf{T}_{3N}$	*N* radar multistatic polarimetric coherency matrix	
$\mathbf{T}$	Jones matrix	
$\boldsymbol{\mathcal{T}}$	axis rotation transformation matrix	
v	velocity of propagation	ms^{-1}
v	radar platform along track velocity	ms^{-1}
v_r	receiver platform along track velocity	ms^{-1}
v_t	transmitter platform along track velocity	ms^{-1}
$\boldsymbol{v}$	vertical polarisation unit vector	
w	trunk width in a tree scattering model	m
$\mathcal{w}$	window (weighting) function	
$\mathbf{w}$	polarisation filter vector	
$\mathbf{Z}$	multi-look averaged covariance matrix	
α	rate of the transmitted chirped ranging pulse	$Hz\ s^{-1}s^{-1}$
α	attenuation constant	$Np\ m^{-1}$
α	average alpha angle of the eigenvectors of the coherency matrix	deg
α	$\cos^{-1}$ of the first entry of the coherence matrix eigenvector	
α_{IF}	interferometric phase factor	$m\ rad^{-1}$
α_{xx}	refection parameter in the small perturbation model	
β	rate of the Doppler induced azimuth chirp	$Hz\ s^{-1}$
β	phase constant	$rad\ s^{-1}$
β	antenna half power beamwidth	rad
β	bistatic angle	rad
δ	depth of penetration	m
$\Delta\phi$	interferometric phase angle	rad
ε	emissivity	dimensionless
ε	polarisation ellipticity angle	rad

ε_P	emissivity at polarisation P		dimensionless
ε	permittivity		Fm^{-1}
ε_c	canopy emissivity		dimensionless
ε_s	soil emissivity		dimensionless
ε_o	permittivity of free space (vacuum)	8.85	pFm^{-1}
ε_r	dielectric constant (relative permittivity)		dimensionless
ε_r'	real part of the complex dielectric constant		dimensionless
ε_r''	imaginary part of the complex dielectric constant		dimensionless
$\boldsymbol{\varepsilon}$	emissivity vector		
γ	propagation constant		m^{-1}
γ	scale parameter of the exponential distribution		
γ	complex polarimetric interferometric coherency		
Γ	reflectivity		dimensionless
Γ_c	canopy reflectivity		dimensionless
Γ_s	soil reflectivity		dimensionless
Γ_P	reflectivity at polarisation P		dimensionless
$\Gamma(n)$	n^{th} Gamma function		
η	wave impedance of free space	377	Ω
η_s	speckle standard deviation		
κ	generalised scattering coefficient or target vector		
κ_a	power absorption coefficient		m^{-1}
κ_e	power extinction coefficient		m^{-1}
κ_s	scattering loss coefficient		m^{-1}
λ	wavelength		m
λ	eigenvalue		
Λ	spatial wavelength of a periodic surface		m
$\boldsymbol{\Lambda}$	diagonal matrix of eigenvalues of the coherency matrix		
μ	permeability		Hm^{-1}
μ_o	permeability of free space	400π	nHm^{-1}
ν	coherence eigenvalue		
θ	incidence angle		rad
ϑ	ground slope angle		rad
Θ	antenna beamwidth, in general		rad
Θ_a	azimuth (along track) beamwidth of the SAR antenna		rad
Θ_v	vertical (across track) beamwidth of the SAR antenna		rad
ρ	radar reflectivity		dimensionless

$\rho(g)$	pixel reflectivity as a function of elevation within its vertical detail		
ρ	Fresnel reflection coefficient		dimensionless
ρ_P	Fresnel reflection coefficient for polarisation P		dimensionless
ρ_H	Fresnel reflection coefficient for horizontal polarisation		dimensionless
ρ_V	Fresnel reflection coefficient for vertical polarisation		dimensionless
ρ_t	trunk reflection coefficient (polarisation unspecified)		dimensionless
ρ_g	ground reflection coefficient (polarisation unspecified)		dimensionless
σ	conductivity		Sm^{-1}
σ	radar cross section		m^2
σ_B	bistatic radar cross section		m^2
σ^o	radar scattering coefficient (often called sigma nought)		m^2m^{-2}
σ^o_o	radar scattering coefficient for vertical incidence		m^2m^{-2}
σ_v	volumetric backscattering coefficient		m^2m^{-3}
$\boldsymbol{\sigma}$	Stefan-Boltzmann constant	$5.67040\text{x}10^{-8}$	$Wm^{-2}K^{-4}$
τ	shape parameter of the Rayleigh distribution		
τ	transmission coefficient		dimensionless
τ_a	compressed azimuth chirp width (duration)		s
τ_r	width of radar ranging pulse		s
ω	radian frequency		rad s^{-1}
ω_o	radar operating (or carrier) frequency		rad s^{-1}
Ω	Faraday rotation angle		rad
$\boldsymbol{\Omega}_{12}$	joint image complex coherency matrix		
ξ	squint angle		rad

Operators and mathematical conventions

operator	meaning
$\mathfrak{E}$	expectation
$\langle\ \rangle$	average value
$\mathcal{R}e$	real part of a complex quantity
Im	imaginary part of a complex quantity
z^*	conjugation of the complex number z
(a,b)	$a<x<b$
[a,b)	$a\leq x<b$
(a,b]	$a<x\leq b$
[a,b]	$a\leq x\leq b$
$\in$	is a member of
$\otimes$	Kronecker product
$\lvert x\rvert$	magnitude or absolute value of x
Δ	small change in
∇	Gradient operator
$\angle$	angle (argument) of a complex number or phasor
$\exp(x)$	e^x
$x(t)\vert_{t_O}$	x evaluated at $t=t_o$
$*$	correlation of two functions
$\bullet$	scalar or dot product of vectors
x	vector or cross product of vectors

CHAPTER 1

THE IMAGING RADAR SYSTEM

1.1 Why Microwaves?

Understanding remote sensing with imaging radar can be more difficult than with optical imaging because the technology itself is more complicated and the image data recorded is more varied. Since there are so many concepts and techniques to be assimilated, this chapter provides an overview of the topic as a framework for the later chapters. It also draws attention to the knowledge assumed on the part of the reader.

First, we should establish why we are interested in radar imaging as a remote sensing modality. A simple answer can be found by examining the wavelength of the radiation used compared with that of the visible and infrared radiation employed in optical remote sensing. Optical imaging technologies operate at wavelengths of the order of 1μm or so – that is a millionth of a metre. Radar imaging, on the other hand, is based on microwaves that have wavelengths of the order of 10cm – approximately 100,000 times as long. With such a disparity in wavelength one would expect that features on the earth's surface would appear differently at microwave than they would optically. That is certainly the case. In many situations the data types are complementary in that what is difficult to discern in one is sometimes more easily discriminated in the other. As a result combined optical and radar data sets feature in geographic information systems.

There is another major difference. While there can be some penetration through media such as water and thin leaves at optical wavelengths, the longer wavelengths of radar can often penetrate vegetation canopies, and even very dry soils. Thus, whereas the imagery recorded optically usually represents the surface elements of the landscape, radar image data is more complex because it often contains volumetric and sub-surface information as well.

At the relatively long wavelengths used for radar imaging surfaces also appear much smoother than at visible and infrared wavelengths so that there is a greater occurrence of mirror like reflections that, at once, can be both helpful and problematic.

Finally, with radar we have control over the properties of the incident energy. That allows a wide variety of data types to be recorded and enables innovative applications such as topographic mapping, landscape change detection and, to a limited extent, three dimensional modelling of the volume detail of a resolution element.

1.2 Imaging with Microwaves

In seeking to form an image with any technology the first consideration is where the energy comes from with which to view the landscape. In the case of optical data it is visible and infrared sunlight, or thermal energy from the earth itself. Although there is a limited amount of microwave energy available from the earth and sun, it is so small that

J.A. Richards, *Remote Sensing with Imaging Radar*, Signals and Communication Technology,
DOI: 10.1007/978-3-642-02020-9_1,

we generally need to provide our own source of incident radiation for most purposes[1]. A microwave transmitter is carried on the remote sensing platform and used to illuminate the earth. Energy scattered back to the platform is received and used to create an image of the landscape. There could be two platforms – one carrying the energy source and the other (or even several) receiving scattered energy. To date most radar remote sensing systems have used the same platform and are called *monostatic*. When two platforms are used the radar system is called *bistatic*, which is now emerging as a significant remote sensing modality.

Microwave energy is just one form of electromagnetic radiation; it is part of a continuous spectrum as shown in Fig. 1.1. The spectrum also includes the visible and infrared energy that is the basis of optical remote sensing. The most significant difference in properties is wavelength. In principle, we could contemplate using any wavelength for imaging the earth's surface, the only real limit being the levels of energy available at the surface.

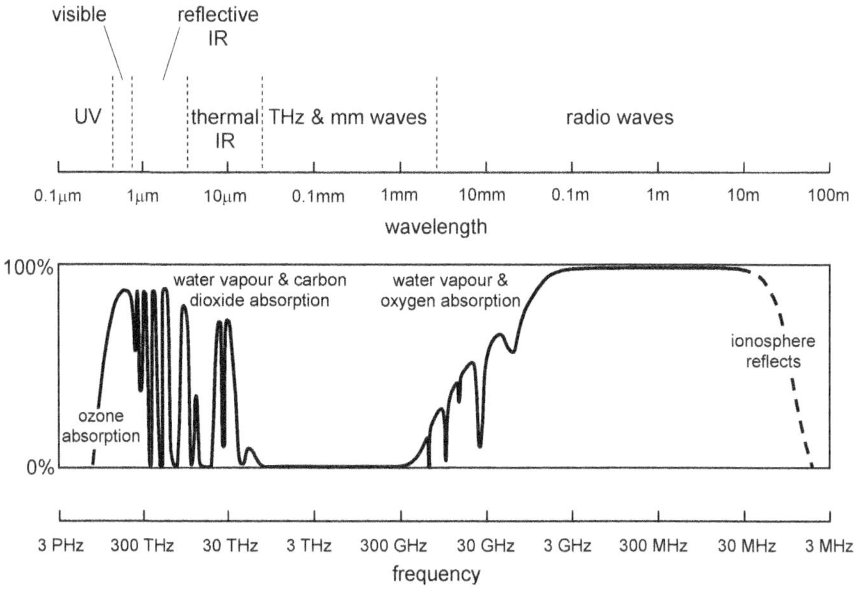

Fig. 1.1. The electromagnetic spectrum and the indicative transmittance of the atmosphere on a path between space and the earth

In Chapt. 2 we will examine specifically the energy levels from the sun over a number of wavelengths of interest. But, as above, we need also to consider a platform carrying its own energy source. We then need to ask whether there are any fundamental limitations to using any particular wavelength range for remote sensing purposes. There is one: the earth's atmosphere is not transparent at all wavelengths. That is very fortunate since the absorption of a significant proportion of the sun's ultraviolet radiation is important for our well being. There is also substantial atmospheric absorption in the far infrared. The

[1] We can use naturally emitted microwave energy for remote sensing but it needs to be collected over very large pixels to give acceptable and practical signal levels.

absorptive characteristics of the atmosphere are quite complex because of its molecular composition. Fig. 1.1 shows atmospheric transmittance as a function of wavelength, covering the range from the ultraviolet up (in wavelength) to the radio wave spectrum. Several aspects are noteworthy. For most of the spectrum the atmospheric constituents of water vapour, oxygen and carbon dioxide selectively block the transmission of electromagnetic energy through the atmosphere. Regions in which there is little absorption are often referred to as atmospheric windows, the most important being in the visible and near infrared region (~0.3-1.3μm), the middle infrared (~1.5-1.8μm, ~2.0-2.6μm, ~3.0-3.6μm, ~4.2-5μm) and the thermal infrared (~7.0-15μm). Below the visible range the ozone content of the upper atmosphere blocks solar radiation. The atmosphere is also essentially closed to radiation for wavelengths beyond the thermal infrared until we encounter the radio wave part of the spectrum[2]. For wavelengths beyond about 3cm the atmosphere is regarded as transparent. For terrestrial applications that applies indefinitely. However, on paths from space to the earth's surface, or vice versa, the region of the atmosphere called the ionosphere, consisting of a weakly but significantly ionised set of layers, reflects electromagnetic energy with wavelengths longer than about 10m or so, both radiated upwards from the earth, or downwards from a space vehicle. For remote sensing purposes we therefore regard the atmosphere potentially to be a problem at those longer wavelengths[3].

1.3 Components of an Imaging Radar System

Fig. 2.1. summarises the technology of radar imaging, depicting the essential system components that need to be understood in developing an overall appreciation of the field. The first consideration is to be able to resolve the field of interest into resolution cells, or pixels. Different principles are used to create resolution in the direction parallel to the motion of the platform (along track or azimuth), and that orthogonal to it (across track or range). We will see that the principle of radar is important for resolving detail across range. Irradiation of the landscape uses pulses of energy; the time they take from transmission to the landscape and back to the radar determines how far away that part of the landscape is. Innovative signal processing techniques will be shown to make high spatial resolutions possible in this dimension. We will also see that it is because of the radar principle that the system has to be side looking.

In the along track direction the motion of the platform relative to the landscape will be seen to give a Doppler change in the frequency of the radiation that is used for illuminating the landscape (just as there is a Doppler change in the frequency of the siren of a passing ambulance). By keeping track of the Doppler shift as the platform passes regions of interest we will see, again, that signal processing methods can be used to achieve very high spatial resolutions in azimuth even from a space borne vantage point – this is where the concept of "synthetic aperture" comes in.

The next important aspect to understand is how a swath of image data can be established, from which images are selected. Although it is, in principle, just a property of the antenna carried on the platform that sets the swath width, we will find that the swath

---

[2] Even though it is accompanied by high atmospheric attenuation there is a growing interest in the use of terrahertz radiation for short distance defence and security applications since it can penetrate dry, non-metallic media (see the Special Issue on T-Ray Imaging, Sensing and Retection, *Proceedings of the IEEE*, vol. 95, no. 8, August 2007.)

[3] For the same reason, transmission to and from a telecommunications satellite has to take place at frequencies in excess of about 50MHz.

is limited by the emergence of ambiguous signals if it is too wide. As a consequence we will consider the application of a principle now referred to as ScanSAR to achieve very wide coverage to the side of the platform.

To use the data meaningfully we need to understand distortions that may have been introduced into the recorded imagery. They can be quite severe and guidance is needed on how radar imaging should be configured to minimise the types of distortion that impact on particular applications.

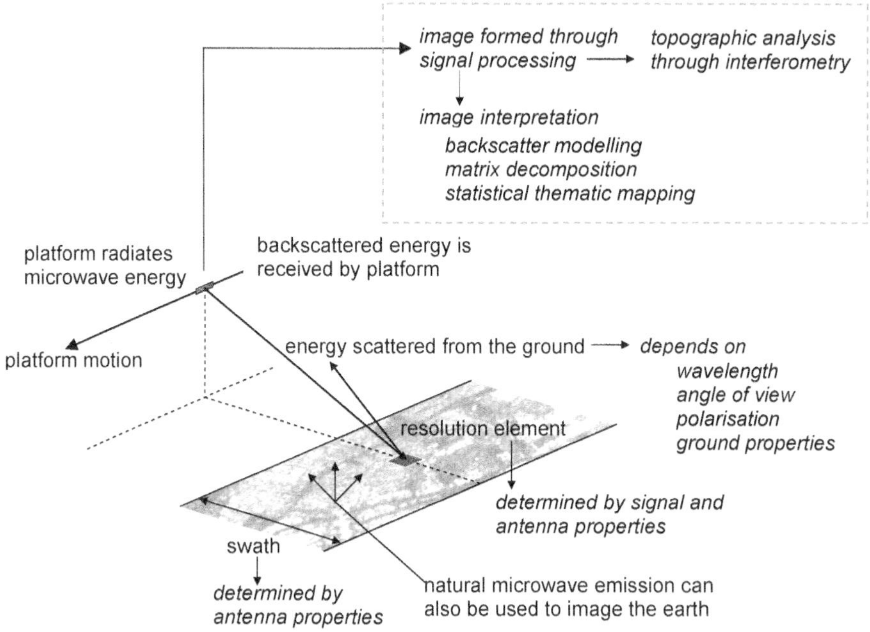

Fig. 1.2. Summary of the essential elements of an imaging radar remote sensing system

Once we get to this stage we will have an understanding of how synthetic aperture radar operates. We next need to understand how the incident radiation scatters from the landscape since the backscattered energy contains information about the properties of the part of the earth's surface being imaged. That is a major consideration since it forms the basis of remote sensing with microwaves. It is a significant study in is own right and will occupy a large part of the treatment of this book. Not only is it important to understand the scattering properties of earth surface materials but it is desirable to be able to model them, since that can be an important step in radar image interpretation.

It is helpful to introduce a little terminology at this point. In remote sensing we generally resolve the scene of interest into pixels. The same is true in radar; however we often use "resolution cell" as a synonym for pixel. Additionally, we will often call the pixel a "target". That comes about because of the heritage of radar technology as a means for detecting discrete objects. Sometimes our pixels will look like discrete targets, such as when they are dominated by a single scattering object like a large tree or a building, or possibly a ship on the surface of the ocean. Most often though they will be composed of a distributed collection of incremental scatterers. Nevertheless we will still loosely use the term "target" when referring to a pixel. We will be a little more precise about the term when we discuss scattering properties.

Complexity is added by the fact that the earth will respond differently for different polarisations and wavelengths of the incident energy, as it will for different angles with which the landscape is viewed. Knowledge of how cover types scatter as a function of these radiation characteristics is essential; an underlying important generalisation will emerge – the radar properties of a region on the earth's surface are dependent on its geometry and its moisture content.

Just as we map the landscape in optical remote sensing using quantitative thematic mapping techniques, so in radar imaging we would like to be able to turn the recorded radar image data into maps of land cover and land use. While, in principle, it is possible to use similar procedures to those employed with optical image data, the nature of radar imagery suggests that it is better to develop methods matched to the properties of radar; this is so important that a separate chapter is devoted to radar image analysis.

A special, and annoying, feature of images that are recorded using the relatively pure electromagnetic radiation in radar, is that they have an overlying speckled appearance. We will see that that is the result of interference of the energy reflected from the many elemental scatterers that occur within a resolution cell (pixel). Speckle needs to be understood, as do means for reducing its impact.

Also, because of the rather pure (or coherent) nature of the energy used in microwave imaging, it is possible to develop intereferometric techniques from which topographic features of the landscape can be derived with very high spatial resolution and with which spatial changes, either short or long term, can be detected and mapped with very high precision.

An associated imaging technique uses the very small naturally emitted microwave energy from the earth as a means for forming images. Although not a radar technique, we provide an overview of that technology because it has particular relevance to oceanographic and soil moisture applications.

So in summary, a comprehensive understanding of imaging radar involves:

1. knowing where the energy comes from, and its properties;
2. knowing how to resolve the scene into pixels;
3. appreciating the scattering properties of earth surface features;
4. understanding the dependence of scattering on system parameters such as the wavelength of the radiation, its polarisation and the angle with which the radiation intersects the earth's surface;
5. knowing how an image is formed;
6. knowing how to interpret the recorded imagery;
7. understanding how thematic mapping can be carried out from radar data; and
8. appreciating special applications of microwave imaging such as interferometry.

1.4 Assumed Knowledge

In the following the background knowledge, particularly mathematical, needed to understand radar imaging is outlined, noting what will be assumed on the part of the reader and where assistance can be found.

1.4.1 Complex Numbers

Understanding the technology of optical remote sensing rarely requires an in-depth knowledge of the properties of the visible and infrared radiation used, apart from its wavelength. With radar, however, understanding the properties of the incident radiation is inescapable, as is its mathematical description.

Although the waveforms used in radar can be characterised using trigonometric functions, it is much more convenient to use a description based on exponential functions. That requires a knowledge of complex numbers. Complex numbers are also convenient descriptors for the earth surface properties encountered at radio wavelengths.

Despite their name, complex numbers are not difficult to understand. For readers without that background, Appendix A provides a summary of the necessary concepts and could be read in conjunction with those sections of the book in which complex arithmetic occurs.

1.4.2 Vectors and Matrices

Understanding radar imaging is considerably simplified by the use of vector and matrix algebra to describe lengthy equations and expressions. Appendix B gives an overview of relevant concepts, and the properties of matrices and vectors that are important in describing imaging radar.

1.4.3 Differential Calculus

A familiarity with differential, and introductory integral, calculus is important for appreciating the development of imaging radar as a remote sensing modality They are less important for understanding the interpretation of radar imagery, provided some developments can be taken for granted. It is nevertheless assumed here that the reader does have an introductory calculus background.

It is not necessary here, however, to understand vector calculus. Even though it is widely used in the theory of electromagnetic wave propagation, our development does not need to use it, except in a very rudimentary way.

1.5 Referencing and Footnotes

Considerable use is made of footnotes to refer to supplementary material and to add comments that are important but perhaps not mainstream. We have also used footnotes to provide citations to published work, rather than the method of numbered end notes more commonly used with scientific and technical books. That decision was made to improve flow by avoiding the distraction of having to go to another part of the text to identify relevant sources. The disadvantage of the footnoting approach for referencing is some repetition: that is minimised by using the now less encountered *ibid* (in the same place immediately before) and *loc cit* (in the place cited), when citations are closely located.

1.6 A Critical Bibliography

Many books on radar remote sensing have appeared in the recent past, especially with multi-polarisation and interferometric developments. It is, however, important to recognise the benchmark books by Ulaby, Moore and Fung that provided the first comprehensive treatment of microwave remote sensing in monograph form and remain valuable to this day. There are three volumes:

F.T. Ulaby, R.K. Moore and A.K. Fung, *Microwave Remote Sensing: Active and Passive, Volume 1 Microwave Remote Sensing and Fundamentals*, Addison-Wesley, Reading Mass., 1981.

F.T. Ulaby, R.K. Moore and A.K. Fung, *Microwave Remote Sensing: Active and Passive, Volume 2 Radar Remote Sensing and Surface Scattering and Emission Theory*, Addison-Wesley, Reading Mass., 1982.

F.T. Ulaby, R.K. Moore and A.K. Fung, *Microwave Remote Sensing: Active and Passive, Volume 3 Volume Scattering and Emission Theory, Advanced Systems and Applications*, Addison-Wesley, Reading Mass., 1986.

Another comprehensive treatment, with chapters covering fundamentals and many application domains, is

F.M. Henderson and A.J. Lewis (Eds), *Principles and Applications of Imaging Radar, Manual of Remote Sensing, 3rd ed, Volume 2*, John Wiley and Sons, N.Y., 1998.

Most of the application chapters in that treatment do not require significant mathematical skills; some of the theory chapters though do appeal to deeper mathematical and electromagnetic knowledge.

A comprehensive coverage of the importance of multi-polarisation radar is given in

F.T. Ulaby and C. Elachi (Eds), *Radar Polarimetry for Geoscience Applications*, Artech House, Norwood Mass., 1990.

It has very good chapters on the fundamental theory, including coordinate systems and polarisation synthesis.

A very readable account of both radar and passive remote sensing is given in

I.H. Woodhouse, *Introduction to Microwave Remote Sensing*, Taylor and Francis, Boca Raton, Florida, 2006.

More recently the following treatment by Massonnet and Souyris gives an excellent overview of the status of radar imaging and particularly of the problems of deriving high quality multi-polarisation imagery for both planimetric and interferometric applications.

D. Massonnet and J-C Souyris, *Imaging with Synthetic Aperture Radar*, EPFL Taylor and Francis, Boca Raton, Florida, 2008.

It also provides good coverage of how radar targets (and pixels) are characterised in a multi-polarisation environment. Its mathematical detail is moderate and comparable to that in this book.

A more detailed, mathematically based, treatment of polarimetric radar is

H. Mott, *Remote Sensing with Polarimetric Radar*, IEEE Press John Wiley and Sons, Hoboken, N.J., 2007.

Its strong focus is on the characterisation of radiation and the scattering properties of targets. Another technical treatment is

B-C Wang, *Digital Signal Processing Techniques and Applications in Radar Image Processing*, John Wiley and Sons, Hoboken, N.J., 2008.

As its title suggests it is very much based on a signal processing treatment of synthetic aperture radar and radar imaging and, like Mott's book, is perhaps a coverage more for the systems specialist than the applications scientist.

The comprehensive, long standing and excellent treatment of optics in

M. Born and E. Wolf, *Principles of Optics, 7th ed.*, CUP Cambridge, 2006

is a must for anyone interested in the theory of polarisation in radar. Although an optics text, the equivalence of light and radio waves as two different forms of electromagnetic radiation means that it is equally applicable to characterising the radiation used in radar. Indeed much of the early theoretical development of multi-polarisation radar is based directly on the coverage in that book.

For the reader interested in an easily read coverage of radar generally, not just for remote sensing, Skolnik is the standard text:

M.I. Skolnik, *Introduction to Radar Systems*, 3rd ed., McGraw-Hill, N.Y., 2001.

While it is written principally for engineers its mathematical detail is not deep and the treatment is very readable even for the non-expert.

For the reader seriously interested in understanding and modelling the scattering of electromagnetic radiation by objects an older, but nevertheless still standard treatment, is the two volume set:

G.T. Ruck, D.E. Barrick, W.D. Stuart and C.K. Krichbaum, *Radar Cross Section Handbook*, Plenum, N.Y., 1970

With the emergence of interest in polarimetric radar in the past decade dedicated books are now appearing that give a level of detail beyond the coverage in this treatment. Lee and Pottier below comprehensively covers the important aspects of polarimetric radar as an imaging tool, and means for data handling and analysis. Cloude focuses on the properties of the electromagnetic waves used to carry radar signals; he then looks at the polarimetric theory of scattering and how target information is found, culminating in the theory and applications of interferometric and polarimetric interferometric synthetic aperture radar.

J-S Lee and E. Pottier, *Polarimetric Radar Imaging: From Basics to Applications*, CRC Press, Taylor and Francis, Boca Raton, Florida, 2009.

S.R. Cloude, *Polarisation: Applications in Remote Sensing*, Oxford University Press, 2009

In the past few years there has emerged an interest in applying bistatic radar concepts to remote sensing. Although bistatic configurations, in which the transmitter and receiver are on separate platforms or in different locations, have been well known for many years in surveillance and similar applications, their use for earth surface mapping has been limited. The classic treatment of bistatic radar will be found in

N.J. Willis, *Bistatic Radar, 2nd ed*, SciTech, Raleigh, NC, 2005

The two recent books edited by Cherniakov give a contemporary account of the field, but also include an excellent background treatment of radar technology in general, although with more classic, rather than remote sensing, applications in mind.

M. Cherniakov (ed), *Bistatic Radar Principles and Practice*, John Wiley and Sons, Chichester, 2007.

M. Cherniakov (ed), *Bistatic Radar Emerging Technology*, John Wiley and Sons, Chichester, 2008.

From time to time it is important to know a little about the propagation of radio waves when dealing with radar. There are many excellent treatments available for those with an engineering or physical sciences background including

J.D. Kraus and D.A. Fleisch, *Electromagnetics with Applications, 5th ed.*, McGraw-Hill, N.Y., 2000.

A more introductory level coverage is given in

J.A. Richards, *Radio Wave Propagation: An Introduction for the Non-Specialist*, Springer, Berlin, 2008.

1.7 How this Book is Organised

The remaining chapters are organised around the components of Fig. 1.2.

Chapter 2 treats the radiation framework for radar, discussing naturally occurring levels of microwave energy and establishing the need for the system to provide its own source of primary energy. The underlying electric and magnetic fields are introduced along with properties such as polarisation, interference and the Doppler effect that are so central to later developments.

Chapter 3 develops the technological basis for radar imaging, showing why a radar technique is needed for resolving the landscape with reasonable spatial resolution in the direction at right angles to the motion of the platform. Means by which resolution is achieved in the along track direction are also covered. While the process by which an image is formed from received radar signals is alluded to in that chapter, details are saved for Appendix D to avoid disrupting the flow of the development of the technology of imaging radar from a user perspective.

The concepts of radar cross section and scattering coefficient are also introduced in Chapt. 3 as the essential descriptors of the properties of a target, or of the earth's surface. The polarisation dependence of target and earth surface scattering is also examined leading to the concept of polarisation synthesis, with which we can see how the landscape would respond to any specified polarisation configuration.

Chapter 4 considers sources of geometric and radiometric error in recorded radar imagery and how they can be "corrected". Calibration devices are also introduced with that material.

Chapter 5 is central to the book. It looks at the scattering characteristics of a range of earth surface materials and features, so that the reader will develop an idea of what can be mapped with radar. Both simple and composite situations are examined, based on an understanding of how surfaces and volumes behave, and how artificial and other strong reflecting structures respond to microwave illumination.

Chapter 6 is devoted to the topic of radar interferometry, building on the concept of interference introduced in Chapt. 2. The emerging area of radar tomography as a means for understanding the structure of a pixel's volume is also included in this treatment.

Chapter 7 re-examines the radar concept. After looking at the possibilities for radar systems involving more than one transmitter and more than one receiver it focuses on recent moves to bi-static radar imaging and its use as a remote sensing tool.

Chapter 8 builds on Chapt. 5 by examining methods for interpreting radar image data. Both qualitative methods, involving the human interpreter, and quantitative methods based on automated recognition methods, are covered. They include procedures specially devised for producing thematic maps from radar data.

An overview of the associated topic of passive microwave imaging is given in Chapt. 9, in which the fundamental concepts are developed and its major benefits are highlighted.

Apart from the appendices already mentioned, others provide data on metric prefixes and coordinate systems.

CHAPTER 2

THE RADIATION FRAMEWORK

2.1 Energy Sources in Remote Sensing

The acquisition of information about features on the earth's surface using remote sensing platforms depends on measuring energy emanating from the region of interest so that an image can be formed. The energy can originate from the earth itself, as a result of its finite temperature, or it can be the reflection of energy incident on the earth's surface from an external source such as the sun. It could also come from an artificial source such as a laser or a generator of some other form of radiant energy carried on an aircraft or space craft platform. Irrespective of the energy source used, the principle is to measure upwelling radiation, usually on a pixel by pixel basis, to help understand and map the earth's surface (and possibly the near sub-surface as seen in Chapt. 5).

It is important to look at expressions that describe the actual energy levels generated by the sun and the earth so we know how much is available from common, natural sources. It is of benefit first, though, to look at the means by which energy propagates outwards in free space from a point source generator. This will help in understanding some of the terminology and units used in microwave remote sensing and will be of value when the technology of imaging radar is examined in Chapt. 3.

The sun radiates its energy approximately uniformly in all directions in space. To this extent it can be called an *isotropic radiator*, even though that term is more usually applied to an idealised point source of energy that radiates equally in all directions. Such a point source is shown in Fig. 2.1. Because we observe the sun from such a large distance we will assume it can be modelled in that manner. Rather than describe its properties in terms of energy, it is more usual to talk about the *rate* at which the radiator can generate energy – i.e. energy per unit time, or *power*. While energy is measured in joules, power is expressed in watts (joules per second). The energy is carried forward by an electromagnetic wave that we will have more to say about later; for the present we will simply say the energy is carried outwards by an expanding wavefront.

Suppose we observe the radiator at a distance R, as indicated. If we want to intercept, or collect, all the power being radiated outwards it would be necessary to enclose the radiator by a sphere. At any given radius R we can imagine that the power radiated is spread out uniformly over the surface area of the sphere. In practice we generally don't try to observe the total power output from the point source, but only that portion being radiated in a given direction (for example in the direction of the earth from the sun). We therefore need to describe how much power is spread over just the part of the sphere in the direction of interest. To be able to do that we introduce the concept of *power density*. If the power being radiated by the point source is P_t as shown in Fig. 2.1 then the power density at distance R from the source is given by dividing the transmitted power by the area of a sphere at that distance:

$$p = \frac{P_t}{4\pi R^2} \quad \mathrm{Wm}^{-2} \tag{2.1}$$

J.A. Richards, *Remote Sensing with Imaging Radar*, Signals and Communication Technology, DOI: 10.1007/978-3-642-02020-9_2,

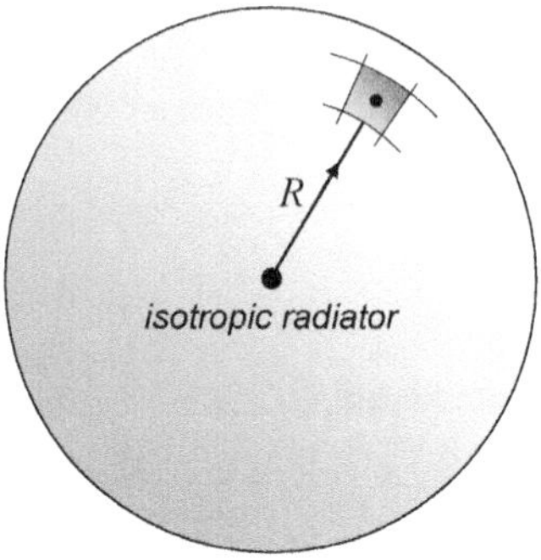

Fig. 2.1. The isotropic radiator and power density produced at a distance R

Note that power density has units of watts per square metre and diminishes as the square of distance from the source, consistent with other inverse square laws found in nature (gravity, sound, electrostatics etc). Knowing the power density at a given distance allows us to determine how much power can be extracted from the spherical wavefront propagating towards us if we intercept the wavefront over a specified cross-sectional area.

A simple analogy to these concepts can be created by considering a light bulb, which is a rough approximation to an isotropic radiator if there is no reflector behind it. A light bulb capable of generating 100W of optical power will radiate that power uniformly in all directions. On the floor, 2.3m below the light, there will be a power density of

$$p = \frac{100}{4\pi(2.3)^2} = 1.5\text{Wm}^{-2}$$

A mirror, with cross sectional area of 0.01m^2 on the floor will intercept and reflect 15mW of optical power.

We can now determine the levels of power or energy available from the sun or earth for imaging purposes. This is based on Planck's radiation law, which describes the power emitted by a so-called *black body* – an ideal emitter and absorber of energy over all wavelengths. For our present purposes the sun and the earth can be regarded approximately as black body radiators.

If a black body is at a temperature T then the so-called *spectral radiant exitance*, or the *spectral power density* (the power per unit of surface area of the body per unit of wavelength) emitted, is

$$M_\lambda = \frac{c_1}{\lambda^5(e^{c_2/\lambda T} - 1)} \text{Wm}^{-2}\mu\text{m}^{-1} \quad (2.2)$$

where

$$c_1 = 2\pi hc^2$$

$$c_2 = hc/k$$

in which

$h = 6.62607\text{x}10^{-34}\text{Js}$ (Planck's constant)

$c = 299.792\text{Mms}^{-1}$ (velocity of light)

$k = 1.38065\text{x}10^{-23}\text{JK}^{-1}$ (Boltzmann's constant) so that

$$c_1 = 3.74176\text{x}10^8\,\text{Wm}^{-2}\mu\text{m}^4$$
$$c_2 = 1.43877\text{x}10^4\,\mu\text{mK}$$

Fig. 2.2 shows the spectral power density emitted by a black body according to (2.2), at different temperatures, plotted as a function of wavelength. The wavelength range chosen covers the ultraviolet through to the so-called thermal infrared range. The dependence of spectral power density on temperature is strong, a feature of importance in thermal remote sensing. Three curves are shown, corresponding first to an ideal black body at the approximate temperature of the sun's surface, secondly to a burning fire on the earth's surface and thirdly to the earth itself at an average temperature of 300K. It is interesting to note that if we wanted to sense fires burning on the earth's surface then we would use an instrument maximally sensitive in the 3-5μm region whereas if we wanted to measure so-called thermal emissions from the earth itself then we would use about 8-12μm.

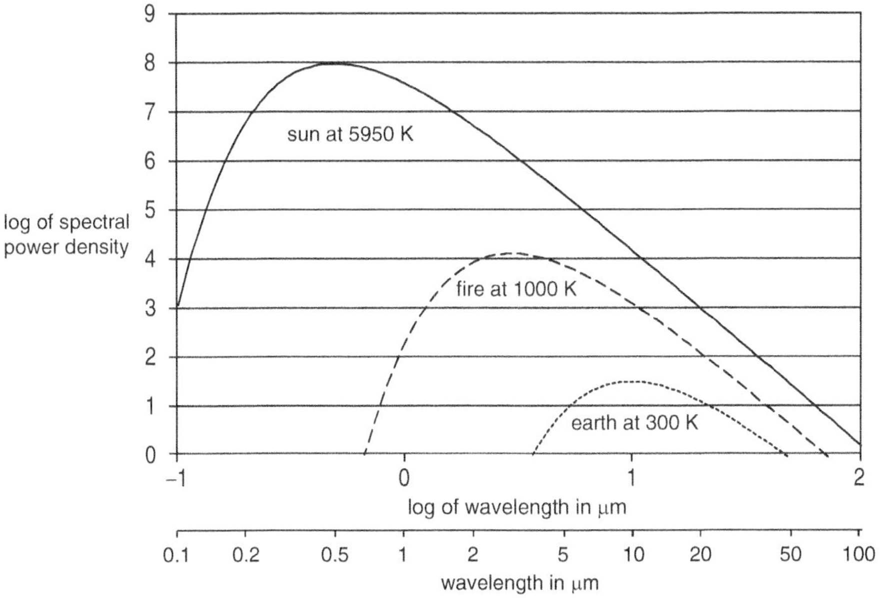

Fig. 2.2. Spectral power density available for ideal black bodies at three different temperatures, computed from (2.2)

Real emitters of radiant energy do not behave as ideal black bodies according to (2.2). Instead, the spectral power density is smaller by a factor ε, referred to as the *emissivity* of the body (or its surface). Emissivity is generally wavelength dependent and is in the range $0<\varepsilon<1$. We will return to emissivity later in the book, but for now we can retain the assumption of ideal black body behaviour.

It is important to recognise that the curves in Fig. 2.2 are those at the surface of the respective black bodies – i.e. they are the power available, per unit of surface area of the body, per unit of wavelength. The ideal solar curve, therefore, does not represent the solar spectral power density at the earth's surface. To find the level of solar energy available at the earth we need to reduce the magnitude of the solar curve as a result of the inverse square law dispersion of solar power density during its passage to the earth. In principle, we do that by computing the total power available from the sun as though it were truly a point source, and by then applying (2.1) using the earth-sun distance for R. That is the same as diminishing the solar power density from (2.2) by the ratio of the squares of the sun's radius and the earth-sun distance, i.e. by the factor

$$\left[\frac{\text{sun radius (Mm)}}{\text{earth sun distance (Mm)}}\right]^2 = \left[\frac{695.5}{149{,}597}\right]^2 = 2.16147\text{x}10^{-5}$$

Fig. 2.3 shows how Fig. 2.2 changes when that correction is made to the solar curve. That explains why sensing the earth at 10-12μm reveals thermal properties of the earth itself, and not those dependent on reflected sunlight, as might have been implied from Fig. 2.2. The corrected solar curve, however, does not take account of the effect of the earth's atmosphere.

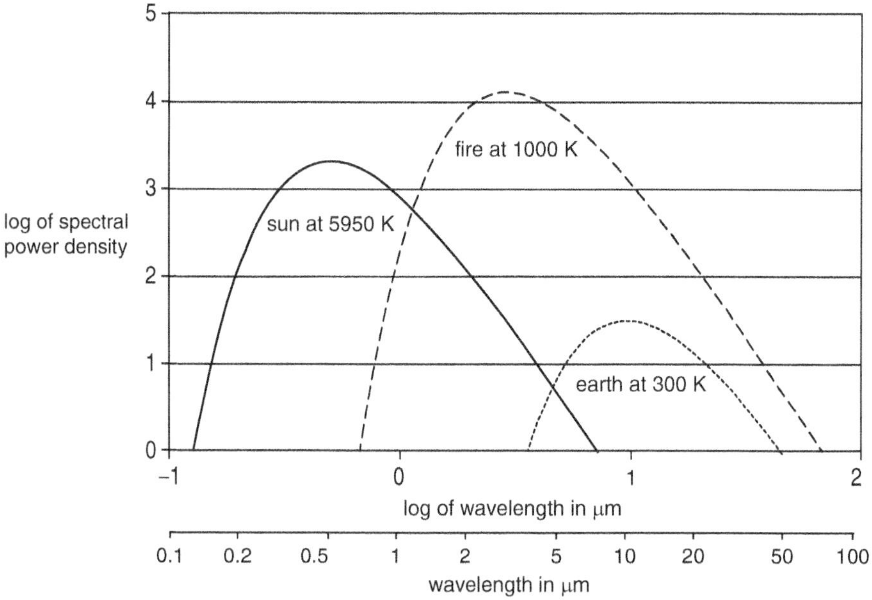

Fig. 2.3. Idealised spectral power densities with the solar curve as it would appear at the top of the earth's atmosphere

2.2 Wavelength Ranges used in Remote Sensing

We have already commented above on the use of sensors operating in the vicinity of 10μm for looking at the surface features that are dependent on the earth's thermal properties. Fig. 2.3 shows that thermal emission from the earth dominates over reflected solar energy for wavelengths in excess of about 5μm with terrestrial thermal emission maximal at about 10μm. The whole field of thermal remote sensing has developed around those observations. Similarly, as noted earlier, if we are interested in imaging very hot objects on the surface, such as active forest fires, sensors operating in the vicinity of 3 or 4μm would be preferred. Line scanners, typically working over the 3-5μm range, have been developed specifically for that purpose.

Fig. 2.3 also shows why typical optical (visible and near infrared) sensors operate in the range of approximately 0.4 to 2.5μm. The sun's output is largest over that regime and the atmosphere allows adequate transmission of energy from the sun to the surface and from the surface to the sensor, apart from a few isolated absorption bands, as illustrated in Fig. 1.1. Moreover, earth surface properties demonstrate good differentiation over those wavelengths[1]. Multispectral and hyperspectral sensors are designed typically to operate within that range.

There are of course many other ranges of wavelength that could be used. In principle, any wavelength could be chosen with which to view the earth and sense its properties. Practical limitations are imposed by the opacity of the atmosphere at many wavelengths, and the availability of suitable technologies. There are, however, few such limitations in the radio wave portion of the electromagnetic spectrum. Energy sources are available and the atmosphere is rarely a problem. Further, one would expect that because of the vast contrast in wavelengths, the properties of the earth would look different from their appearance at optical wavelengths. That is partly the motivation for adopting radar as a remote sensing modality.

While lower radio frequencies (longer wavelengths) could in principle be used, it turns out that the most interesting wavelengths to employ in terms of sensing surface features are those in the microwave portion of the electromagnetic spectrum. This will become clearer when scattering properties are considered in Chapt. 5.

The microwave energy of interest in remote sensing is largely in the range of about 300MHz to about 20GHz. Because of control of the radio spectrum by international treaties, the frequencies used tend to be quite specific within that range. Fig. 2.4 shows the operating frequencies and equivalent wavelengths for a number of past and current radar remote sensing programs. The frequency and wavelength of radiation are related by the very useful formula

$$f(\mathrm{MHz}) = \frac{300}{\lambda(\mathrm{m})} \tag{2.3}$$

The particular bands of frequency and wavelength indicated for different radar remote sensing programs in Fig. 2.4 are usually described by the letter designations as shown.

2.3 Total Available Energy

The energy output from a black body described by (2.2) is expressed per unit of wavelength. Often it is important to know the total available energy over a given range of wavelengths. This can be obtained by integrating (i.e. summing) the spectral power density over the wavelength range (λ_1, λ_2) of interest:

$$M = \int_{\lambda_1}^{\lambda_2} M_\lambda d\lambda \quad \mathrm{Wm}^{-2}. \tag{2.4}$$

If the power density over *all* wavelengths is of interest it can be shown that

[1] See P.H. Swain and S.M. Davis, *Remote Sensing: The Quantitative Approach*, McGraw-Hill, N.Y., 1978

$$M = \int_0^\infty M_\lambda d\lambda = \boldsymbol{\sigma} T^4 \ \mathrm{Wm}^{-2} \tag{2.5}$$

which is known as the Stefan-Boltzmann radiation law, where

$$\boldsymbol{\sigma} = 5.67040\mathrm{x}10^{-8} \ \mathrm{Wm}^{-2}\mathrm{K}^{-4}$$

is the Stefan-Boltzmann constant. As an interesting application of this law consider the total power density available at the top of the earth's atmosphere from solar radiation.

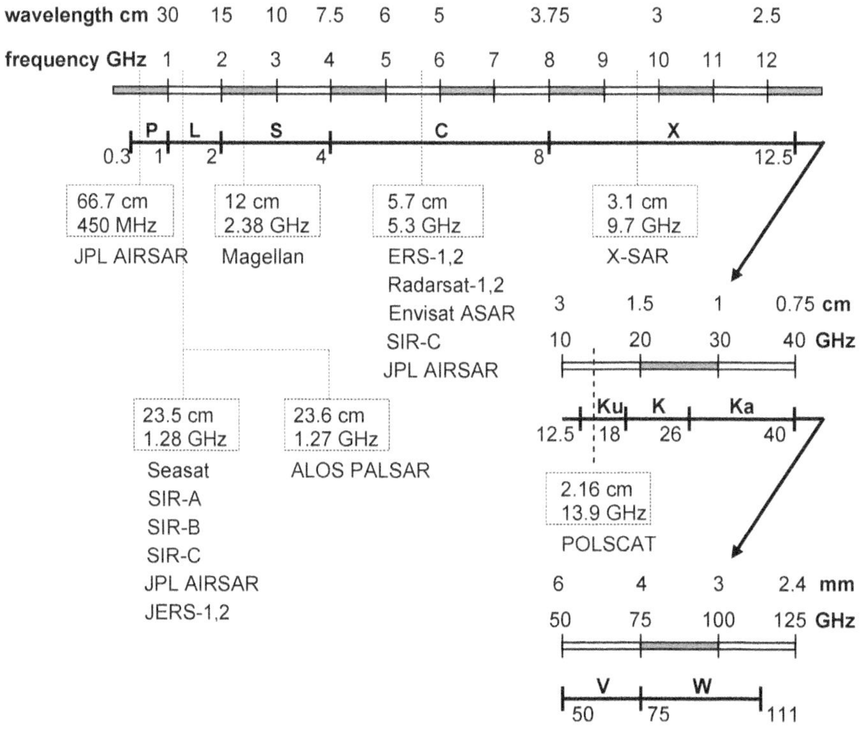

Fig. 2.4. Wavebands used in radar remote sensing; there is some uncertainty about the specification of the bottom of V band

The power density at the surface of the sun (at an assumed temperature of 5950K) is, from (2.5)

$$M = 5.67040\mathrm{x}10^{-8}\mathrm{x}(5950)^4 = 71 \ \mathrm{MWm}^{-2}$$

Multiplying this figure by the surface area of the sun allows its power output to be determined. Assuming this emanates from an equivalent point source which radiates isotropically then the power density produced at the earth is found by dividing the sun's power output by the surface area of the sphere which has the sun-earth distance as its radius or, as above, by multiplying the figure for M by the square of the ratio of the sun's radius to the sun-earth distance. This gives the solar power density at the top of the earth's atmosphere as

$$M_e = 71\text{x}10^6\text{x}2.16147\text{x}10^{-5} = 1.53\text{kWm}^{-2}$$

This would be the earth surface solar power density in the absence of any absorption by the atmosphere, and assuming that the Sun acts as an ideal black body radiator in the sense required by Planck's law. The solar curve of Fig. 2.3 departs from ideal black body behaviour as observed at the earth's surface because of selective absorption by atmospheric constituents[2] and the sun's composition.

The actual solar power density at the earth is known as the *solar constant* and has the value of 1.37kWm^{-2}. It differs from the value computed above for a number of reasons including, first, that the correct temperature to use in the computation of Planck's law depends on wavelength and, secondly, because solar emission at different wavelengths comes from differing portions of the sun's outer layers. By using an average sun temperature of 5800K, Schott[3] obtains a value of 1.39kWm^{-2} for the solar constant.

2.4 Energy Available for Microwave Imaging

Consider now the level of microwave energy available from the earth to see if it can be used for imaging purposes. Even though the infinite wavelength range in Planck's law implies there is also microwave energy available from the sun, we will concentrate on energy emanating from the earth itself because we see in Fig. 2.3 that beyond about 10μm the energy available at the earth's surface from sunlight will be significantly below that from the earth itself.

Fig. 2.5 shows the black body radiation curve for the earth at 300K extended out to microwave wavelengths. Given that the curve has its maximum at about 10μm, we are interested now in that portion well out into the tail of the distribution.

The shortest wavelength of interest in remote sensing using imaging radar will be seen later to be typically 0.03m (i.e. 3x10^{4}μm). For a radiator at 300K, the exponent in the denominator of (2.2) then has the value

$$c_2 / \lambda T = \frac{1.43877\text{x}10^4}{3\text{x}10^4\text{x}300} \approx 1.6\text{x}10^{-3}$$

so that we can approximate the exponential by the first two terms of its Taylor series

$$e^{c_2/\lambda T} \approx 1 + c_2 / \lambda T$$

This allows (2.2) to be approximated as

$$M_\lambda = \frac{c_1 T}{c_2 \lambda^4} = \frac{aT}{\lambda^4} \text{MWm}^{-2}\mu\text{m}^{-1} \qquad (2.6)$$

[2] See Fig. 3.6 of P.N. Slater, *Remote Sensing: Optical and Optical Systems*, Addison-Wesley, Reading Mass., 1980.

[3] See J. R. Schott, *Remote Sensing: The Image Chain Approach*, OUP, N.Y., 1997.

where $$a \approx 2.6\text{x}10^4\,\text{Wm}^{-2}\mu\text{m}^3\text{K}^{-1}$$

Equation (2.6) is called the Rayleigh-Jeans law (or the Rayleigh Jeans approximation to Planck's law) which can be now be used to assess the microwave spectral power density levels from the earth.

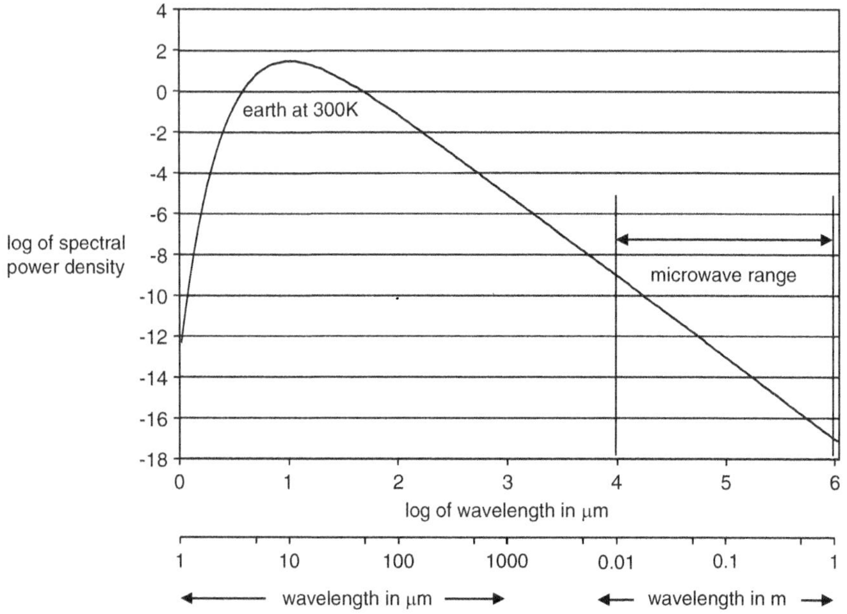

Fig. 2.5. The portion of the earth's Planck radiation curve relevant to microwave frequencies

Consider the microwave power density emanating from the earth's surface at 300K over the wavelength range 0.03m to 0.3m, which is a much larger range than would be used for a single band in microwave imaging. The power density is given by integrating (2.6) over that range of wavelengths:

$$\begin{aligned} M &= \int_{0.03m}^{0.3m} M_\lambda d\lambda \\ &= aT \int_{0.03m}^{0.3m} \lambda^{-4} d\lambda \\ &= -\frac{780x10^{-14}}{3} \lambda^{-3}\Big|_{0.03m}^{0.3m} \\ &= 96.2\text{nWm}^{-2} \end{aligned}$$

in which we have used a=2.6x10^{-14}Wm^{-2}m^{3}K^{-1}, which expresses wavelength in metres.

This is a very small power density indeed, particularly considering that it was computed over such a broad range of wavelengths; in a 100MHz bandwidth around 10GHz the figure is about 3nWm^{-2}, while the available power density is 29pWm^{-2} at

1GHz in a 100MHz bandwidth. These figures are so small that the earth's surface can be considered "dark" at microwave frequencies, in much the same way that the earth is dark at night visually, in the absence of sunlight. To see at night a torch (or flashlight) is used – in other words an artificial source of energy is employed to irradiate the landscape. The same principle can be used at microwave frequencies, day or night. A generator of microwave radiation is carried on board an aircraft or spacecraft and used to irradiate the earth's surface so that image data can be gathered at those wavelengths. The image is constructed by observing the microwave energy scattered back to the platform, as depicted in Fig. 2.6. This is the basis of radar remote sensing developed further in Chapt. 3 and is referred to as an active remote sensing technique. Although we came to this approach by assessing the very low levels of natural terrestrial microwave emissions, being able to irradiate the surface using an artificial energy source gives more control over imaging parameters and methodologies, the importance of which will be seen later when looking at the microwave response of the earth's surface.

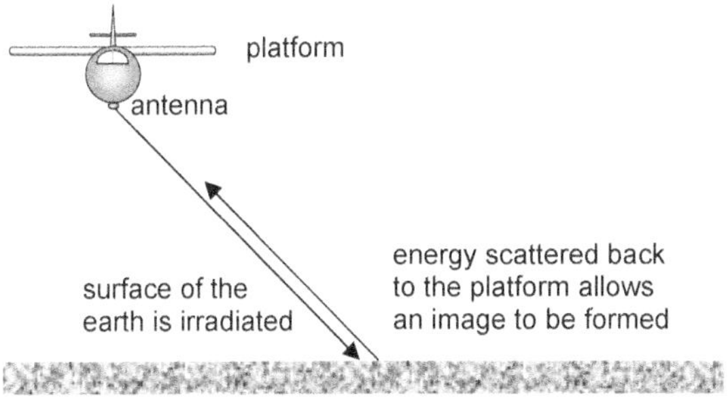

Fig. 2.6. The fundamental arrangement for active microwave remote sensing

2.5 Passive Microwave Remote Sensing

Although the Earth is essentially "dark" at microwave wavelengths there is nevertheless some energy emitted, as demonstrated above. In order to obtain enough power to measure, it is necessary to observe over a large enough area of the earth's surface. Consequently, passive microwave imaging is possible provided large pixel sizes are used. The study of passive microwave remote sensing is the topic of Chapt. 9.

2.6 The Atmosphere at Microwave Frequencies

Just as with optical remote sensing that employs visible and infrared wavelengths, care must be taken with the choice of wavelength in radar remote sensing to ensure that atmospheric properties do not interfere with the imaging process. Atmospheric scattering and attenuation is much less a problem at microwave as was noted in Chapt. 1 and as can be seen in more detail in Fig. 2.7, which shows attenuation by common atmospheric constituents in the range of wavelengths commonly used for radar remote sensing. Note that there is no appreciable effect until wavelengths become smaller than about 1cm. Since most radar imaging is carried out with wavelengths no shorter than 3cm (X band) atmospheric effects can generally be ignored. As noted in Fig. 1.1 though, at very low radar frequencies the ionosphere can be a problem. While remote sensing radars would not be operated at those frequencies for which the ionosphere appears opaque (about 10MHz and lower) an effect known as Faraday rotation can be problem at about L band (1GHz) and lower. The free electrons in the ionosphere coupled with the earth's magnetic field can cause the plane of polarisation of a wave[4] passing through the ionosphere to be rotated. That effect is treated in Sect. 3.24.

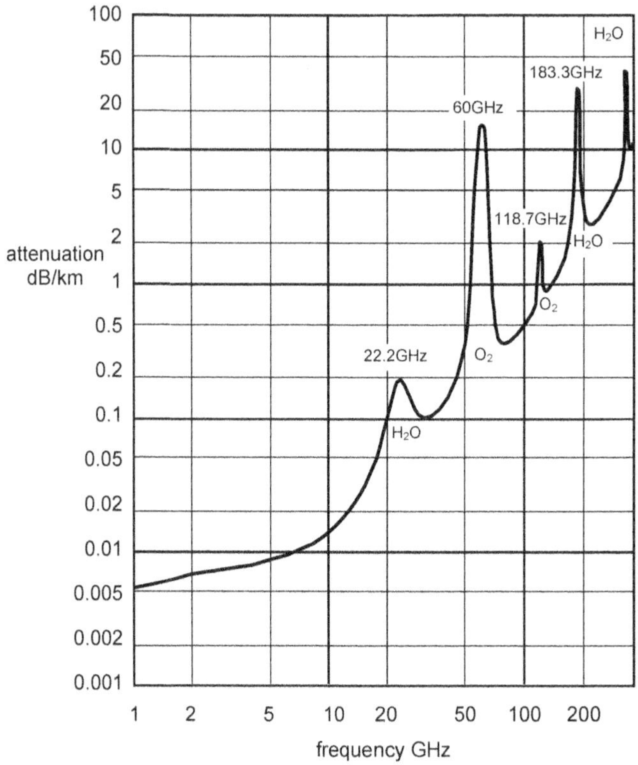

Fig. 2.7. Attenuation of microwave radiation by atmospheric constituents (from J.A. Richards, *Radio Wave Propagation: An Introduction for the Non-Specialist*, Springer, Berlin, 2008)

The fine droplet size in most clouds means they do not significantly scatter microwave energy at the wavelengths used for remote sensing. While optical energy cannot penetrate clouds to any appreciable extent, rendering imaging through clouds at visible and infrared wavelengths largely impossible, one of the great benefits of radar imaging is that clouds are, for all intents and purposes, transparent.

Rainfall can be a problem, but only for very short imaging wavelengths and when the rainfall is particularly heavy as can be assessed from Fig. 2.8.

[4] See Sect. 2.8

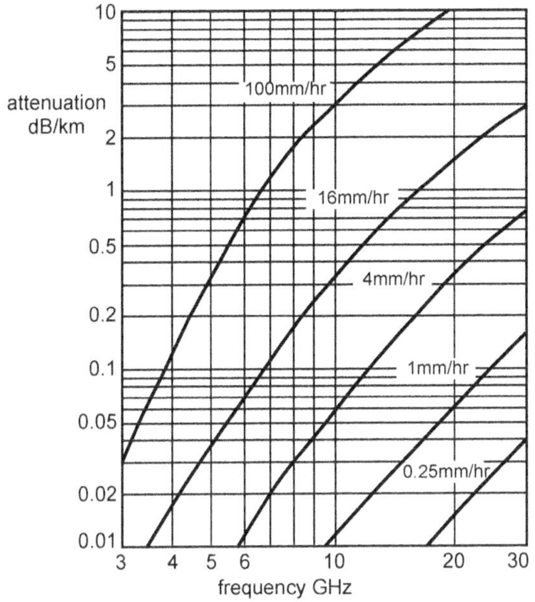

Fig. 2.8. Effect of rainfall on microwave propagation (from J.A. Richards, *Radio Wave Propagation, An Introduction for the Non-Specialist*, Springer, Berlin, 2008)

2.7 The Benefits of Radar Remote Sensing

Since clouds and other atmospheric constituents do not interfere with detection, and thus imaging, at the wavelengths used for radar remote sensing, and since the platform carries its own primary energy source, radar imaging can be carried out at any time of day and under any weather conditions, unless there is particularly severe rainfall and very short wavelengths are used. In general, though, radar imaging is thought of as an all-weather, all-hours technology.

Furthermore, since the wavelengths used with radar are about four or five orders of magnitude longer than those employed in optical remote sensing, quite different properties of earth cover types can be detected at microwave. It will be seen in Chapt. 5 that radar scattering is determined largely by geometric properties, such as shape and surface roughness, and by moisture content. Also, depending on the wavelength employed and the moisture content of the near earth surface being imaged, it is sometimes possible to image beneath the surface. It is certainly possible to image below vegetation canopies at longer wavelengths. As a result of all of these effects, radar image data can capture a different set of properties of the region being imaged than is the case for optical data. It thus finds its own applications, but is particularly valuable when used in association with optical imagery.

2.8 Looking at the Underlying Electromagnetic Fields

Our development of radar as an imaging modality so far has been based on understanding levels of power and power density. That will remain the case for much of our treatment. Nevertheless, it is important to understand some of the properties of the electric and magnetic fields that carry the power to and from the earth's surface.

In Fig. 2.6 the power radiated towards the earth travels as an expanding wavefront similar to that depicted in Fig. 2.1. If the transmitting antenna were isotropic then the wavefront would be spherical; a real antenna will shape that somewhat although the power density still diminishes with the inverse square of distance. Irrespective of the antenna used, when we are well away from the platform the wavefront appears planar as shown in Fig. 2.9, with the electric field propagating forward as a plane wave as depicted. Often we show the wave either by the ray that points in the direction of propagation or as a sinusoid in the plane of the wave, as illustrated. It is important to note in the figure that the vertical dimension is the strength of the electric field and not a vertical movement in space. It is a field that oscillates in amplitude vertically in the case that has been drawn – we describe that as a wave that has *vertical polarisation*. If we stood at a point in space and observed the wave passing us we would see the electric field strength alternating in a sinusoidal fashion vertically. We could also draw the wave oscillating in the horizontal direction – it is then said to have *horizontal polarisation*.

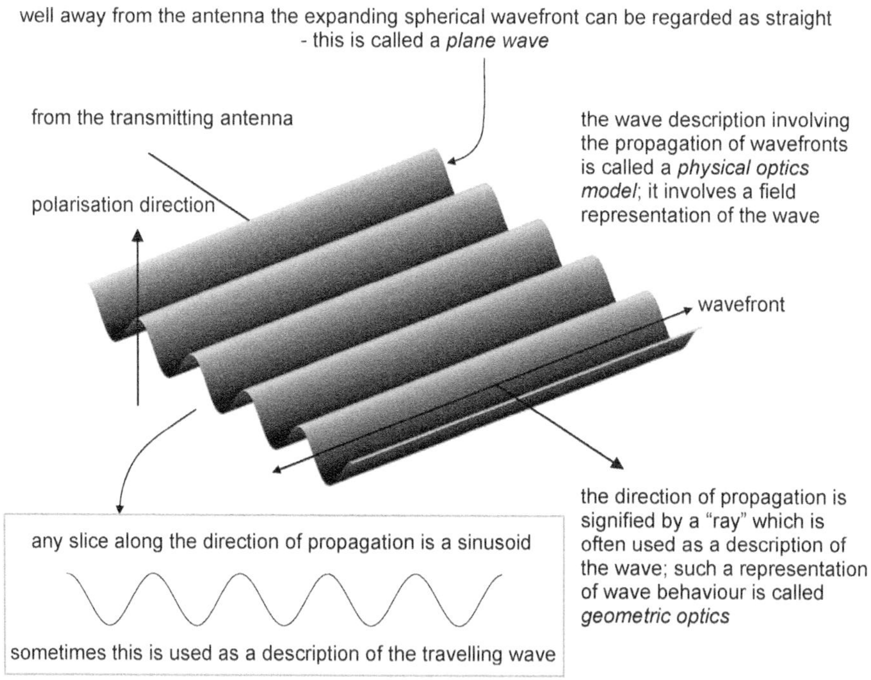

Fig. 2.9. A propagating plane wave; the vertical dimension is field strength and not a vertical distribution or displacement

Power, or power density, is carried forward as the result of both an electric field and a magnetic field that oscillate at right angles to each other and to the direction of propagation as illustrated by the field vectors (indicating polarisation – i.e. the plane of polarisation) in Fig. 2.10. It is therefore called a *transverse electromagnetic* (*TEM*) *wave*. There is an important relationship between the two field vectors and the direction of propagation – they follow the right hand screw rule. If a screwdriver is aligned as though it were to drive a screw in the direction of propagation then it would have to move from the direction of the electric to the direction of the magnetic field vector in doing so – i.e. in a clockwise sense when viewed from behind.

The electric and magnetic fields have four properties that we will need to consider from time to time. They are the frequency at which they oscillate (corresponding to the wavelength of the radiation being used), their amplitudes, their relative phase angles and directions in which they point in space. We can write them (with their units) as

electric field	$\mathbf{E}=E\mathbf{e}$	Vm^{-1} (volts/metre)
magnetic field	$\mathbf{H}=H\mathbf{h}$	Am^{-1} (amperes/metre)

in which **e** and **h** are vectors of unit magnitude that point in the direction of the respective field vector – i.e. in the direction of oscillation. Seldom in this treatment will we need to consider those unit vectors explicitly since we will normally know quite well the spatial orientations of the fields. From what we said above about the fields themselves, **e** and **h** are at right angles to each other and to the direction of propagation. We use a bold faced entry (**E**) if we infer the complete description of a field, whereas the un-bolded version (E) lacks reference to the direction in which it points spatially, but encompasses the other three properties – these are called the vector magnitudes.

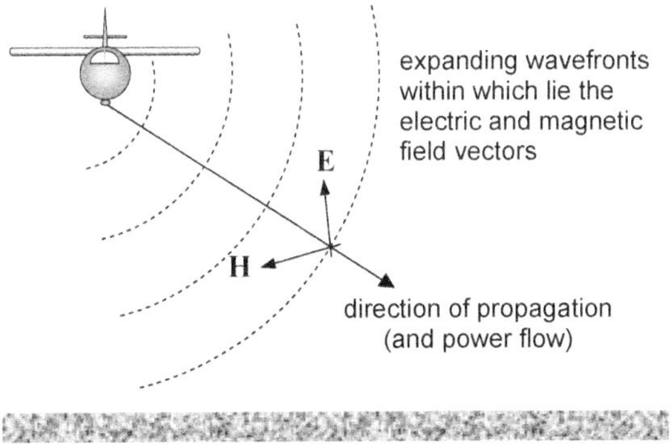

Fig. 2.10. The power radiated to the landscape is carried by electric and magnetic field vectors at right angles to each other and to the direction of propagation

There are two ways of writing the magnitudes of the vectors to describe the other properties, either as sinusoids explicitly

$$E = E_o \cos(\omega t + \phi_e)$$
$$H = H_o \cos(\omega t + \phi_h)$$

or in their sometimes more convenient complex exponential form[5]

$$E = E_o \exp j(\omega t + \phi_e)$$
$$H = H_o \exp j(\omega t + \phi_h)$$

in which E_o and H_o are the amplitudes of the fields and ϕ_e and ϕ_h are their phase angles. In free space the electric and magnetic fields will be in phase with each other so the phase angles are the same. The complete bracketed arguments are generically called the *phases* of the respective sinusoids.

The radian frequency ω is related to the commonly used measure of frequency f by

$$\omega = 2\pi f \text{ rads}^{-1}, \text{ with } f \text{ expressed in hertz (Hz)}$$

The period of the sinusoid (the duration of one period of its oscillation) measured in seconds is

$$T = \frac{1}{f}$$

and the relationship between frequency and wavelength is

$$f = \frac{c}{\lambda}$$

in which c is the velocity of light. This leads to (2.3) when the appropriate units are used. From the last two expressions we have $\lambda = cT$; thus if we observed a wave travelling past us the frequency of oscillation will be determined by the wavelength and speed of propagation.

If we take the product of the amplitudes of the two fields we see that the units are VAm^{-2} or watts per square metre, which are precisely the units of power density. We may thus equate

$$p_p = EH \qquad \text{Wm}^{-2} \tag{2.7a}$$

where p_p is called the peak power density, which we will simplify shortly. In free space E and H are not independent. The are related via the *impedance of free space* η:

$$E = \eta H$$

η has the value of $120\pi \approx 377\Omega$. Because of this dependence it is normal in remote sensing, when needing to appeal to a field as against a power description, just to talk about the electric field, knowing that the magnetic field can be described if needed. Thus, the peak power density at a given point where the electric field strength is E will be given by

$$p_p = \eta E^2 \tag{2.7b}$$

We don't often use peak power density in practice. Because E is a function of time p_p will also fluctuate with time. We are more interested in average power quantities including average power density, which is given by the average value of p_p found from

[5] See Appendix A for a brief review of how complex numbers can be used in this manner.

$$p = \frac{1}{T}\int_0^T p_p(t)dt = \frac{\eta}{T}\int_0^T E^2(t)dt = \frac{1}{2}\eta E_o^2 = \eta E_{rms}^2 \tag{2.7c}$$

in which $E_{rms} = E_o/\sqrt{2}$ is called the root mean square value (rms) of the field amplitude. Whenever we describe electric (and magnetic) field strength in applications it is normally understood that we are talking about rms values. When we write expressions for fields, especially in the exponential form, the amplitudes will be in rms form.

The field expressions above show their time variations at a given point in space (since they are written without a distance term). If we wish to show complete expressions for the magnitudes, including how the waves propagate, we need to incorporate a dependence on position R in their phases in the following manner

$$E = E_o\cos(\omega t - \beta R + \phi_e)$$
$$H = H_o\cos(\omega t - \beta R + \phi_h)$$

or

$$E = E_o\exp j(\omega t - \beta R + \phi_e)$$
$$H = H_o\exp j(\omega t - \beta R + \phi_h)$$

in which β is called the *phase constant* (measured in radians per metre); it is often also written as the *wave number k*, which is in all respects equivalent. In free space[6]

$$\beta = \frac{\omega}{c} = \frac{2\pi f}{\lambda f} = \frac{2\pi}{\lambda} \tag{2.8}$$

To see that the wave is actually travelling in the positive R direction we lock ourselves onto a point of constant phase and get carried along with the wave, much as a surf board rider gets carried by a water wave by sitting at an equivalent point of constant phase. This is illustrated in Fig. 2.11. We have, at any given point on, say, the electric field wave

$$\omega t - \beta R + \phi_e = \phi = \text{constant}$$

therefore

$$R = \frac{1}{\beta}(\omega t - \phi + \phi_e)$$

from which the velocity of the wave is

$$v = \frac{dR}{dt} = \frac{\omega}{\beta} = c$$

which is positive in the positive R direction and equal to the velocity of light. Thus, when there is a negative sign in front of βR in the field expression the wave travels in the positive R direction.

[6] See J.A. Richards, *Radio Wave Propagation: An Introduction for the Non-Specialist*, Springer, Berlin, 2008.

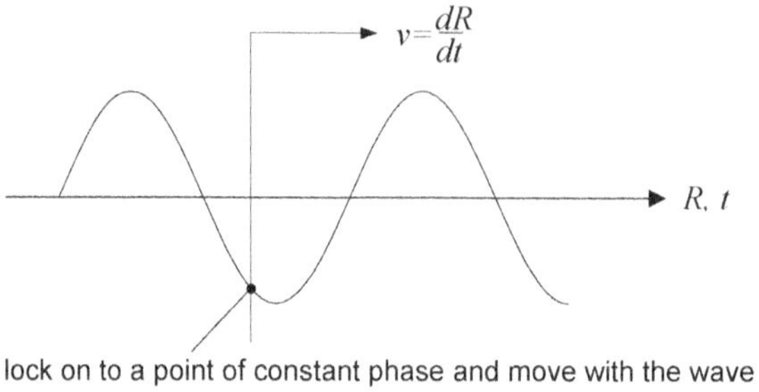

Fig. 2.11. Demonstrating the velocity and direction of travel of a sinusoidal field component

Figure 2.12 summarises the conventions of nomenclature used in describing travelling electromagnetic fields that we adopt in this book.

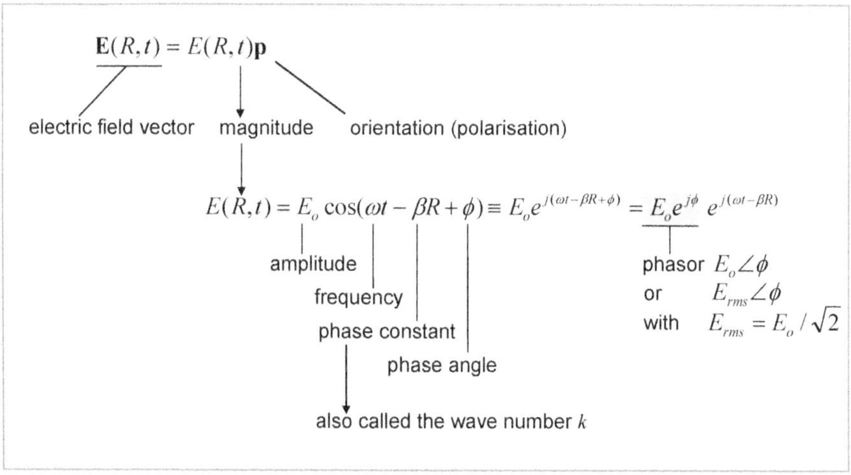

Fig. 2.12. Summary of the nomenclature conventions used with electromagnetic fields; although expressed as an electric field, magnetic field vectors are described in the same way

2.9 The Concept of Near and Far Fields

In radar remote sensing we assume the transmitted and scattered radiation propagates as transverse electromagnetic (TEM) waves. However near the antenna, and similarly in the near vicinity of a radar target, that is not the case. We have to be a certain distance away from each before we can assume the fields are TEM and thus a simple view of propagation can be used. When we can assume TEM behaviour we say we are in the *far field*; otherwise we are in the *near field* of the antenna or target. To develop an understanding of where the transition to far field behaviour occurs it is instructive to consider the simplest of all antennas, the so-called short dipole. Fig. 2.13 shows the geometry of a short dipole in which distance and direction out from the antenna is described by the radial vector **r**. Any point in space is described by the spherical coordinates (r,ϕ,θ). The complete set of field components generated about the short dipole is

$$E_r = \frac{Ae^{j(\omega t-\beta r)}\cos\theta}{2\pi\varepsilon_o}\left(\frac{1}{cr^2}+\frac{1}{j\omega r^3}\right) \tag{2.9a}$$

$$E_\theta = \frac{Ae^{j(\omega t-\beta r)}\sin\theta}{4\pi\varepsilon_o}\left(\frac{j\omega}{c^2 r}+\frac{1}{cr^2}+\frac{1}{j\omega r^3}\right) \tag{2.9b}$$

$$H_\phi = \frac{Ae^{j(\omega t-\beta r)}\sin\theta}{4\pi}\left(\frac{j\omega}{cr}+\frac{1}{r^2}\right) \tag{2.9c}$$

where A is a constant and the exponential terms describe propagation outwards from the dipole. We will say more about that term later, but it is not important in this discussion. Equations (2.9) show that there are transverse components (θ,ϕ) of the magnetic and electric fields. There is also a radial electric field component (r) – i.e. *in* the direction of propagation. Note however that it has a stronger inverse dependence on distance than the transverse components so that if the distance is sufficiently large it disappears. This is demonstrated by letting r go large in (2.9a-c) to give

$$E_r = 0$$

$$E_\theta = \frac{j\omega Ae^{j(\omega t-\beta r)}\sin\theta}{4\pi\varepsilon_o c^2 r}$$

$$H_\phi = \frac{j\omega Ae^{j(\omega t-\beta r)}\sin\theta}{4\pi cr}$$

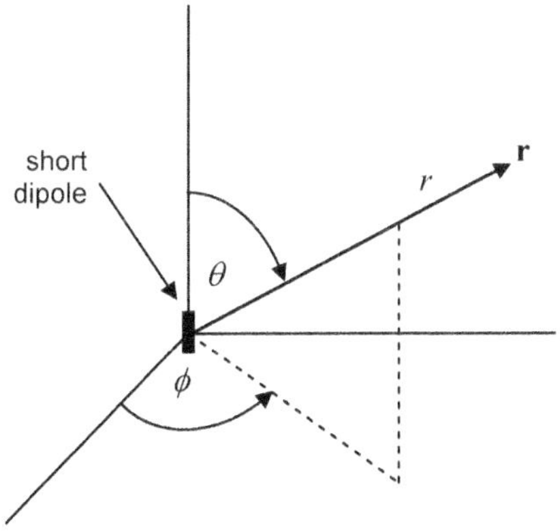

Fig. 2.13. The short dipole

Thus for large distances the wave is TEM – transverse electromagnetic. These equations describe the *far field* of the antenna. The far fields are inverse distance dependent and the treatment in this book, based on simple power and power density relationships, is valid. In contrast, closer to the antenna (2.9a-c) are needed to describe the *near field* for this simple structure. The transition from near to far field for the short monopole is said to occur when the inverse distance terms in (2.9b,c) are equal to the inverse distance squared terms, assuming that any inverse cubic terms are then negligible. Therefore the near field/far field transition for this particular case is when

$$\left|\frac{\omega}{cr}\right| = \left|\frac{1}{r^2}\right|$$

which gives $r \approx \lambda/6$ for the short monopole. Usually we can assume we are in the far field when we are a few wavelengths from the antenna or scatterer.

2.10 Polarisation

The orientations of the electric and magnetic field vectors shown in Fig. 2.10, which define the *polarisation* of the propagating wave, are not strictly important in terms of the propagation of radiation in free space. However, when the radiation strikes the ground the response of surface materials can be different for different orientations of the vectors. Therefore, we need a convention to describe the directions in which the field vectors point. Because the magnetic field is always at right angles to the electric field it sufficient to concentrate on the polarisation of electric field alone. Fig. 2.14 shows an incoming ray interacting with a surface.

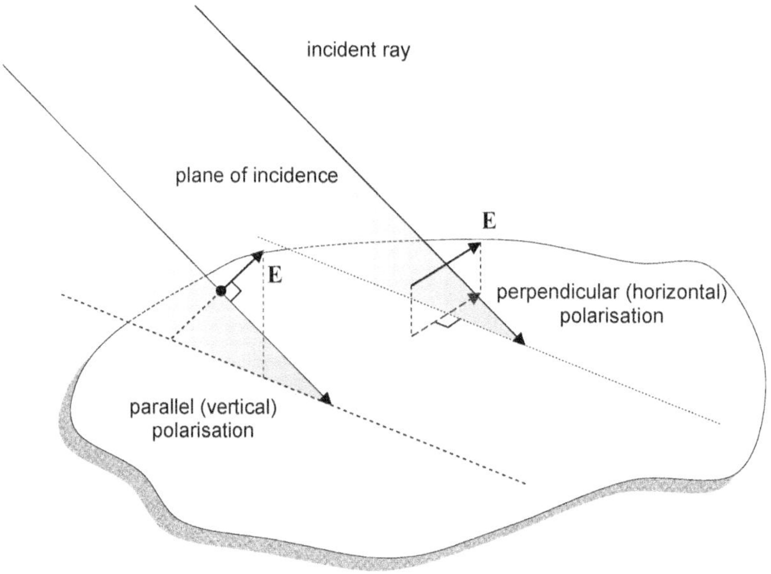

Fig. 2.14. Definitions of polarisation with respect to the plane of incidence and the surface

We define the plane of incidence as that which is at right angles to the surface and which contains the ray. By reference to the plane of incidence we define two types of polarisation of the electric field; both are illustrated in Fig. 2.14. If the electric field vector lies in the plane of incidence then the field is said to have *parallel polarisation*, whereas if it is at right angles to the plane of incidence it is said to have *perpendicular polarisation*. Note that a perpendicularly polarised wave is horizontal to the earth's surface. In remote sensing it is therefore more often called *horizontal polarisation*. Even though not strictly correct, parallel polarisation is similarly referred to as *vertical polarisation*. Remember that the plane of polarisation is that in which the electric field vector oscillates sinusoidally as described in the expressions above and in (2.10, 2.11) below.

Imagine now that the ray travelling towards the earth's surface has a polarisation that is neither horizontal nor vertical but can be resolved into vertical and horizontal components, as shown in Fig.2.15. We can write the field vector as

$$\mathbf{E} = E_H \boldsymbol{h} + E_V \boldsymbol{v} \tag{2.10}$$

where $\boldsymbol{h}$ and $\boldsymbol{v}$ are unit length vectors that point in the respective directions as reminders of the horizontal and vertical directional components of the field.

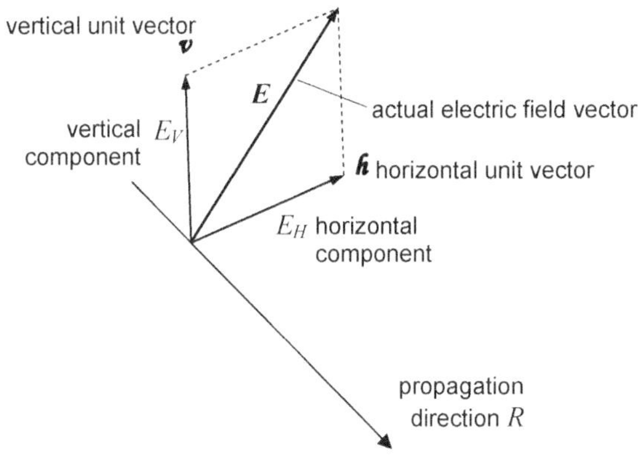

Fig. 2.15. Resolution of an electric field into its horizontal and vertical components

Most generally, the components' magnitudes can be written

$$E_H = a_H \cos(\omega t - \beta R) \tag{2.11a}$$

$$E_V = a_V \cos(\omega t - \beta R + \delta) \tag{2.11b}$$

in which a_H and a_V are the amplitudes of the two components, R is the direction of propagation and, for generality, δ is a *phase difference* between them. From here on we assume that a_H and a_V are explicitly the rms values of the field amplitudes.

Fig. 2.16 shows the two components plotted as functions of time at a given position in space to illustrate the significance of the phase difference. Note that the time origin is on the left of the figure. If the waves were travelling towards a target or spot on the ground the left hand sides of the plots would arrive first.

Imagine those two sinusoids were the horizontal and vertical components shown in Fig. 2.15. Let us now look at what might be happening to the actual electric field vector with time if we viewed it as it approaches us – in other words we are looking back at the approaching wave from the position of the arrowhead in Fig. 2.15. In Fig. 2.17 we have shown the situation we would observe at three different times if curve A in Fig. 2.16 represents the vertical component and curve B the horizontal, and the phase angle δ is positive and equal to 90°. As seen, the effect of the time dependence of the two components, and the phase difference between them, is that the actual electric field vector traces out a clockwise circular path around the direction of propagation. While that is happening the wave is also travelling forward because of the ωt-βR arguments in (2.11). The field vector propagates forward in a corkscrew fashion, which we don't see when we view the field along the direction of propagation.

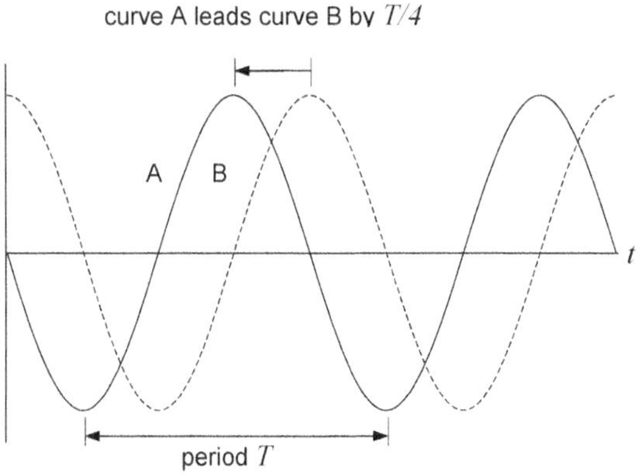

Fig. 2.16. Illustrating phase difference; in this case curve A has a positive phase difference (leads) with respect to curve B

If the phase angle in (2.11b) were negative then the vertical component would lag behind the horizontal, and the total field vector would rotate in the anti-clockwise direction when viewed from the surface upon which the field is incident. Viewed from behind that would be how a screwdriver would rotate when driving a screw in the direction of propagation; consequently that is referred to as *right circular polarisation.* In the former case of the total vector rotating clockwise when viewed in approach, the effect from behind would emulate a left handed screwdriver. It is then called *left circular polarisation.*

Pure circular polarisation only occurs when the two components have the same amplitude and the phase difference between them is 90° (or a quarter of a period). In the most general case the approaching figure shown in Fig. 2.17 would be an ellipse. Again there will be *left elliptical polarisation* and *right elliptical polarisation* depending on the sign of the phase angle between the components. Circular polarisation is a special case. Another special case is simple *linear polarisation*; that occurs when there is no phase angle between the components. Their relative amplitudes will determine the orientation of the actual field vector, which then oscillates in amplitude along that spatial direction.

We can actually derive the equation of the ellipse around which the field vector moves in the most general case. Expand (2.11b) as

$$E_V = a_V\left[\cos(\omega t - \beta R)\cos\delta - \sin(\omega t - \beta R)\sin\delta\right]$$

From (2.11a)

$$\cos(\omega t - \beta R) = \frac{E_H}{a_H} \quad \text{and} \quad \sin(\omega t - \beta R) = \sqrt{1 - \frac{E_H^2}{a_H^2}}$$

so that

$$E_V = a_V \left[\frac{E_H}{a_H} \cos\delta - \sqrt{1 - \frac{E_H^2}{a_H^2}} \sin\delta \right]$$

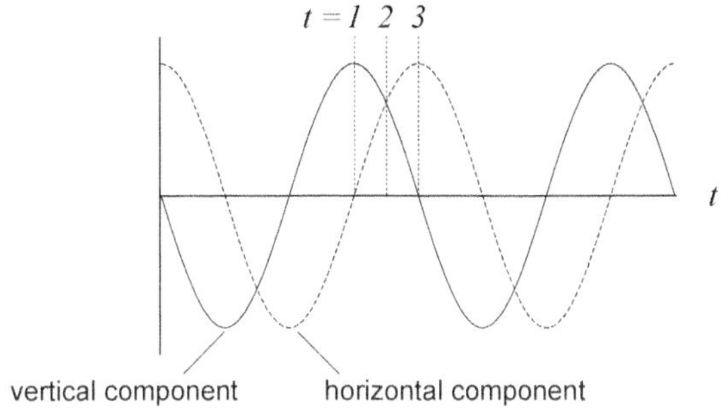

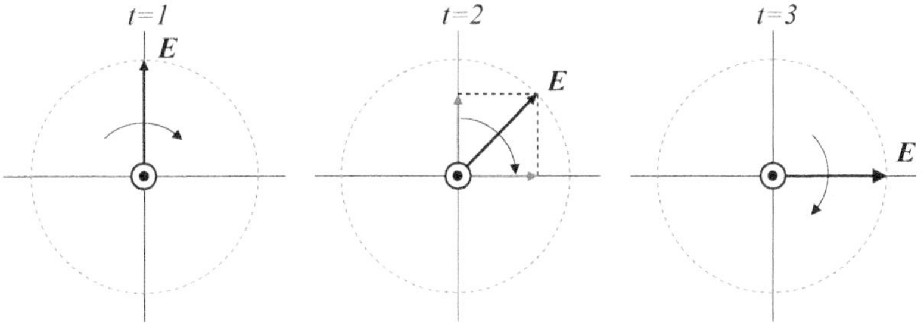

Fig. 2.17. Demonstrating how the actual field vector rotates in transmission as a result of the time phase difference between its two components; in this case the vertical component has a positive (leading) phase angle with respect to the horizontal giving a clockwise rotation of the field vector viewed from the front as indicated by the small dot and circle at the origin (representing the arrowhead in Fig. 2.15); when viewed from behind this is left circular polarisation

Rearranging we have

$$\frac{E_H}{a_H}\cos\delta - \frac{E_V}{a_V} = \sqrt{1 - \frac{E_H^2}{a_H^2}} \sin\delta$$

which on squaring leads to

$$\left(\frac{E_V}{a_V}\right)^2 + \left(\frac{E_H}{a_H}\right)^2 - 2\frac{E_H E_V}{a_H a_V}\cos\delta = \sin^2\delta \tag{2.12}$$

which is the equation of an ellipse in the variables E_V and E_H, centred on the origin, as shown in Fig. 2.18. Note that the ellipse is inscribed in a rectangle, parallel to the field axes, of dimensions $2a_H$, $2a_V$. It can also be shown that[7] $a_H^2 + a_V^2 = e^2 + f^2$, where e and f are respectively the semi-minor and semi-major axes of the ellipse.

This ellipse is the general version of the circles shown in Fig. 2.17. The field vector indicated rotates around the ellipse as determined by the phase angle δ between the components. If $\delta=\pm 90^\circ$, (2.12) reduces to the equation of an ellipse with axes parallel to the field components (see Fig. 2.19 following). If, in addition $a_V = a_H$ then the ellipse degenerates into a circle. If $\delta=0$, (2.12) reduces to the equation of a straight line through the origin with slope a_V/a_H, since

$$\left(\frac{E_V}{a_V}\right)^2 + \left(\frac{E_H}{a_H}\right)^2 - 2\frac{E_H E_V}{a_H a_V}\cos\delta = \left(\frac{E_V}{a_V} - \frac{E_H}{a_H}\right)^2 = 0$$

Likewise if $\delta = \pi$, (2.12) reduces to

$$\left(\frac{E_V}{a_V} + \frac{E_H}{a_H}\right)^2 = 0$$

which is the equation of a straight line through the origin with slope $-a_V/a_H$.

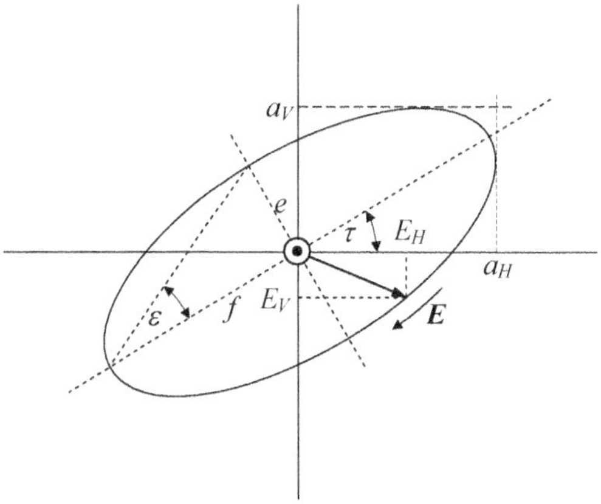

Fig. 2.18. The polarisation ellipse as the locus of the approaching electric field vector; again the small circle and point in the centre is meant to represent the tip of the directional arrow head pointing out of the page

There are two properties of the ellipse that relate directly to the polarisation state of the radiation. The first is its *ellipticity* or *eccentricity* which describes how different it is from a circle or, at the other extreme, a straight line. The other is its *tilt* or *inclination* with respect to the horizontal. Tilt is described by the angle τ and eccentricity can be described by the so-called axial ratio f/e, or more often by the angle ε when the term ellipticity is

[7] M. Born and E. Wolf, *Principles of Optics*, 7th ed., Cambridge University Press, Cambridge, 2006.

used. As is to be expected they are related to the phase difference δ and the relative amplitudes a_H and a_V. If we define

$$\alpha = \tan^{-1}\frac{a_V}{a_H} \tag{2.13}$$

then it can be shown that[8]

$$\tan 2\tau = \tan 2\alpha \cos\delta \tag{2.14}$$

and

$$\sin 2\varepsilon = \sin 2\alpha \sin\delta \tag{2.15}$$

which are explicit relationships between properties of the wave (a_H, a_V, δ) and those of the polarisation ellipse (τ, ε).

Since a_V and a_H are positive (because they are simply amplitudes) then α is positive so that the sign of ε follows the sign of δ. As a result, ε is positive for left elliptical polarisation and negative for right elliptical polarisation. Note that the range of ε is -45° to $+45^\circ$, which goes between the extremes of right to left circular polarisations; linear polarisation occurs when ε=0. From Fig. 2.18 we can see that the range of τ will be -90° to $+90^\circ$.

In radar remote sensing we most frequently encounter linearly polarised systems, but a knowledge of elliptical polarisation is important to gain most insight from polarisation synthesis radar treated in Sect. 3.22. Some circularly polarised radars are also encountered in practice.

2.11 The Jones Vector

The electric field described by (2.10) and (2.11) can be re-written using the parameters of the polarisation ellipse in the following manner. We commence by expressing the field in the convenient exponential form

$$\mathbf{E} = a_H \exp j(\omega t - \beta R)\boldsymbol{h} + a_V \exp j(\omega t - \beta R + \delta)\boldsymbol{v} = \{a_H\boldsymbol{h} + a_V e^{j\delta}\boldsymbol{v}\}\exp j(\omega t - \beta R) \tag{2.16}$$

in which recall that the unit vectors $\boldsymbol{h}$ and $\boldsymbol{v}$ are simply convenient reminders of the horizontal and vertical directions of the respective field components. From the previous section we know that this field rotates around the polarisation ellipse of Fig. 2.18. Taking out the common factor $\sqrt{a_H^2 + a_V^2}$, which is the total amplitude of the wave, gives

$$\mathbf{E} = \sqrt{a_H^2 + a_V^2}\{\cos\alpha\boldsymbol{h} + \sin\alpha\boldsymbol{v}e^{j\delta}\}\exp j(\omega t - \beta R) \tag{2.17}$$

in which $\alpha = \tan^{-1}\frac{a_V}{a_H}$ from (2.13).

It is now common to replace the representation using unit vectors by one in which the two components of the field are the two elements of a column vector; this is just another useful way of writing the composite field. The first entry represents the magnitude of the horizontal component and the second the magnitude of the vertical component. We also

[8] M. Born and E. Wolf, *loc cit.*

drop the exponential factor in time since we know it applies to all fields. We are then left with

$$\mathbf{E} = Ae^{\zeta}\begin{bmatrix}\cos\alpha \\ \sin\alpha e^{j\delta}\end{bmatrix} = Ae^{\zeta}\mathbf{E}_J \tag{2.18}$$

in which A is the amplitude from (2.17) and ζ is the absolute phase corresponding to $-j\beta R$, resulting from any propagation path that is important to consider. Sometimes we ignore that term along with the time exponential. While in radar the full expression in (2.18) is sometimes referred to as the Jones vector, we will call $\mathbf{E}_J$ the *Jones vector* unless there is a particular reason to include the amplitude and total phase terms; this is the version most commonly adopted in optics.

Now consider the polarisation ellipse lying parallel to the axes as in Fig. 2.19. From that geometry we have $\alpha=\varepsilon$ and, by comparison to Fig. 2.18, $\tau=0$. From (2.14) that requires $\delta = \pm 90^{\circ}$ if α is non zero. This makes $e^{j\delta} = \pm j$ so that (2.18), for the ellipse parallel to the axes, becomes

$$\mathbf{E} = Ae^{\zeta}\begin{bmatrix}\cos\varepsilon \\ \pm j\sin\varepsilon\end{bmatrix} = Ae^{\zeta}\mathbf{E}_J$$

It's awkward having the $\pm$ sign in the vector since we must keep in mind the corresponding sign of δ. Alternatively, we can "absorb" the sign into the eccentricity ε. When the second element of the Jones vector is $+j\sin\varepsilon$ we have left elliptical polarisation. We accommodate that by agreeing that positive eccentricity corresponds to left elliptical polarisation. Negative eccentricity then corresponds to right elliptical polarisation, giving the negative sign in the last expression because sine is an odd function. That is the same as making δ negative in (2.11b). Adopting that convention for eccentricity the Jones vector for the ellipse lying parallel to the axes is just simply

$$\mathbf{E} = Ae^{\zeta}\begin{bmatrix}\cos\varepsilon \\ j\sin\varepsilon\end{bmatrix} = Ae^{\zeta}\mathbf{E}_J \tag{2.19}$$

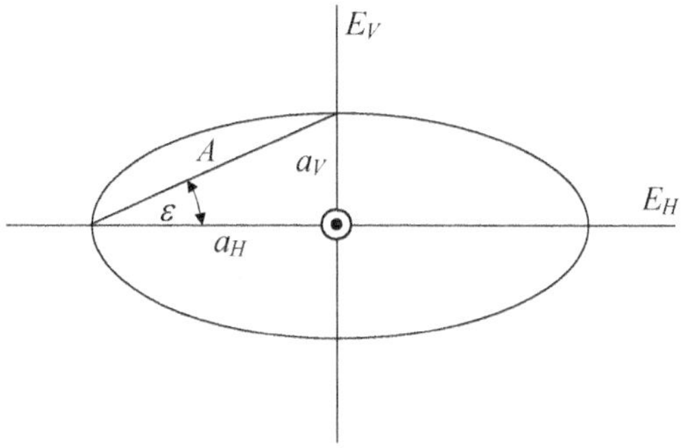

Fig. 2.19. Polarisation ellipse used in the derivation of the Jones vector

Consider some examples of special Jones vectors. For a horizontally polarised wave $a_V=0$, so that $\varepsilon=0$. The Jones vector is then as shown in the first entry of Table 2.1. For vertical polarisation ε=90° but the situation needs a little care to handle. We return to (2.18) and put δ=0 because the relative phase angle of the vertical field compared to the horizontal has no meaning in expressions like (2.16) when there is no horizontal component. Thus $e^{j\delta}=1$ so that with ε=90° (2.18) yields the entry in Table 2.1.

Now consider left circular polarisation for which $a_H = a_V$ and thus $\varepsilon = 45^o$ giving $\sin\varepsilon = \cos\varepsilon = 1/\sqrt{2}$. The Jones vector then becomes as seen in Table 2.1. To get the Jones vector for right circular polarisation we choose $\varepsilon = -45^o$. That is equivalent to reversing the sign of the horizontal component of the field, which is the same as adding 180° to its phase and thus making it lead the vertical component by 90°.

The Jones vector of (2.19) was derived on the basis of the polarisation ellipse lying parallel to the axes in Fig. 2.19. We can transform the vector so that it applies to the more general case of the inclined ellipse of Fig. 2.18 by rotating the axes clockwise by the inclination angle.

Fig. 2.20 shows two sets of axes within which the same electric field vector is described. The XY set is rotated anti-clockwise from the HV set by an angle τ. The two descriptions of the field are related by the matrix transformation:

$$\begin{bmatrix} E_X \\ E_Y \end{bmatrix} = \begin{bmatrix} \cos\tau & \sin\tau \\ -\sin\tau & \cos\tau \end{bmatrix} \begin{bmatrix} E_H \\ E_V \end{bmatrix} = \mathbf{\mathcal{T}}(\tau) \begin{bmatrix} E_H \\ E_V \end{bmatrix} \tag{2.20}$$

In order to apply (2.19) to the inclined polarisation ellipse of Fig 2.18, it is necessary to rotate the axes of Fig. 2.19 clockwise by the angle τ. That involves the inverse of the transformation matrix of (2.20) (or reversal of the sign of τ) which is readily shown to be

$$\mathbf{\mathcal{T}}^{-1}(\tau) = \begin{bmatrix} \cos\tau & -\sin\tau \\ \sin\tau & \cos\tau \end{bmatrix} \tag{2.21}$$

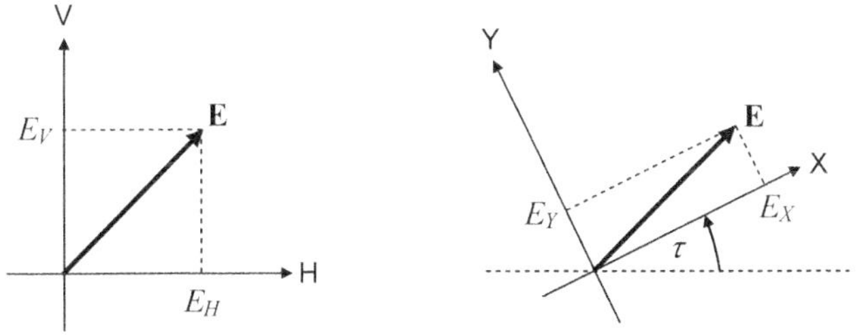

Fig. 2.20. Transforming the axis description of an electric field vector

Thus the electric field vector of (2.19) for the most general case is

$$\mathbf{E} = Ae^{\zeta} \begin{bmatrix} \cos\tau & -\sin\tau \\ \sin\tau & \cos\tau \end{bmatrix} \begin{bmatrix} \cos\varepsilon \\ j\sin\varepsilon \end{bmatrix} \tag{2.22}$$

which is the required (full) description of the field in terms of the parameters of the polarisation ellipse. This allows us to generate Jones vectors for polarisation configurations not describable by (2.19). For example, a linearly polarised wave with an inclination of ±45° will have Jones vectors given by rotating a horizontally polarised wave by ±45° using (2.22), generating the results shown in Table 2.1.

Table 2.1 Some common Jones vectors

horizontal polarisation	$\begin{bmatrix}1\\0\end{bmatrix}$	from (2.19)
vertical polarisation	$\begin{bmatrix}0\\1\end{bmatrix}$	from (2.18) and (2.16)
right circular polarisation	$\frac{1}{\sqrt{2}}\begin{bmatrix}1\\-j\end{bmatrix}$	from (2.19)
left circular polarisation	$\frac{1}{\sqrt{2}}\begin{bmatrix}1\\j\end{bmatrix}$	from (2.19)
+45° linear polarisation	$\frac{1}{\sqrt{2}}\begin{bmatrix}1\\1\end{bmatrix}$	from (2.22)
-45° linear polarisation	$\frac{1}{\sqrt{2}}\begin{bmatrix}1\\-1\end{bmatrix}$	from (2.22)

2.12 Circular Polarisation as a Basis Vector System

In the previous sections we have represented a travelling wave as a combination of horizontal and vertical components, as in (2.10) and (2.11). It is also possible to choose right circularly polarised and left circularly polarised field components as the basis with which to express any general electric field:

$$\mathbf{E} = E_L\boldsymbol{\ell} + E_R\boldsymbol{r} \equiv E_H\boldsymbol{h} + E_V\boldsymbol{v} \tag{2.23}$$

$\boldsymbol{\ell}$ and $\boldsymbol{r}$ are unit vectors that signify left circular and right circular polarisation respectively and E_L and E_R are the corresponding magnitudes of the field components. The unit vectors rotate around the unit circle in their respective directions carrying the relevant field magnitudes with them.

A purely left circularly polarised wave will have E_R=0. In that case the resultant field is a vector that travels around the circle of Fig. 2.17 as a special case of Fig. 2.18. From Sect. 2.10 we know that that will happen if the horizontally and vertically polarised components have the same amplitude and the vertical field component leads (has a positive phase angle with respect to) the horizontal component by 90°. Using exponential notation we can thus write

$$E_L\boldsymbol{\ell} = \exp j(\omega t - \beta R)\boldsymbol{h} + \exp j(\omega t - \beta R + \pi/2)\boldsymbol{v}$$

in which we have assumed unit amplitudes. Note that E_L will have the same dependence on time and position as its two components, which we can remove as a common factor, leaving

$$|E_L|\boldsymbol{\ell} = \boldsymbol{h} + e^{j\pi/2}\boldsymbol{v} = \boldsymbol{h} + j\boldsymbol{v}$$

Since $\boldsymbol{\ell}$, $\boldsymbol{h}$ and $\boldsymbol{v}$ are unit vectors this last expression requires $|E_L| = \sqrt{2}$ giving

$$\boldsymbol{\ell} = \frac{1}{\sqrt{2}}(\boldsymbol{h} + e^{j\pi/2}\boldsymbol{v}) \equiv \frac{1}{\sqrt{2}}(\boldsymbol{h} + j\boldsymbol{v}) \tag{2.24a}$$

It is interesting to compare this with the Jones vector for left circular polarisation in Table 2.1. In a similar manner a right circularly polarised wave will have its unit vector expressible as

$$\boldsymbol{r} = \frac{1}{\sqrt{2}}(\boldsymbol{h} - j\boldsymbol{v}) \tag{2.24b}$$

Using (2.24a,b) in (2.23) gives

$$\begin{aligned}\mathbf{E} &= E_L\frac{1}{\sqrt{2}}(\boldsymbol{h} + j\boldsymbol{v}) + E_R\frac{1}{\sqrt{2}}(\boldsymbol{h} - j\boldsymbol{v})\\ &= \frac{1}{\sqrt{2}}(E_L + E_R)\boldsymbol{h} + \frac{j}{\sqrt{2}}(E_L - E_R)\boldsymbol{v}\end{aligned}$$

so that

$$\begin{aligned}E_H &= \frac{1}{\sqrt{2}}(E_L + E_R)\\ E_V &= \frac{j}{\sqrt{2}}(E_L - E_R)\end{aligned}$$

These last two expressions can be written in matrix form:

$$\begin{bmatrix}E_H\\E_V\end{bmatrix} = \frac{1}{\sqrt{2}}\begin{bmatrix}1 & 1\\ j & -j\end{bmatrix}\begin{bmatrix}E_L\\E_R\end{bmatrix} \tag{2.25}$$

This indicates how the linear field components can be computed from the circular components. By inverting the matrix in (2.25) we can find the circular components in terms of the linear components:

$$\begin{bmatrix}E_L\\E_R\end{bmatrix} = \frac{1}{\sqrt{2}}\begin{bmatrix}1 & -j\\ 1 & j\end{bmatrix}\begin{bmatrix}E_H\\E_V\end{bmatrix} \tag{2.26}$$

Note from (2.24a,b) that we can write

$$\boldsymbol{h} = \frac{1}{\sqrt{2}}(\boldsymbol{r} + \boldsymbol{\ell}) \tag{2.27a}$$

$$\boldsymbol{v} = \frac{j}{\sqrt{2}}(\boldsymbol{r} - \boldsymbol{\ell}) \qquad (2.27b)$$

which demonstrate that a horizontally polarised wave is made up of right and left circularly polarised waves starting in phase (and contra-rotating). A vertically polarised wave is made up of the two contra-rotating circular components starting in anti-phase. The j in (2.27b) is a time phase term common to both components, advancing them by 90° and thus causing the vertical component to have a value of unity at t=0.

2.13 The Stokes Parameters, the Stokes Vector and the Modified Stokes Vector

The Stokes parameters provide a very convenient means by which to describe the power density relationships in an electromagnetic wave in radar, whether it be the wave used to irradiate the earth's surface or that which is scattered. For a single frequency signal (monochromatic) such as we assume to be the case for radar remote sensing, they are defined by[9]:

$$s_0 = a_H^2 + a_V^2 \qquad (2.28a)$$

$$s_1 = a_H^2 - a_V^2 \qquad (2.28b)$$

$$s_2 = 2a_H a_V \cos\delta \qquad (2.28c)$$

$$s_3 = 2a_H a_V \sin\delta \qquad (2.28d)$$

The first parameter s_0 is equal to the amplitude squared – or intensity – of the actual field vector shown in Fig. 2.18 and thus from (2.7) is directly proportional to the power density being carried by the wave. The second parameter indicates whether the wave is more horizontally than vertically polarised, while the third and fourth indicate the ellipticity of the wave's polarisation; in particular, if δ=0 we have linear polarisation and s_3=0. If δ=90°, s_2=0 and the polarisation ellipse will be aligned vertically and horizontally. It will be circular if the magnitudes of the vertical and horizontal components are also equal. It is straightforward to demonstrate that

$$s_0^2 = s_1^2 + s_2^2 + s_3^2 \qquad (2.29)$$

Often the Stokes parameters are collected together into a column vector called the *Stokes vector*:

$$\mathbf{s} = \begin{bmatrix} s_0 \\ s_1 \\ s_2 \\ s_3 \end{bmatrix} = \begin{bmatrix} a_H^2 + a_V^2 \\ a_H^2 - a_V^2 \\ 2a_H a_V \cos\delta \\ 2a_H a_V \sin\delta \end{bmatrix} = \begin{bmatrix} |E_H|^2 + |E_V|^2 \\ |E_H|^2 - |E_V|^2 \\ 2\mathcal{R}e(E_H^* E_V) \\ 2\mathcal{I}m(E_H^* E_V) \end{bmatrix} = I_0 \begin{bmatrix} 1 \\ \cos 2\tau \cos 2\varepsilon \\ \sin 2\tau \cos 2\varepsilon \\ \sin 2\varepsilon \end{bmatrix} \qquad (2.30)$$

in which I_0=s_0, the total power density, or intensity, of the wave.

Equation (2.30) gives alternative expressions for the Stokes vector. The rightmost version shows how the vector can be expressed in terms of the two principal angles of the polarisation ellipse; they follow from (2.40a) and (2.40b) below. The next from the right

[9] See M. Born and E. Wolf, *loc cit.*

shows how the Stokes vector is derived from the vector components of the electric field. The last two parameters in this form need to be written in terms of the complex phasors that describe the fields. Note from the exponential form of the two field components that

$$E_H = a_H \exp j(\omega t - \beta R) \quad E_V = a_V j(\omega t - \beta R + \delta)$$

so that we have $$E_H^* E_V = a_H a_V e^{j\delta} = a_H a_V (\cos\delta + j\sin\delta)$$

giving $$\mathcal{R}e(E_H^* E_V) = a_H a_V \cos\delta \quad \text{and} \quad Im(E_H^* E_V) = a_H a_V \sin\delta$$

Interestingly, the Stokes vector can be generated in the following manner which is useful when we come to look at scattering of radiation from the earth's surface and polarisation synthesis in Chapt. 3. Express the Stokes vector as the product of a matrix **R**, and a vector **g** which contains the four possible complex products of the two components of the field:

$$\mathbf{s} = \mathbf{Rg} \tag{2.31}$$

with $$\mathbf{R} = \begin{bmatrix} 1 & 1 & 0 & 0 \\ 1 & -1 & 0 & 0 \\ 0 & 0 & 1 & 1 \\ 0 & 0 & -j & j \end{bmatrix} \tag{2.32}$$

and $$\mathbf{g} = \begin{bmatrix} E_H E_H^* \\ E_V E_V^* \\ E_H E_V^* \\ E_V E_H^* \end{bmatrix} = \begin{bmatrix} |E_H|^2 \\ |E_V|^2 \\ E_H E_V^* \\ E_V E_H^* \end{bmatrix} \tag{2.33}$$

Sometimes the vector **g** is written in the Kronecker product form[10] (the last three elements are re-ordered)

$$\mathbf{g}_c = \mathbf{E} \otimes \mathbf{E}^* = \begin{bmatrix} E_H E_H^* \\ E_H E_V^* \\ E_V E_H^* \\ E_V E_V^* \end{bmatrix}$$

in which $\mathbf{E} = \begin{bmatrix} E_H \\ E_V \end{bmatrix}$; $\mathbf{g}_c$ is called the *coherency vector*. Sometimes its elements are written in the form of the *wave coherency matrix*[11]

[10] See Sect. B.3 in Appendix B

[11] See M. Born and E. Wolf, *loc cit*, D. Massonnet and J-C Souyris, *Imaging with Synthetic Aperture Radar*, Taylor and Francis, Roca Baton, Florida, 2008 and H. Mott, *Remote Sensing with Polarimetric Radar*, IEEE Press John Wiley and Sons, Hoboken, N.J., 2007.

$$\mathbf{J}=\begin{bmatrix} E_H E_H^* & E_H E_V^* \\ E_V E_H^* & E_V E_V^* \end{bmatrix}.$$

To use the coherency vector in (2.31) requires the matrix **R** in (2.32) to be re-expressed

$$\mathbf{R}_c=\begin{bmatrix} 1 & 0 & 0 & 1 \\ 1 & 0 & 0 & -1 \\ 0 & 1 & 1 & 0 \\ 0 & j & -j & 0 \end{bmatrix}$$

The *modified Stokes vector* $\mathbf{s}_m$ uses the intensities of the two orthogonal power components as its first two elements rather than the total power and the difference of the orthogonal power components; the second and third component remain the same:

$$\mathbf{s}_m=\begin{bmatrix} s_0 \\ s_1 \\ s_2 \\ s_3 \end{bmatrix}=\begin{bmatrix} a_H^2 \\ a_V^2 \\ 2a_H a_V \cos\delta \\ 2a_H a_V \sin\delta \end{bmatrix}=\begin{bmatrix} |E_H|^2 \\ |E_V|^2 \\ 2\mathcal{R}e\left(E_H^* E_V\right) \\ 2Im\left(E_H^* E_V\right) \end{bmatrix}=I_0\begin{bmatrix} 1+\cos 2\tau\cos 2\varepsilon \\ 1-\cos 2\tau\cos 2\varepsilon \\ \sin 2\tau\cos 2\varepsilon \\ \sin 2\tau \end{bmatrix} \tag{2.34}$$

2.14 Unpolarised and Partially Polarised Radiation

Equations (2.30) and (2.34) imply that the wave is of a single frequency and has a well defined phase difference between its components. That is often not the case in nature. Generally we can assume that the radiation sources used for illuminating the landscape in radar remote sensing are sufficiently pure to be regarded as polarised in the manner treated in the previous sections. However the radiation about us is often unpolarised, or only partially so, such as from an incandescent room light. That can easily be checked using a pair of polarising sunglasses. Rotating the sunglasses shows no variation in transmitted light intensity; the intensity would vary if the light were polarised. Similarly the sunlight that irradiates the landscape is largely unpolarised. The fact that polarising sunglasses will show the light reflecting from roadways and other landscape features as partially polarised has to do with how those features differentially reflect light of different polarisations rather than anything to do with the polarisation of the sunlight.

Radar energy backscattered from the landscape will often be polarised, but if it comes from random scattering media or time-varying scatterers, such as the surface of the ocean, that will not be the case.

If we were to observe, in two orthogonal transverse axes, an approaching wave that is totally unpolarised we would see the two amplitudes fluctuating randomly and without any relationship between them – in other words the amplitude variations would be uncorrelated. In addition, the relative phase between the components would be totally random – further reinforcing the lack of correlation. The wave however still carries energy (power density) so it is worth exploring the Stokes parameters in such a situation. The first parameter, as we have seen, is related to the power density, or intensity, of the wave since it is the sum of the two amplitudes squared. Given that the amplitudes are fluctuating in this unpolarised case we need to look at their averages over time. We write the averages as

$$\langle a_H^2 \rangle \quad and \quad \langle a_V^2 \rangle$$

where the angular brackets are the symbol used for time averaging. The squares of the amplitudes are employed because we are interested in power related quantities. If we took the averages of the amplitudes of the fields themselves then they most likely would be zero if the fields were randomly varying with time.

If the two orthogonal components – in say our H and V directions described above – are totally random, such that there is no preferred polarisation then their average squared amplitudes will be the same. Therefore the first Stokes parameter of (2.28a) is

$$s_0 = 2\langle a_H^2 \rangle$$

while the second Stokes parameter of (2.28b) will be zero. Similarly, since the relative phase is random, one can reason that the third and fourth Stokes parameters (2.28c and d) are meaningless (the time average of the trigonometric function of a randomly varying angle will be zero). Thus $s_2=s_3=0$. Therefore, the Stokes vector for unpolarised radiation is

$$\mathbf{s}^{unpol} = \begin{bmatrix} 2\langle a_H^2 \rangle \\ 0 \\ 0 \\ 0 \end{bmatrix} \tag{2.35}$$

Radiation can also be partially polarised resulting from the addition of unpolarised and polarised components. In that case the Stokes vectors can be added. No weighting coefficients or mixing parameters are needed in the sum since the field amplitudes themselves take account of the mix of polarised and unpolarised fields. Thus if we have

$$\mathbf{s}^{pol} = \begin{bmatrix} s_0^{pol} \\ s_1 \\ s_2 \\ s_3 \end{bmatrix} \text{ and } \mathbf{s}^{unpol} = \begin{bmatrix} s_0^{unpol} \\ 0 \\ 0 \\ 0 \end{bmatrix} \text{ then } \mathbf{s}^{total} = \begin{bmatrix} s_0^{pol} + s_0^{unpol} \\ s_1 \\ s_2 \\ s_3 \end{bmatrix} = \begin{bmatrix} s_0^{total} \\ s_1 \\ s_2 \\ s_3 \end{bmatrix} \tag{2.36}$$

We can use this last expression to define the degree of polarisation $\mathcal{P}$ of a wave as the power density carried by the polarised part of the field as a proportion of the total power density of the combination:

$$\mathcal{P} = \frac{\text{power density of the polarised part}}{\text{total power density}} = \frac{s_0^{pol}}{s_0^{pol} + s_0^{unpol}} = \frac{\sqrt{s_1^2 + s_2^2 + s_3^2}}{s_0^{total}} \tag{2.37}$$

2.15 The Poincaré Sphere

A very interesting geometric summary of the Stokes parameters and the state of polarisation of a wave emerges from the realisation that (2.29) is the equation of a sphere in the s_1, s_2, s_3 coordinate space. This is shown in Fig. 2.21. Named after Poincaré, who first described it in 1892, the sphere has a radius of s_0 and its surface is the locus of all possible polarisation states. Given that the polarisation of a wave can be described by the amplitudes of its two orthogonal components a_H and a_V, and their relative phase δ, or alternatively by the angles of the polarisation ellipse ε and τ and the wave intensity s_o, it is reasonable to expect that the geometry of the sphere, and in particular the coordinates of a particular point on its surface, should be expressible in sets of those parameters. Fig. 2.22 shows the sphere with that information added, as derived in the following.

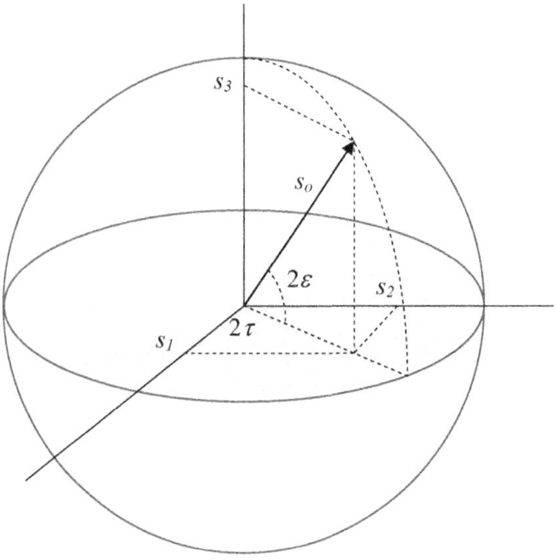

Fig. 2.21. The Stokes parameters plotted spherically

It can be shown that[12]

$$\sin 2\varepsilon = \frac{2a_H a_V}{a_H^2 + a_V^2}\sin\delta = \frac{s_3}{s_0} \quad \text{thus } s_3 = s_0 \sin 2\varepsilon \tag{2.38}$$

$$\tan 2\tau = \frac{2a_H a_V}{a_H^2 - a_V^2}\cos\delta = \frac{s_2}{s_1} \quad \text{thus } s_2 = s_1 \tan 2\tau \tag{2.39}$$

Substituting these relationships into (2.29) gives

$$s_0^2 = s_1^2 + s_1^2 \tan^2 2\tau + s_0^2 \sin^2 2\varepsilon$$

---

[12] H. Mott, *loc cit.*

so that we have $s_0^2(1-\sin^2 2\varepsilon) = s_1^2(1+\tan^2 2\tau)$ i.e. $s_0^2\cos^2 2\varepsilon = \dfrac{s_1^2}{\cos^2 2\tau}$

Thus $$s_1 = s_0\cos 2\varepsilon\cos 2\tau \tag{2.40a}$$

giving, from (2.39), $$s_2 = s_0\cos 2\varepsilon\sin 2\tau \tag{2.40b}$$

which are the forms given in the last column of (2.30). With (2.38), (2.39) and (2.40) we now have equations that describe the coordinates of a point on the Poincaré sphere in terms of the angles of the polarisation ellipse and the wave intensity s_0 as seen in Fig. 2.22. Also shown are points on the ellipse corresponding to particular wave polarisations.

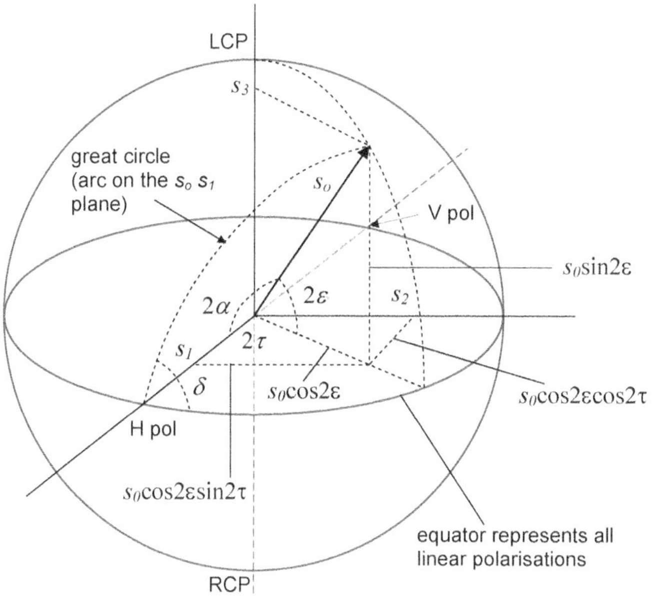

Fig. 2.22. The Poincaré sphere, showing the relationship between the Stokes parameters, the angles of the polarisation ellipse and the parameters of the horizontal and vertical components of the electric field.

Two further angles are marked on the sphere: the first is the angle δ between the equator and the arc joining the tip of the intensity vector to the s_1 axis, which is in fact the phase difference between the wave's horizontal and vertical components in (2.11). The second is the angle 2α, which is between the intensity vector and the s_1 axis – shown also as the angle subtended by the great circle. As we saw in the discussion on the polarisation ellipse in Sect. 2.10, this angle is defined in (2.13). With these two angles we have a description of the point on the Poincaré sphere in terms of the amplitudes and relative phase angle of the original components of the wave, as well as the earlier description in terms of the angles of the polarisation ellipse.

From (2.36) we can see for partially polarised radiation that

$$s_1^2 + s_2^2 + s_3^2 < s_0^2$$

which lies *within* the Poincaré sphere rather than on its surface. Therefore partially polarised waves are described by points inside the sphere; the origin represents the case of fully unpolarised radiation.

2.16 Transmitting and Receiving Polarised Radiation

Electromagnetic waves are launched into free space using an antenna; likewise they are received using an antenna. In the case of radar the same antenna is very often used for both purposes, as we will see in Chapt. 3. The polarisation state of the wave launched is determined by the properties of the transmitting antenna, particularly its orientation around the line of sight of the ray. By appropriately orienting the antenna we can launch vertically polarised or horizontally polarised radiation, or any linear polarisation in between. Some special antennas are designed to launch circularly (or elliptically) polarised radiation.

The polarisation state of the antenna used to receive radiation needs to match that of the wave itself if the received signal is to be maximised. As illustrated in Fig. 2.23, a vertically oriented, or polarised, antenna will receive maximum signal from a vertically polarised wave, but nothing if the wave is polarised horizontally. For any other polarisation of the wave some signal will be received – proportional to the vertical projection of the wave onto the antenna. A convenient means by which to describe such a projection, and which is used extensively in radar, is to employ the scalar or dot product of two vectors, one of which describes the polarisation state of the receiving antenna, with the other describing the polarisation of the radiation incident on that antenna. This is developed in the following for an arbitrarily oriented antenna and wave polarisation.

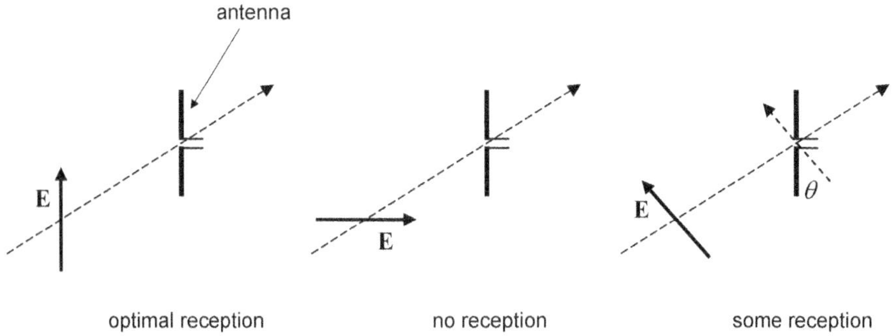

Fig. 2.23. Illustrating how the relative alignment of the field polarisation and the antenna affects reception; the antenna shown here is called a vertical dipole

The electric field vector incident on a receiving antenna can be represented as shown in Fig. 2.15. In that particular case we have chosen two components of the electric field vector that are horizontal and vertical. We could just as easily have described the electric field in terms of a unit vector that actually points in the direction of the field itself so that we could write

$$\mathbf{E} = E\mathbf{p} \tag{2.41}$$

in which E is the magnitude of the field and $\mathbf{p}$ is that unit vector.

Now define a new unit vector $\mathbf{p}^{ra}$ that aligns with the polarisation of the receiving antenna. If ϕ is the angle between the polarisations of the field and antenna then the projection of the field onto the antenna – i.e. the component of field aligned with the antenna – is $E\cos\phi$. This is classically derived from the *dot* or *scalar* product of the vectors which, for two vectors **A** and **B** at an angle of ϕ to each other, is defined as

$$\mathbf{A}.\mathbf{B} = AB\cos\phi \tag{2.42}$$

This is shown in Fig. 2.24, in which the projection of one vector on the other is evident. Note in passing that **A.B**=**B.A**. The component of the incoming electric field that will be detected (received) by the antenna is given by

$$E_r = \mathbf{E}.\mathbf{p}^{ra} = E\mathbf{p}.\mathbf{p}^{ra} \tag{2.43a}$$
$$= E|\mathbf{p}||\mathbf{p}^{ra}|\cos\phi = E\cos\phi \tag{2.43b}$$

since the two polarisation vectors have unit magnitudes.

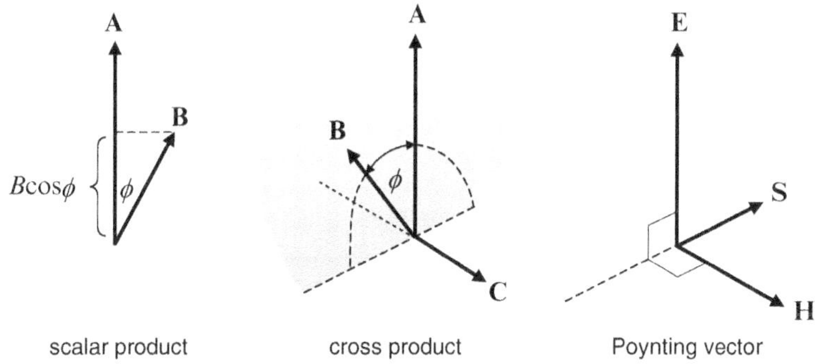

Fig. 2.24. The scalar and cross products, and the Poynting vector shown diagrammatically

We can write the dot product of (2.42) in a different way that will be helpful when we look at polarisation synthesis in Chapt. 3. Write **A** and **B** in terms of their horizontal and vertical components[13]:

$$\mathbf{A} = a_H\boldsymbol{h} + a_V\boldsymbol{v}$$
$$\mathbf{B} = b_H\boldsymbol{h} + b_V\boldsymbol{v}$$

Since $\boldsymbol{h}$ and $\boldsymbol{v}$ are at right angles to each other and have unit magnitudes (2.42) shows

[13] Any pair of orthogonal directions could be chosen for this illustration; we have used those commonly employed when describing electromagnetic radiation, particularly in radar.

$$\boldsymbol{h}.\boldsymbol{h} = 1,\ \boldsymbol{v}.\boldsymbol{v} = 1 \text{ and } \boldsymbol{h}.\boldsymbol{v} = \boldsymbol{v}.\boldsymbol{h} = 0$$

Now look at the dot product of **A** and **B**:

$$\begin{aligned}\mathbf{A}.\mathbf{B} &= (a_H\boldsymbol{h} + a_V\boldsymbol{v}).(b_H\boldsymbol{h} + b_V\boldsymbol{v}) \\ &= a_H b_H \boldsymbol{h}.\boldsymbol{h} + a_H b_V \boldsymbol{h}.\boldsymbol{v} + a_V b_H \boldsymbol{v}.\boldsymbol{h} + a_V b_V \boldsymbol{v}.\boldsymbol{v} \\ &= a_H b_H + a_V b_V\end{aligned}$$

Thus the dot product can also be expressed as the sum of the products term by term of the components of the relevant vectors. If we write the vectors in column form:

$$\mathbf{A} = \begin{bmatrix} a_H \\ a_V \end{bmatrix} \text{ and } \mathbf{B} = \begin{bmatrix} b_H \\ b_V \end{bmatrix}$$

then we see that either of $\mathbf{A}^T\mathbf{B}$ or $\mathbf{B}^T\mathbf{A}$ yields the same result[14]. We therefore have an alternative expression for the dot product that we will use in matrix-vector calculations:

$$\mathbf{A}.\mathbf{B} = \mathbf{A}^T\mathbf{B} = \mathbf{B}^T\mathbf{A}$$

Note the particular case of the magnitude of a vector:

$$|\mathbf{A}|^2 = \mathbf{A}.\mathbf{A} = \mathbf{A}^T\mathbf{A}$$

There is another product of two vectors that is important in electromagnetic wave propagation but which we will use infrequently here. It is called the *cross* or *vector* product and is defined by

$$\mathbf{C} = \mathbf{A}\text{x}\mathbf{B} = AB\sin\theta\mathbf{c} \tag{2.44}$$

The result is a vector; **c** is a unit vector that describes its orientation in space. It points at right angles to both **A** and **B** as illustrated in Fig. 2.24. The specific direction of **C** is that given by a right handed screwdriver when being turned in the direction from **A** to **B**. In this case the order of the vectors in the formula is important.

A celebrated application of the cross product in electromagnetism is in the definition of the *Poynting vector*:

$$\mathbf{S} = \mathbf{E}\text{x}\mathbf{H} \quad \text{Wm}^{-2} \tag{2.45}$$

The Poynting vector is at right angles to both the electric and magnetic field vectors and has units equivalent to power density. It also points in the direction of propagation! Since the electric and magnetic fields are at right angles to each other, the magnitude of the Poynting vector using (2.44) is the product of the magnitudes of the electric and magnetic field strengths. Thus the magnitude of the Poynting vector is exactly that of power density in (2.7a). The benefit of (2.45) is that the propagation direction is defined explicitly by the cross product definition.

[14] See Appendix B for a summary of vectors and matrices.

Return now to looking at the field received by an antenna. In Fig. 2.23 we considered the case of linear polarisation. Forming the scalar product of the antenna polarisation vector and the polarisation vector of the incoming wave, as in (2.43a), is an operation that applies in general, including for the case of elliptical polarisation. For example an antenna designed to radiate right circularly polarised radiation can, from (2.24b), be described by the polarisation vector

$$\mathbf{p}^{ra} = \frac{1}{\sqrt{2}}(\boldsymbol{h} - j\boldsymbol{v})$$

A right circularly polarised wave launched by the antenna will be written

$$E\mathbf{p} = \frac{1}{\sqrt{2}}(\boldsymbol{h} - j\boldsymbol{v})$$

There is an important subtlety though with describing a right circularly polarised wave propagating towards the antenna: to make the wave appear as right polarised when propagating in the negative r direction it has to rotate in the opposite sense than when travelling forward[15]. That means the vertical component needs to lead the horizontal so that the field incident on the antenna must be written

$$E\mathbf{p} = \frac{1}{\sqrt{2}}(\boldsymbol{h} + j\boldsymbol{v})$$

giving as the received field

$$E\mathbf{p}.\mathbf{p}^{ra} = \frac{E}{2}(\boldsymbol{h}.\boldsymbol{h} + \boldsymbol{v}.\boldsymbol{v}) = \frac{E}{2}(1+1) = E\,.$$

In contrast, if the returning field were left circularly polarised then the field received on the right circularly polarised antenna is

$$E\mathbf{p}.\mathbf{p}^{ra} = \frac{E}{2}(\boldsymbol{h}.\boldsymbol{h} - 2j\boldsymbol{h}.\boldsymbol{v} - \boldsymbol{v}.\boldsymbol{v}) = \frac{E}{2}(1+0-1) = 0$$

With elliptical polarisation it is often convenient to describe the wave by its Stokes vector, since that fully specifies the polarisation. We can also define a Stokes vector for the antenna which is the polarisation configuration it is optimised to receive (or, alternatively, the polarisation state of a wave it would generate if it were used as a transmitting antenna). We will use this property explicitly in Chapt. 3 in the topic on polarisation synthesis.

If the same antenna were used for transmission and reception, as in radar, how can the polarisation state of the received field be different from that of the antenna that launched it? That can occur when the transmitted field scatters from earth surface features; the field scattered back to the antenna will have a polarisation state that is related to the properties of the scattering medium.

---

[15] This is a good example of how careful we need to be with coordinate conventions when dealing with multi-polarisation radar. In general, if we reverse the propagation direction we need to conjugate the vertically polarised component in elliptical polarisation: See D. Massonnet and J-C Souyris, *Imaging with Synthetic Aperture Radar*, Taylor and Francis, Roca Baton, Florida, 2008.

2.17 Interference

We now come to a further property of electromagnetic radiation of major importance in radar studies. If two sinusoidal signals of the same frequency are received simultaneously they can interfere with each other; the result is dependent on the time phase angle between the sinusoids.

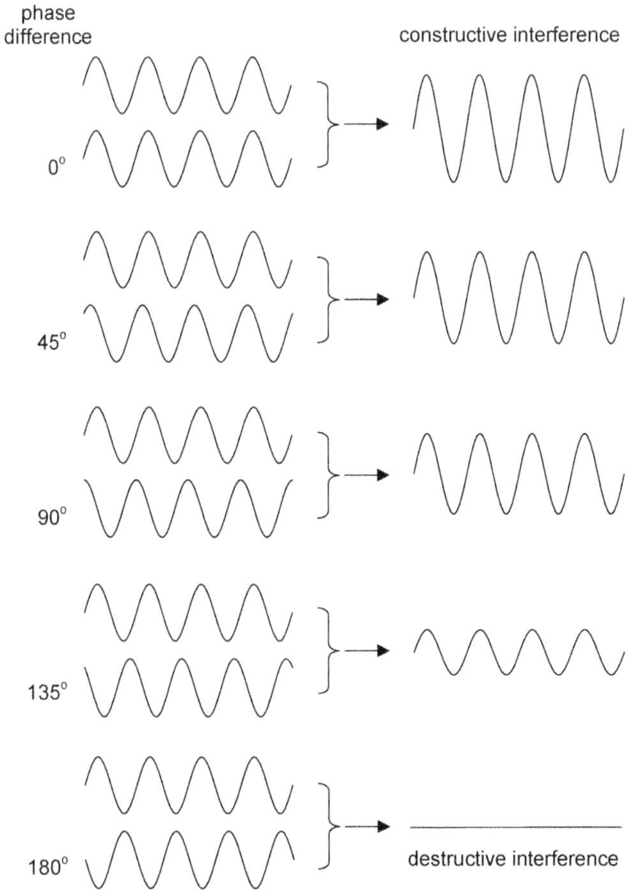

Fig. 2.25 Demonstration of the interference of two sinusoids with varying phase differences

Fig. 2.25 shows examples of two sinusoidal signals adding as the phase difference between them is altered. As seen, if they are in phase (i.e. there is no mutual phase difference) the resultant is a straight addition of the signals. They reinforce each other in what is known as *constructive interference*. If they are totally out of phase (i.e. there is a phase difference of 180° between them) then they cancel in a process called *destructive interference*. For other phase differences there will be neither full reinforcement nor full cancellation.

The results of Fig. 2.25 are easy to demonstrate mathematically. If the two sinusoidal signals are $\cos\omega t$ and $\cos(\omega t+\phi)$, where ϕ is the phase difference, then their sum is

$$\cos\omega t+\cos(\omega t+\phi)=2\cos\left[\omega t+\frac{\phi}{2}\right]\cos\frac{\phi}{2}$$

If $\phi=0$ then the result is the simple sum and if $\phi=180^{\circ}$ then the result is total cancellation. We will meet interference frequently with radar, sometimes with travelling waves.

Signals with different frequencies can also interfere leading to a process called *beating*, provided the frequencies are not too different. That is also easily demonstrated if we have two signals $\cos\omega t$ and $\cos(\omega+\alpha)t$ with α small; here we have not added any phase difference. Adding the signals gives

$$\begin{aligned}\cos\omega t+\cos(\omega t+\alpha)&=2\cos(\omega t+\frac{\alpha}{2})\cos\frac{\alpha}{2}\\&\approx 2\cos\omega t\cos\frac{\alpha}{2}\end{aligned}$$

which is shown plotted in Fig. 2.26. As seen, the result of interference between two sinusoids of slightly different frequencies is a sinusoid at the major frequency multiplied in amplitude by a slower sinusoid of half the frequency difference.

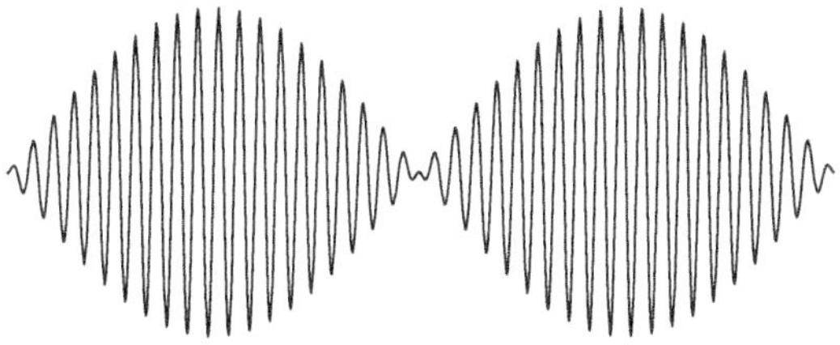

Fig. 2.26. Beating caused by the addition of two sinusoids with a 5% difference in frequency

2.18 The Doppler Effect

The Doppler effect, in which the frequency of a sinusoid is affected by the relative velocity of the generator and receiver of the signal, is central to synthetic aperture radar. We will meet it most often in the signal scattered from the landscape or reflected from a discrete object. Here we will develop it in the more classical situation illustrated in Fig. 2.27 of a moving transmitter approaching and passing a stationary receiver, first with the receiver being in line with the transmitter's velocity vector. Define the (time varying) distance between the receiver and transmitter as $x(t)$, measured in the positive direction to the right of the receiver as drawn. Time t is defined to be zero when the transmitter is at the position of the receiver and positive to the right. At a given distance between the transmitter and receiver of X_o the corresponding time is $t=\frac{X_0}{v}$ in which v is the velocity of the transmitter towards the receiver. That could be the distance when the signal is just noticed by the receiver. At the general time t, $x(t)=vt$.

Suppose the signal being radiated is sinusoidal of the form $\cos\omega t$. The signal arrives at the transmitter after a time delay of $t_D = x(t)/c$ where c is the velocity of light. Thus the received sinusoid can be written

$$\cos\omega(t+t_D) = \cos[\omega t + 2\pi x(t)/\lambda] \text{ since } \omega/c = 2\pi f/c = 2\pi/\lambda$$

Substituting for $x(t)$ we see that the received signal is

$$\cos(\omega t + 2\pi vt/\lambda) = \cos(\omega + 2\pi v/\lambda)t \tag{2.46}$$

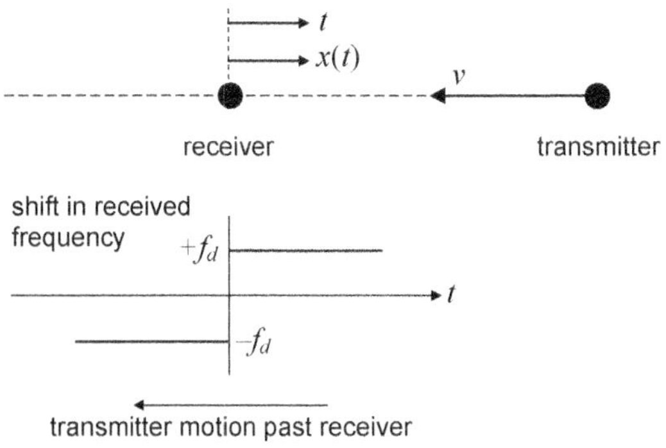

Fig. 2.27. Transmitter moving towards and past a receiver, and the Doppler change in received frequency

The frequency of the received signal is the coefficient of t in this last expression, consisting of the transmitted frequency adjusted by the so-called Doppler component:

$$\omega_r = \omega + \omega_d \text{ with } \omega_d = 2\pi v/\lambda \tag{2.47a}$$

We would generally write this in the normal form with frequency expressed in Hz

$$f_r = f + f_d \text{ with } f_d = v/\lambda. \tag{2.47b}$$

Thus the frequency of the received sinusoid is up-shifted because of the velocity of the transmitter towards the receiver. Once the transmitter passes the receiver the sign of the Doppler shift reverses because the sign of $x(t)$ reverses for t negative.

For this example the change in Doppler frequency as the transmitter passes the receiver is instantaneous as indicated in Fig. 2.27. In practice the transmitter is more likely to pass by the receiver at a distance Y_o as illustrated in Fig. 2.28. The change in frequency is then more gradual as the following analysis demonstrates. This is more complicated but gives a result that is closer to the situation we will encounter with radar in the next chapter.

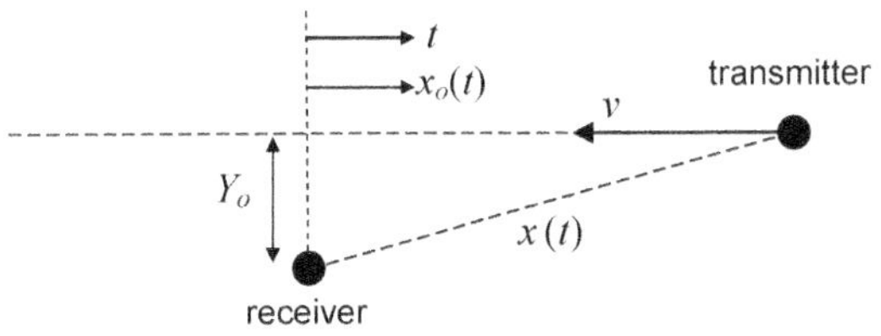

Fig. 2.28. Transmitter moving past a receiver separated by a small distance at broadside

The distance between the transmitter and receiver at any given time t is

$$x(t) = \sqrt{Y_o^2 + (vt)^2} \tag{2.48}$$

In this case we cannot easily handle the situation as we did the simpler case of Fig. 2.27. Instead we note that the received sinusoid arrives with a phase delay $\phi(t)$ given by 2π times the distance expressed as a fraction of a wavelength:

$$\phi(t) = 2\pi \frac{x(t)}{\lambda} \tag{2.49}$$

In the simpler case of (2.46), the phase delay was $2\pi \frac{vt}{\lambda}$. Comparing this with (2.46) and (2.47a) we can induce that the Doppler frequency component is the time derivative of the phase delay. Applying that principle to (2.49) shows that the Doppler frequency shift for the more general case is

$$\omega_d = \frac{d\phi(t)}{dt} = \frac{2\pi}{\lambda}\frac{dx(t)}{dt}$$

so that from (2.48)

$$f_d = \frac{1}{\lambda}\frac{dx(t)}{dt} = \frac{v^2}{\lambda}\left[Y_o^2 + (vt)^2\right]^{-0.5} t \tag{2.50}$$

This is plotted in Fig. 2.29 for a transmitter platform flying at 1000kms^{-1} radiating at 300MHz, with receiver offsets of 0m, 100m, 200m and 300m.

Some interesting special cases can now be considered. First, if Y_o=0 (2.50) reduces to (2.47b). Secondly, suppose $vt \ll Y_o$ – i.e. the transmitter is close to passing the receiver. In this case it is better to return to the expression for $x(t)$ in (2.48) and note that it can be approximated

$$x(t) = Y_o + \frac{v^2}{2Y_o}t^2$$

so that the Doppler frequency shift is

$$f_d = \frac{1}{\lambda}\frac{dx(t)}{dt} = \frac{v^2}{\lambda Y_o}t$$

This tells us that the received frequency varies linearly about the transmitted frequency when the transmitter is passing close by the receiver (i.e. near broadside). That can be seen in the plots of Fig. 2.29.

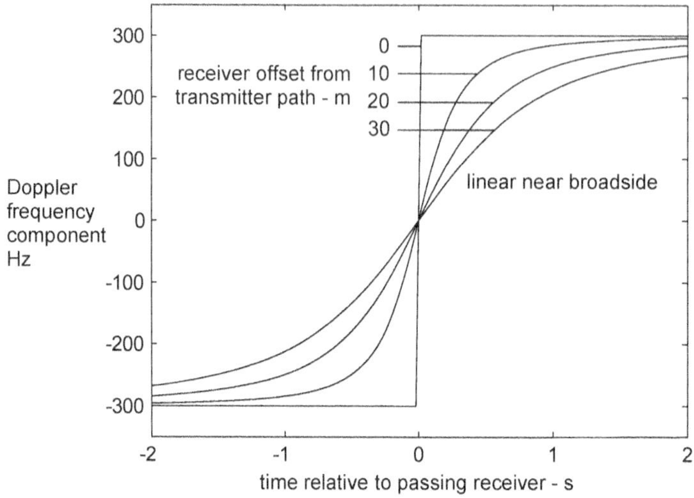

Fig. 2.29. Doppler frequencies for a range of receiver offsets, as the transmitter passes from right to left; time is measured to the right

CHAPTER 3

THE TECHNOLOGY OF RADAR IMAGING

This chapter provides the technical basis for imaging radar. It is broken into two parts, the first of which develops the system itself, showing how spatial resolution can be obtained and how an image is formed. The second section focuses on the target – which can be a discrete entity such as a house or a calibration device, or a distributed medium such as a vegetated pixel. Means by which the target can be described mathematically are derived by building on the radiation framework given in Chapt. 2.

PART A: THE SYSTEM

As shown in Chapt. 2, the levels of naturally occurring microwave energy are almost negligible. Although they can be measured, they are small enough to permit an artificial source of irradiation to be used. The essential radar remote sensing instrument consists therefore of both a transmitter and a receiver of energy at the wavelength of interest. Such an arrangement is called *active*, in contrast to *passive* remote sensing instruments which use the sun or the earth itself as a primary source of energy.

3.1 Radar as a Remote Sensing Technology

The transmitter and receiver in a remote sensing microwave imaging system can be located separately; that configuration is called *bistatic* and is depicted in Fig. 3.1a. Alternatively, the transmitter and receiver can be co-located, often sharing the same antenna to radiate and receive the energy scattered back to the platform[1]. That is referred to as *monostatic* (Fig 3.1b) and to date has been the most commonly encountered arrangement in remote sensing. As might be expected the characteristics of the earth's surface can appear quite differently in monostatic and bistatic systems. We will concentrate on the monostatic configuration in this chapter. Bistatic radar is treated in Chapt. 7.

Consider now how a monostatic active microwave system might be used to form an image of a region on the ground. Conceptually, the easiest approach would be to use a very narrow beam of microwave energy and scan it across the earth's surface normal to the motion of the aircraft or spacecraft, just as optical multispectral scanners acquire data in strips orthogonal to the motion of the platform. Unless the wavelength is very small, and the required resolution coarse, this turns out to be an impractical approach. Instead, active microwave remote sensing is based on the principles of radar in order to achieve practical spatial resolutions.

---

[1] Antennas are reciprocal devices in that the same antenna can be used to transmit or to receive radiation.

J.A. Richards, *Remote Sensing with Imaging Radar*, Signals and Communication Technology,
DOI: 10.1007/978-3-642-02020-9_3,

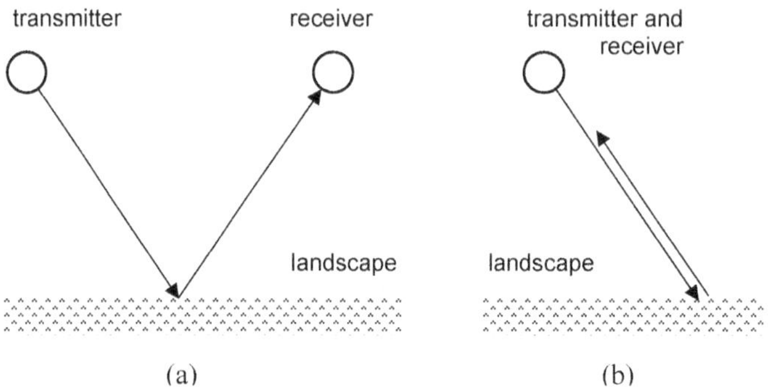

Fig. 3.1. (a) Bistatic and (b) monostatic microwave imaging systems

Consider a transmitter of energy mounted on an aircraft or spacecraft. Using a suitable antenna the energy is radiated to the side of the platform during flight, as shown in Fig. 3.2; the reason for sidewards radiation will become clear shortly. The properties of the antenna are chosen so that the energy radiates over quite a broad beam to the side as shown. This will be seen later to define the swath width of the image data recorded. In contrast, the antenna beam in the direction parallel to the platform motion is usually narrow, as indicated. This is related to the resolution of the system in the direction along the flight path. When the radiated energy reaches the ground some of it will be scattered back towards the platform where it is received and measured. This measured level of scattered energy is related to the properties of the region of earth being irradiated and is the focus of Part B following and of Chapt. 5.

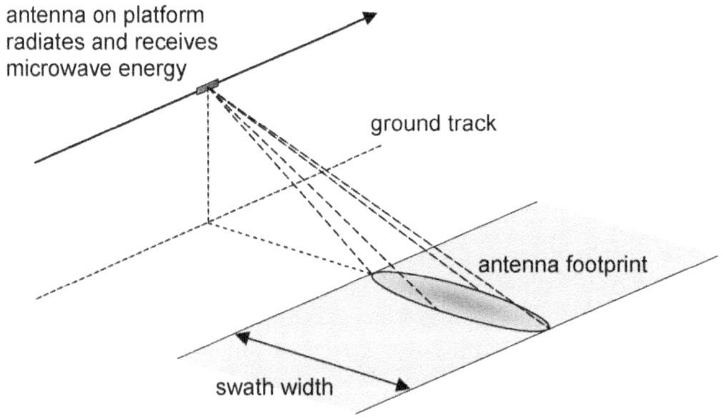

Fig. 3.2. Radar imaging geometry

Since the antenna is fixed and its pattern broad there is no spatial discrimination across the swath. To provide resolution in this dimension, referred to as the *range direction*, the classic principle of radar is employed. That involves transmitting the energy in the form of short bursts or pulses of radiation at the operating frequency of the radar, rather than transmitting continuously. The pulses, of the form shown simplistically in Fig. 3.3a, travel to the ground at the velocity of light c of 300Mms^{-1}; they then scatter and return to the platform at the same speed. Thus, portions of ground closer to the platform (in the *near* swath) give rise to echoes – returned pulses – that arrive at the receiver earlier than those that scatter from further out in the swath (the *far* swath). That is illustrated in Fig. 3.4 which shows just the envelopes (outline shapes) of the pulses rather than their internal details. By using a pulse, the received signals are separable in time in such a way as to allow the strip of terrain to be resolved spatially across the swath.

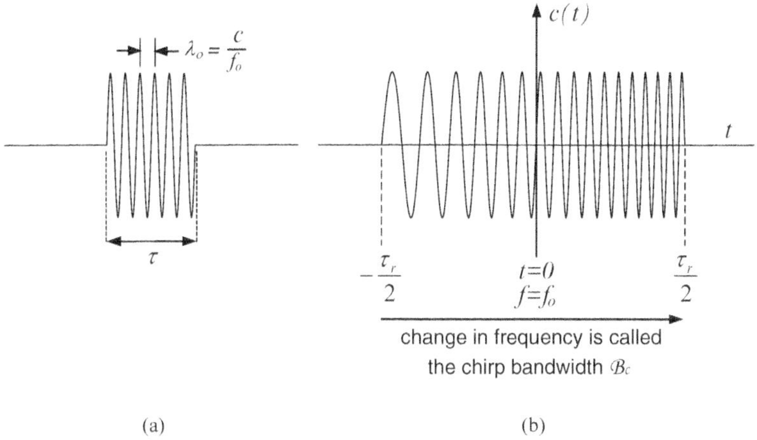

Fig. 3.3. (a) simple and (b) chirped pulses for use in radar systems

The transmitted pulses, commonly called ranging pulses, are repeated at a rate called the *pulse repetition frequency* (*prf*). This is synchronised with the forward velocity of the platform so that contiguous strips of terrain across the swath are irradiated pulse by pulse as shown in Fig. 3.5.

There is an important constraint on *prf*. If all the echoes from a particular transmitted pulse are not received before the next pulse is transmitted then a range ambiguity occurs. We would not know whether some echoes are from targets close in, or are reflections from targets much further away as a result of the previous pulse (such as might happen if the next transmitted pulse in Fig. 3.4 occurred in time between reflections B and C). Effectively, therefore, the highest usable *prf* is bound by the largest *slant range* of the system; slant range is measured along the direct line from the radar to a point on the ground, rather than along the ground. The maximum usable *prf* is considered in Sect. 3.7.

3.2 Range Resolution

Consider two closely-spaced spots on the ground that give rise to radar echoes as shown in Fig. 3.6. The ability to separate those targets in the signal received by the radar is determined by whether the returning pulses are distinguishable. If the targets are too close their echoes will overlap and they cannot be separated in the received signal.

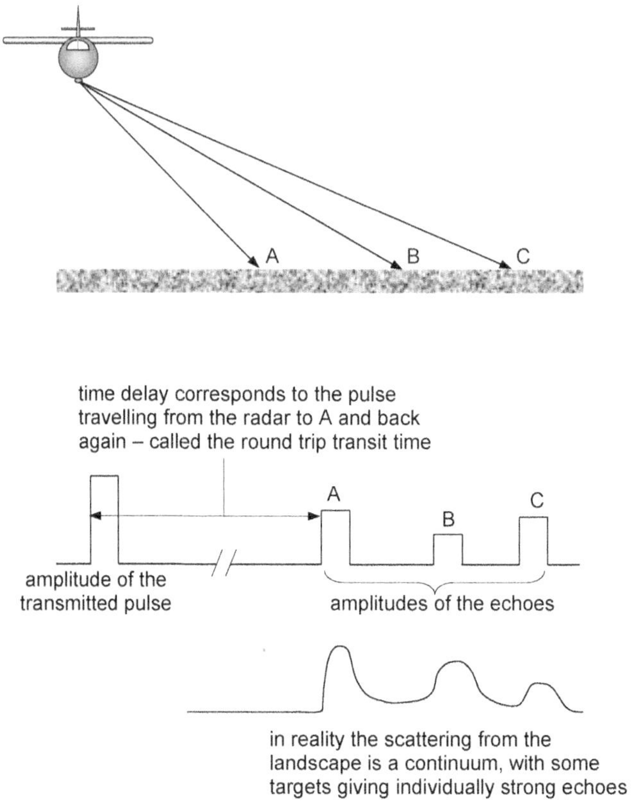

Fig. 3.4. Resolution of targets spatially by time resolution of the received echoes; the pulses shown correspond to the envelopes of the type of pulse shown in Fig. 3.3a

If the targets are Δr apart in slant range as depicted in Fig. 3.6 then the difference in time between their echoes on reception will be $\Delta t = 2\Delta r c^{-1}$ since the pulses travel to and from the ground. We are unable to resolve in time better than the width τ of the pulses so that the lower limit on Δt is τ; the corresponding limit of spatial resolution Δr in the slant range direction is

$$r_r = \frac{c\tau}{2} \quad \text{m} \tag{3.1}$$

which is called *slant range resolution.*

As users of radar we are more interested in how well we can resolve targets along the ground, rather than in the slant direction. If we assume that the angle at the ground with which the beam of radiation is incident locally is θ, called the *incidence angle*, then the spatial resolution in what is commonly called the (ground) range direction is

$$r_g = \frac{c\tau}{2\sin\theta} \quad \text{m} \tag{3.2}$$

This is termed *ground range resolution*.

The angle at the platform measured with respect to the vertical (nadir) is called the *look angle* and is a system design parameter. The incidence angle at the ground will be the same as the look angle if the surface is horizontal and earth curvature can be ignored, normally the case for aircraft altitudes. For spacecraft platforms earth curvature normally makes the look and incidence angles different from each other by a few degrees.

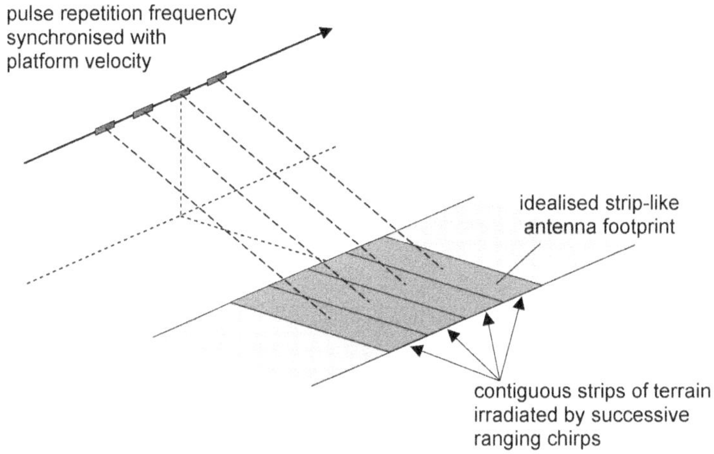

Fig. 3.5. Transmitting successive ranging pulses synchronously with the platform velocity so that contiguous strips of terrain are irradiated

Several important implications can be drawn from (3.1) and (3.2).

1. There is no spatial resolution if $\theta = 0$ – directly under the platform. That explains why the system has to be side looking. Aircraft imaging radars of this type have often been called *side looking airborne radars* (SLAR).
2. The slant and ground range resolutions are independent of the altitude of the platform.
3. Ground range resolution is a function of incidence angle, so that it will vary across the swath shown in Fig. 3.2. It is best in the far swath where θ is largest and poorest in the near swath where θ is smallest. That is opposite to the effect experienced with optical sensors which have their best resolution closest to the platform, just as we can see more detail in the near range when we look out the window of an aircraft.
4. If the antenna radiated to both sides of the aircraft and a single receiver were used there would be a right-left ambiguity in the received signal. That can be circumvented using two antennas and receivers, but most often that is not the case. In practice imaging radars usually radiate only to one side of the platform.

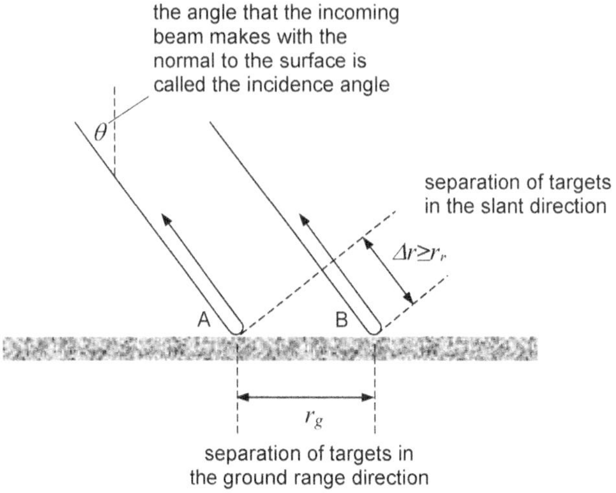

Fig. 3.6. Geometry for computing range resolutions

3.3 Pulse Compression Radar

Consider a simple calculation involving (3.2). Suppose τ=10μs and θ=30°. Then the ground range resolution will be 3000m, which would be considered far too coarse for most remote sensing purposes[2]. Given that the incidence angle is fixed by the location of the platform and the position on the ground being imaged, we can often do little about θ. The only way to improve resolution therefore is to narrow the transmitted pulse. For example, if the pulse were 100ns in duration (i.e. "width") then the resolution becomes 30m, which is much better.

The problem with narrowing the pulse is that the energy it carries is reduced; that limits the sensitivity of the radar, making it harder to detect weaker targets. The energy carried by a pulse is proportional to the product of its duration and the square of its amplitude. If we were to narrow it in pursuit of higher spatial resolution we could, in principle, restore its energy by increasing its amplitude. There is a limit, however, set by the ability of the transmitting circuits to handle pulses of large amplitude without damaging their electronic components. Thus, continuing to narrow the pulse to obtain better spatial resolution is not the answer and a better solution has to be found.

The answer to this problem is amazingly simple; it also provides the groundwork for how we will achieve good resolution in the direction of platform motion, which we have not really mentioned yet. That involves transmitting a long pulse, as shown in Fig. 3.3b, but within which the frequency is swept in a linear fashion with time as indicated. Such a pulse is referred to as a *chirp*, which would be the sort of sound one could imagine if listening to an audio signal swept from a low to a higher frequency. Mathematically, we write the chirp waveform as

---

[2] Since the velocity of light is 300Mms^{-1} the pulses travel 300m in one direction in 1μs, which is a convenient figure to remember in radar applications. Sometimes radar engineers define the *two-way* radar range as 150m/μs. In imperial units it is sometimes helpful to know that the velocity of light is approximately 1 foot per nanosecond.

$$c(t) = p(t)\cos(\omega_o + 0.5at)t = p(t)\cos(\omega_0 t + 0.5at^2)$$
$$= p(t)\cos(2\pi f_o t + 0.5\alpha t^2) \qquad (3.3)$$

in which a and α are referred to as the *chirp rate*, in rad.s^{-1}s^{-1} and Hzs^{-1} respectively, and $p(t)$ is a unit amplitude pulse centred on t=0 that is zero outside the range $-\tau_r/2<t<\tau_r/2$. Note that at $t=0$, the instantaneous frequency of the chirp is ω_o, or f_o as appropriate.

The advantage of transmitting the chirp waveform is that, on reception, it can be compared against a replica of itself using the operation of correlation, the result of which is a compressed pulse with its centre located very precisely in time. That is depicted in Fig. 3.7, showing that the correlation outcome is a very narrow function, albeit with side lobes (bunches), centred on the origin of the chirp replica. That narrow pulse is effectively the ranging pulse used to achieve range resolution.

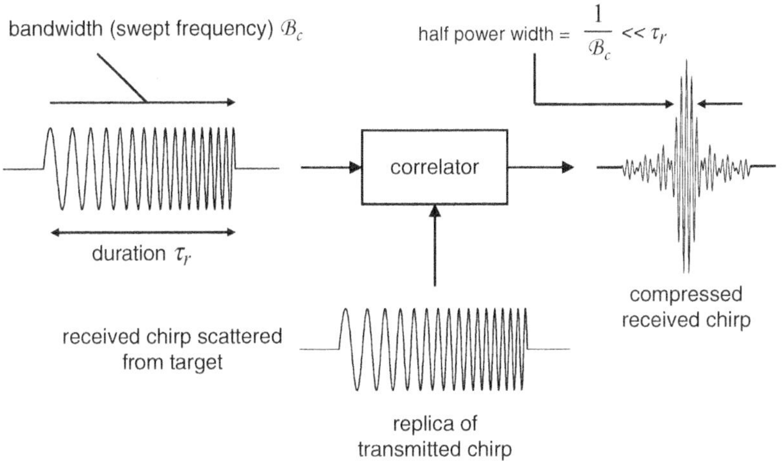

Fig. 3.7. Using the process of correlation to compressed the received chirp.

Assume the received chirp is identical to the replica. That does not happen in practice because of the scattering properties of the earth's surface and the addition of noise, but the assumption is helpful here to allow us to understand some important concepts. The result of the correlation process is, to a very good approximation, given by

$$z(t) = \cos(2\pi f_o t)\,\mathrm{sinc}(\pi \mathcal{B}_c t) \qquad (3.4)$$

The sinc function is the sine of its argument divided by the argument itself, i.e.

$$\mathrm{sinc}\, x = \frac{\sin x}{x}$$

Its half power width is equal to the reciprocal of the chirp bandwidth which is the range of frequencies over which the chirp sweeps and is thus given by $\mathcal{B}_c = \alpha\tau_r$. Substituting this compressed value for the pulse width into (3.1) and (3.2) shows that the slant and ground range resolutions in a pulse compression radar system are:

slant range resolution $$r_r = \frac{c}{2\mathcal{B}_c} \quad \text{m} \tag{3.5a}$$

ground range resolution $$r_g = \frac{c}{2\mathcal{B}_c \sin\theta} \quad \text{m} \tag{3.5b}$$

The side lobes of the sinc function seen in Fig. 3.7 can be a problem since they are large enough, in principle, to be mistaken for smaller targets. In practice, measures are taken to reduce the side lobes, as discussed in Appendix D. Fig. 3.8 shows how the correlation-based compression process allows closely spaced targets to be resolved.

(a)

(b)

(c)

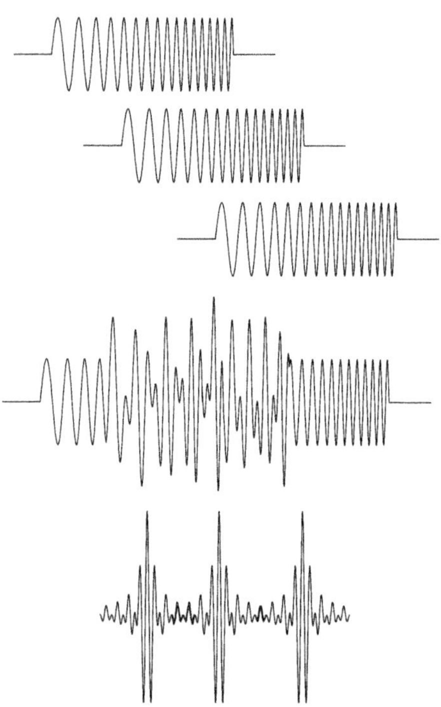

Fig. 3.8. (a) Overlapping echoes from three targets closely spaced in range, (b) the composite signal received by the radar and (c) the outcome of pulse compression (correlation) showing how the targets can be resolved (in practice a greater degree of compression than that illustrated here would be achieved)

The first free-flying satellite radar remote sensing mission was Seasat, launched in 1978. Its ranging chirps were 33.9μs in duration with a bandwidth of 19MHz. On compression following reception their equivalent widths were reduced to 53ns, the reciprocal of 19MHz. Thus the 33.9μs pulse was compressed by a factor of 640! With an incidence angle of 20°, that gives a ground range resolution of 23m (the actual value in practice was about 25m).

3.4 Resolution in the Along Track Direction

We now need to see how the radar provides spatial resolution in the direction parallel to the platform motion. That is referred to as the along-track or *azimuth* direction. The term azimuth is unusual here but commonly used. It is so chosen since it is the direction orthogonal to the range direction; in air traffic control radars azimuth motion is rotational about the radar axis, which makes the term more meaningful.

In Fig. 3.2 the available resolution in the along-track direction is set by the along-track beamwidth of the antenna. For an antenna of length l_a in the along-track direction, large compared with a wavelength, the angular beamwidth subtended by the antenna is given from antenna theory by

$$\Theta_a = \frac{\lambda}{l_a} \quad \text{rad} \tag{3.6}$$

Therefore the along-track dimension of the antenna footprint, which defines the azimuth resolution for this simple system, will be

$$r_a = \frac{\lambda}{l_a} R_o \quad m \tag{3.7}$$

where R_o is the slant distance from the platform to the ground at the point at which the azimuth resolution is being considered. Since this expression depends on R_o, the resolution depends on platform altitude (and any variations during flight), and on position across the swath.

Suppose we have an aircraft system operating at 10GHz (0.03m wavelength) with a 3m long antenna. Then for a slant range to the ground of 2000m the along track resolution will be 20m, which is acceptable although not exceptional for aircraft altitudes. If the same system were to be placed on a spacecraft at 1000km altitude then the azimuth resolution will be no better than 10km (assuming the slant range is not too different from the platform altitude in this case), which is not acceptable. For longer wavelength radar, often needed in practice, the situation will be even worse. Clearly a better method is needed for achieving acceptable azimuth resolutions. The solution adopted is called *synthetic aperture radar* (SAR) since it gives the appearance of synthesising a very long antenna (also called an aperture) as developed in the following section.

3.5 Synthetic Aperture Radar (SAR)

The method adopted to achieve acceptable azimuth resolution at spacecraft altitudes is to synthesise an apparently long antenna by making use of the forward linear motion of the space platform. This is depicted in Fig. 3.9; the length of the synthetic aperture is defined by the time that a particular spot on the ground is irradiated by the radar. To increase the duration of irradiation a very broad beam in azimuth is needed which, from (3.7), suggests that a very short antenna in the along track direction should be used.

While this is an attractive concept it does lead to complexities when forming images from the radar echoes as discussed in the next section. Undertaking that analysis however leads to a quite remarkable result, viz. that the *azimuth resolution* obtainable with SAR is

$$r_a = \frac{l_a}{2} \text{ m} \tag{3.8}$$

where recall l_a is the length of the antenna carried on the spacecraft, measured in the along track direction. This indicates that the azimuth resolution is independent of slant range, and thus platform altitude, and independent of operating wavelength. Since ground range resolution is also height independent a SAR can, in principle, operate at any altitude with no variations in resolution. Consequently, spaceborne operation is acceptable. Because of the benefits of altitude independence and high resolution, SAR technology is also often used with aircraft based imaging radars.

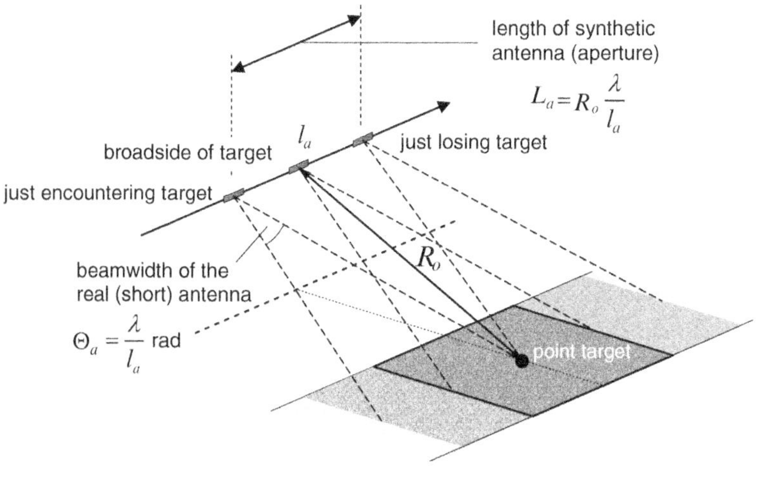

Fig. 3.9. The concept of using the platform motion to synthesise an effectively long antenna; the footprint of the real antenna on the ground is shown as rectangular for simplicity

In contrast to real aperture (SLAR) systems described by (3.7), for SAR (3.8) shows the azimuth resolution depends directly (and not inversely) on the physical antenna length. This is an amazing result since it says that improvement in azimuth resolution can be made by reducing the antenna length. The penalty in doing so will be an increase in signal processing demand as seen in Appendix D.

3.6 The Mathematical Basis for SAR

Consider a slant range projection of the geometry of Fig. 3.9, shown in Fig. 3.10. We define the vehicle's position along its track by the coordinate x; it has its origin broadside of a point target and is positive when the platform is prior to broadside. Likewise we define the time origin at broadside so that t is also positive before broadside is encountered. The platform's along track velocity is v ms^{-1}.

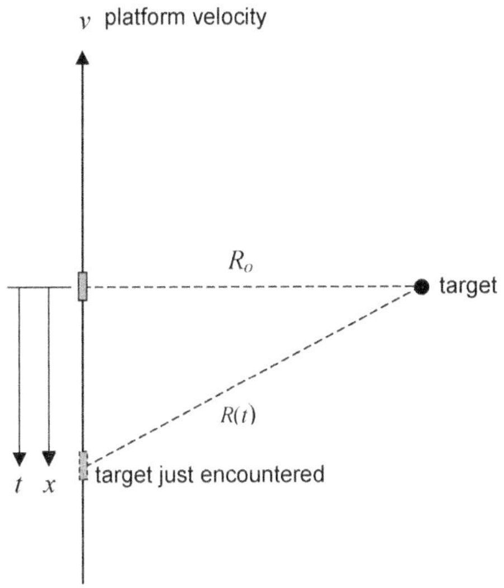

Fig. 3.10. Slant plane view (containing the velocity vector and R_o) of the platform passing a point target

Let the slant range to the target be described by $R(t)$. From Fig. 3.10 this is seen to be

$$R(t) = \sqrt{R_o^2 + x^2} = R_0\left[1 + \left(\frac{vt}{R_o}\right)^2\right]^{1/2}$$

Typically we can assume $\frac{vt}{R_o} << 1$ so that the square root can be approximated, to give

$$R(t) = R_o + \frac{(vt)^2}{2R_o}$$

To make the following development less complicated imagine the signal transmitted from the platform is a simple sinusoid of the form $\cos\omega_o t$. That ignores the chirp modulation but is a helpful approximation that does not significantly affect the result to be generated. With this simplification the signal received back at the (moving) platform after reflection from the target will be of the form $\cos\omega_o(t + t_D)$ in which t_D is the time taken for the two way trip and is given by

$$t_D = \frac{2R(t)}{c} = \frac{2}{c}\left\{R_o + \frac{(vt)^2}{2R_o}\right\}$$

Noting that $\omega_o / c = 2\pi f_o / c = 2\pi / \lambda$ the received signal is

$$\cos\left\{\omega_o t + \frac{4\pi R_o}{\lambda} + 2\pi\frac{v^2t^2}{\lambda R_o}\right\} = \cos[\omega_o t + \phi_R(t)] = \cos\phi_T(t) \tag{3.9a}$$

in which the phase delay $\phi_R(t)$ is the result of the two way travel between the platform and target and $\phi_T(t)$ is the total phase angle of the received signal. We can also derive the phase delay directly as

$$\phi_R(t) = 2\text{x}2\pi.\frac{R(t)}{\lambda} = \frac{4\pi R_o}{\lambda} + \frac{2\pi(vt)^2}{\lambda R_o} \tag{3.9b}$$

which is twice the distance from the platform to the target, expressed in wavelengths, and multiplied by 2π to convert the result to an angle, in radians.

The instantaneous frequency of a sinusoid is the first time derivative of the total phase angle[3], so that

$$\omega = \frac{d\phi_T(t)}{dt} = \omega_o + \frac{d\phi_R(t)}{dt}$$

which from (3.9a) or (3.9b) gives

$$\omega = \omega_o + \frac{4\pi v^2}{\lambda R_o}t = \omega_o + bt \tag{3.10}$$

Thus the received signal has a frequency variation induced on it as a result of the motion of the platform. This is the classical Doppler shift experienced with moving platforms, treated more generally in Sect. 2.18. It shows that the carrier frequency of the received signal is higher than transmitted when the platform is ahead of broadside (the frequency is up-shifted just as the siren of an approaching ambulance appears) and is downshifted after broadside (as will be the siren of the ambulance when receding). The parameter b is called the Doppler rate and is given by

$$b = \frac{4\pi v^2}{\lambda R_o} \quad \text{rad.s}^{-1}\text{s}^{-1} \tag{3.11a}$$

or, if expressing frequency in hertz, by

$$\beta = \frac{b}{2\pi} = \frac{2v^2}{\lambda R_o} \quad \text{Hzs}^{-1} \tag{3.11b}$$

Consider a typical value for β. For the JERS-1 mission the slant range is approximately 720km and the satellite orbital velocity is 6.883kms^{-1}. The operating wavelength is 0.235m, corresponding to a frequency of 1.275GHz. This gives β= 5760Hzs^{-1}.

This signal commences when the platform first acquires the target and stops when the target is lost, having travelled a distance equal to the real azimuth beamwidth on the ground; during this period the Doppler modified signal appears as a chirp. In this case it is of decreasing frequency, as against the rising chirp illustrated in Fig. 3.3b. It can, however, still be compressed using the same approach as for range compression – by correlating it against a replica of itself. That is generally done off-line after all the echoes have been received for a given region of terrain, as outlined in Appendix D.

We saw from the development leading to (3.5) that the half power width of the compressed chirp after correlation is the inverse of the chirp bandwidth. In the current analysis the chirp bandwidth is $\mathcal{B}_c=\beta T_a$ where T_a is the time over which the azimuth chirp

[3] The frequency of a sinusoid is defined by the rate of angular rotation of the vector $\mathcal{R}e^{j\phi(t)}$ on the complex plane that is used to generate the sinusoid (see Appendix A). The angular velocity of the vector is $d\phi(t)/dt$. For the simple case of $\cos\omega t$, which we can write as $\cos\phi=\mathcal{R}e(e^{j\phi})$ with $\phi=\omega t$, then $d\phi/dt=\omega$.

exists – in other words, while the point target is visible to the radar. This is equal to L_a/v where L_a is the azimuth footprint of the antenna on the ground. Therefore the width of the compressed azimuth chirp in seconds is

$$\tau_a = \frac{1}{\mathcal{B}_c} = \frac{1}{\beta T_a} = \frac{v}{\beta L_a} = \frac{\lambda R_o}{2vL_a} \tag{3.12}$$

The time duration of the compressed chirp can be turned into spatial resolution in the azimuth direction by multiplying it by the platform velocity:

$$r_a = \frac{\lambda R_o}{2L_a} \tag{3.13}$$

The azimuth antenna footprint is given as the beamwidth of the antenna from (3.6) multiplied by the slant range R_o:

$$L_a = \frac{\lambda R_o}{l_a} \tag{3.14}$$

where l_a is the length of the real, physical antenna. Combining (3.13) and (3.14) leads to the remarkable result of (3.8)! L_a in (3.14) is explicitly the length of the synthetic aperture. It is the large, apparent antenna length that gives rise to the fine azimuth resolution.

Note that we can express the azimuth resolution as

$$r_a = \frac{v}{\mathcal{B}_c} \equiv \frac{l_a}{2} \tag{3.15}$$

a form that is important when considering the ScanSAR mode of operation in Sect. 3.9.

This derivation of SAR azimuth resolution was based on transmitting a continuous sinusoid that gets transformed into a chirp as a result of the finite acquisition time of the target. It ignores the fact that the transmitted signal is not continuous but a series of ranging chirps repeated at the pulse repetition frequency needed to acquire adjacent strips across the swath as the platform moves forward. How does that affect our derivation? When each of those ranging chirps is received by the radar it is compressed according to the material of Sect. 3.3 to become a sinc pulse as illustrated in Fig. 3.7. Those sinc pulses are extremely narrow and have a centre frequency f_o. The motion induced azimuth chirping just discussed represents a Doppler shift imposed on the centre frequency of the compressed ranging chirps. For a typical *prf* of, say, 1500 chirps per second and a time over which the target is visible of about 5s, 4500 chirps are reflected from the target and received at the radar as the platform passes by, during which time the azimuth induced Doppler effect is imposed.

We could regard those 4500 echoes, after compression, simply as very narrow finite time samples of a continuous sinusoid on which is induced the azimuth chirping above; thus the complete echo history for the target is just a set of samples of a waveform of frequency f_o, modulated linearly at the rate β Hzs^{-1}. Provided those samples represent a fair (un-aliased) model of the continuous sinusoid then this description is acceptable.

It is instructive now to do a simple calculation. The antenna on the Seasat satellite was 10.74 m long, which means that the azimuth resolution should have been 5.37m. In fact the actual resolution was 25m, about 4 times coarser, which matches the ground range resolution of about 25m as seen in Sect. 3.3. Why are the theoretical and actual pixel sizes different in azimuth? Sets of four pixels in the azimuth dimension were averaged to give the 25mx25m pixels in the final image product. That averaging helps to reduce significantly the influence of what is known as speckle, discussed in Sects. 4.3.1-4.3.3.

Although the averaging is done slightly differently, as outlined in Appendix D, the result is the same. The number of pixels averaged is called the number of "looks" in the language of SAR. Thus Seasat image data used "four look averaging".

3.7 Swath Width and Bounds on Pulse Repetition Frequency

The width of the image swath recorded in a radar remote sensing system is determined principally by the "vertical" beamwidth of the antenna. The antenna is made small in its vertical (or across track) dimension so that a large beamwidth, and thus swath, is obtained. Even though (3.6) was employed for calculating azimuth beamwidth it can also be used to compute the vertical beamwidth Θ_v of the antenna if l_a is replaced by the vertical antenna dimension l_v

$$\Theta_v = \frac{\lambda}{l_v} \text{ rad}$$

The swath width will therefore be approximately

$$S = \frac{\Theta_v R_o}{\cos\theta} = \frac{\lambda R_o}{l_v \cos\theta} \text{ m} \tag{3.16}$$

in which R_o is the slant range at mid swath and θ is the incidence angle at mid swath. This can be seen from the geometry of Fig. 3.12. If R_o=800km, λ=0.235m, θ=23° and the antenna is l_v=2.16m in the across track direction, then the (ideal) swath width will be 95 km, close to the actual value of 100km for the Seasat satellite. It is significant to recognise that these calculations have all been based on a flat earth model. For a curved earth a larger swath would result. The actual swath width in practice is usually smaller than the value determined from the antenna beamwidth, being governed instead by the number of range samples (pixels) actually recorded by the particular radar system within the available antenna beam.

We are now in the position to understand the limits on the pulse repetition frequency used for the ranging pulses. As noted at the end of Sect. 3.1 the *prf* needs to be synchronised with the velocity of the platform so that adjacent range lines are contiguous, or at least do not have spaces between them. The width of a range line is the azimuth resolution of the system r_a. If the platform velocity is v then we need to transmit one ranging pulse every r_a/v seconds for there to be no gaps in the coverage[4]. That, with (3.8), gives a minimum *prf* of

$$prf_{\min} = \frac{v}{r_a} = \frac{2v}{l_a} \tag{3.17}$$

[4] From a signal analysis point of view that means we are sampling the scene in azimuth on the assumption that there are no spatial frequency components with periods shorter than two azimuth resolution cells. That is called Nyquist rate sampling; if sampling is carried out slower than the Nyquist rate we incur a form of distortion known as aliasing.

in which l_a is the (azimuth) length of the antenna.

The upper bound on *prf* is set by the need to ensure that the returns from the far edge of the swath from one ranging pulse do not overlap with those from the near edge of the swath from the next ranging pulse. If S is the swath width, and the incidence angle does not vary significantly across the swath, then from Fig. 3.11 the additional two way distance to the far swath edge relative to the near swath edge is approximately $2S\sin\theta$, where θ is the mid swath incidence angle. To avoid the range ambiguity just mentioned resulting from transmitting too quickly, the upper bound on *prf* is[5]

$$prf_{\max} = \frac{c}{2S\sin\theta} \tag{3.18a}$$

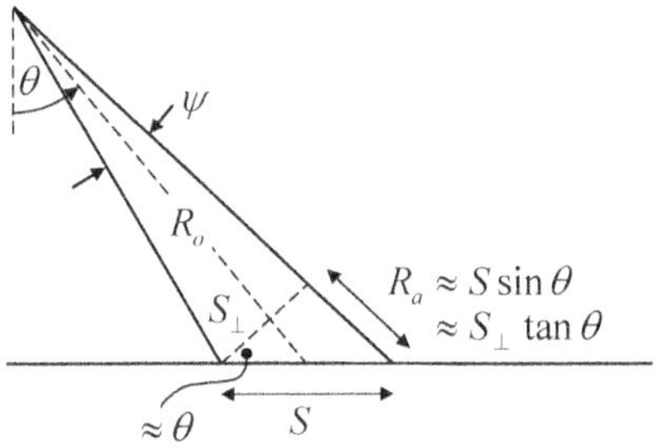

Fig. 3.11. Computing the maximum pulse repetition frequency to avoid range ambiguity

We can alternately express this in terms of the dimension $S_\perp$ shown in the figure, which in turn is a function of the vertical beamwidth of the antenna Θ_v, itself being dependent on the vertical dimension of the antenna l_v:

$$prf_{\max} = \frac{c}{2S_\perp \tan\theta} = \frac{c}{2\Theta_v R_o \tan\theta} = \frac{l_v c}{2R_o\lambda\tan\theta} \tag{3.18b}$$

Bringing the two constraints on *prf* together we have

$$\frac{v}{r_a} \le prf \le \frac{c}{2S\sin\theta} \tag{3.19}$$

The limiting condition is when all three terms in this last expression are equal, which gives

$$\frac{r_a}{S} = 2\frac{v}{c}\sin\theta \tag{3.20}$$

This is a critically important equation since, for a given mid swath incidence angle, it says that there is a direct relationship between azimuth resolution and achievable swath width.

[5] Multi-polarisation radars operate with only half the swath for a given *prf*, see Sect. 3.23.

This forms the basis of our consideration of ScanSAR in the next section. Note from (3.17) and (3.18b) we can also write

$$\frac{2v}{l_a} \leq prf \leq \frac{l_v c}{2R_o \lambda \tan\theta}$$

which, along with $c/\lambda = f$, can be re-arranged to give

$$l_v l_a \geq \frac{4vR_o \tan\theta}{f}$$

$l_v l_a$ is the area or aperture of the antenna, so that the last expression can be written

$$\text{antenna area} \geq \frac{4vR_o \tan\theta}{f} \tag{3.21}$$

which is another fundamental radar equation; it acts as a constraint on the minimum antenna size (but not its individual dimensions).

Finally, recall from the end of Sect. 3.6 that several looks in azimuth are usually averaged to reduce speckle in the image. The achievable azimuth resolution $l_a/2$ in (3.8) is therefore degraded by the number of looks used. If N_L is the number of looks then (3.17) can be expressed

$$prf_{\min} = \frac{v}{r_a} = \frac{2v}{N_L l_a} \tag{3.22}$$

Using this in (3.20) gives

$$\frac{r_a}{S} = 2\frac{v}{c} N_L \sin\theta \tag{3.23}$$

3.8 The Radar Resolution Cell

We can now envisage the landscape resolved into discrete cells, or pixels, defined by the ground range and azimuth resolutions of the system as shown in Fig. 3.12. The number of cells across the swath, and the number of ranging lines recorded for a given region determine the size of a radar image in pixels.

3.9 ScanSAR

Most imaging radars have swath widths of about 50km-100km, which are often too narrow for many mapping and monitoring applications, especially over wide, relatively homogeneous fields such as the ocean. Suppose we want to design a system with a swath of, say, 300km. Noting (3.23) what then is the best achievable azimuth resolution? To determine that we need to know typical values of the other parameters. Suppose we use v=7.5kms^{-1} and θ=23°, typical of Seasat and ERS. Then choosing a swath of 300km limits r_a to about 6m for a one look system or to about 24m for a four look system (the usual image product), which does not look too bad. Equation (3.23) though is derived on the basis of an ideal set of conditions. In particular it assumes that the azimuth beam pattern of the synthetic aperture cuts off sharply at the edges of the resolution cell and that the returning ranging pulses are well enough defined that we can apply the ambiguity criterion of (3.20) exactly.

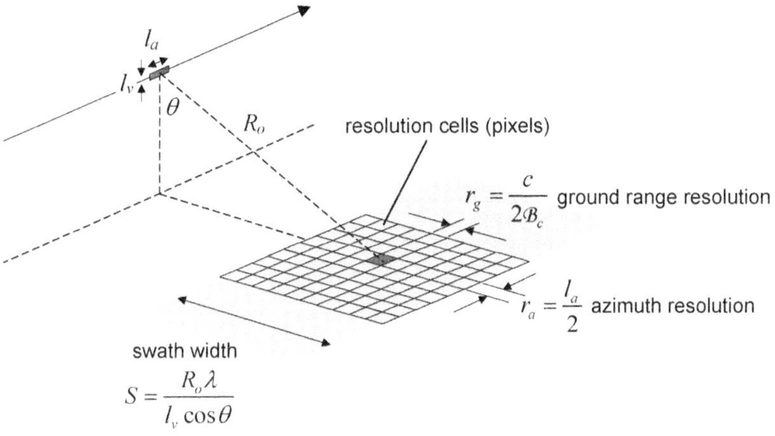

Fig. 3.12. Resolution of the image field into resolution cells, defined by the ground range and azimuth resolutions for a single look image

To give a margin of safety in design so that any system non-idealities don't lead to azimuth or range ambiguities, the minimum *prf* is generally chosen a bit higher (say by 50%) than the value given by (3.17) and the maximum *prf* is generally chosen to be a bit lower (say by 50%) than the value given by (3.18). The net effect of those safety margins can be accommodated by including a factor k in (3.23):

$$\frac{r_a}{S} = 2k\frac{v}{c}N_L \sin\theta \tag{3.24}$$

k clearly has a minimum value of 1; a value of 3 would give a reasonable design margin in most cases. Using 3 adjusts the above azimuth resolutions for a 300km swath to 18m for one look and 72m for four looks. Again, these are not necessarily bad figures, especially for oceanographic applications. However, consider how long physically the antenna has to be. For an 18m one look azimuth resolution the along track dimension of the antenna from (3.8) needs to be 36m! That is too big for orbiting on a spacecraft and for ensuring good manufacturing tolerances. Radarsat 2 is able to image with a 500km swath, which would mean a 60m antenna; yet the imaging is done with a 15m antenna azimuth dimension. Clearly there must be another approach.

The ScanSAR principle[6] is used to provide wide swath imaging, with reasonable spatial resolution and practical antenna sizes. ScanSAR relies upon breaking the imaging process up into blocks, both in the along track and across track directions as shown in Fig. 3.13.

[6] See R.K. Moore, J.P. Claassen and Y.H. Lin, Scanning spaceborne synthetic aperture radar with integrated radiometer, *IEEE Transactions on Aerospace and Electronic Systems*, vol. AES-17, no. 3, May 1981, pp. 410-421, and K. Tomiyasu, Conceptual performance of a satellite borne, wide swath synthetic aperture radar, *IEEE Transactions on Geoscience and Remote Sensing*, vol. GE-19, no. 2, April 1981, pp. 108-116.

We have shown N_s blocks in azimuth over the distance of the equivalent synthetic aperture and, for simplicity, the same number of blocks across the desired swath width.

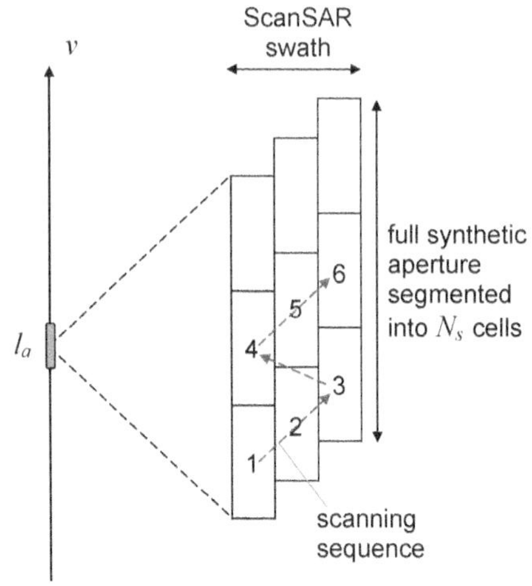

Fig. 3.13. The use of scanning cells to construct a wide swath with a practical antenna using the ScanSAR principle

The antenna carried on the platform is capable of being steered electronically (and therefore extremely quickly) from one block to the next in a sequence such as that indicated in the figure. While it is dwelling in one block – often called a scanning cell – the normal SAR process applies: ranging chirps are transmitted to resolve the scene across the scanning cell (whose edges are defined by the real vertical beamwidth of the antenna) and the Doppler history of the signal in azimuth is used to provide azimuth resolution. However, since the full azimuth chirp bandwidth is not now used for compression the achievable azimuth resolution is poorer by the factor of the number of scanning cells in azimuth; thus from (3.15)

$$r_a = \frac{v}{(\mathcal{B}_c / N_s)} = \frac{N_s l_a}{2} \tag{3.25a}$$

For wide swath operation this poorer azimuth resolution is generally not a problem for the types of application envisaged. If, in addition, we average N_L resolution cells in azimuth for speckle reduction then for a given antenna length, the actual achievable azimuth resolution is

$$r_a = \frac{v}{(\mathcal{B}_c / N_s N_L)} = \frac{N_L N_s l_a}{2} \tag{3.25b}$$

Note that the ambiguity constraint of (3.24) applies within each of the sub-swaths of Fig. 3.13. Also note that (3.24) already incorporates the number of looks in azimuth so when using that expression, (3.25a) is the corresponding formula for the azimuth resolution. For a given azimuth resolution specified by (3.25a) the maximum total achievable swath is

$$S_{scansar} = N_s S = \frac{N_s}{N_L} r_a \frac{v}{2kv\sin\theta} \tag{3.26}$$

Consider some typical values. Suppose v=7.5kms^{-1}, N_L=2 and θ=20°. Suppose, further, we wish to achieve a 90m azimuth resolution with a 15m antenna, then what overall swath is available? Equation (3.25b) tells us that we need to have N_s=6 scanning blocks. Using (3.26), and choosing a safely factor of 4 to be conservative, we then find that the overall ScanSAR swath available to us is approximately 400km. In contrast, if we tried to achieve the same swath width with a conventional SAR system (3.24) shows that the azimuth resolution would be 48m. Being a 2 look system that requires a 48m antenna, which is impractical.

This example has been very simple and has ignored a number of system related factors concerned with transmitter power, receiver noise, antenna efficiency and earth curvature. Nevertheless it serves to demonstrate that segmenting the swath into a number of individual scanning cells allows wide swaths to be achieved while maintaining practical antenna sizes.

As might be expected the processing of ScanSAR data is more complex than with conventional SAR because of the need to join the scanned cells, but this penalty is manageable given the wide swath benefit that results.

3.10 Squint and the Spotlight Operating Mode

If the antenna beam does not point exactly to broadside the radar is said to have *squint*. Squint can occur inadvertently as a result of platform yaw or because of the rotation of the earth during imaging, or intentionally in (military) applications where the platform needs to maintain a safe distance from the area being imaged. Squint is also often an operating feature of bistatic radar, as seen in Chapt. 7. As expected, not only will squint lead to geometric distortion, particularly since the range lines are not orthogonal to the flight line, but the Doppler history in azimuth will be changed. The centre of the azimuth chirp will be displaced from the broadside position and the Doppler bandwidth will be reduced, leading to drop in azimuth resolution as demonstrated below. It also leads to a coupling between the azimuth and range coordinates which can increase the problem of range walk, outlined in Appendix D, and which has to be corrected in image formation. Range resolution is not significantly affected by squint.

Figure 3.14 shows a slant plane view with the radar antenna squinting forward by an angle ξ. In order to see its effect on azimuth resolution it is sufficient to determine the azimuth bandwidth under squint conditions because that determines resolution as seen in (3.15). We will find the bandwidth by identifying the motion induced Doppler component on the carrier frequency at the start and the end of the period that the point target is in view.

As with radar without squint the distance from the radar to a point target is given by

$$R(t) = \sqrt{R_o^2 + x^2} = R_o\sqrt{1 + \frac{(vt)^2}{R_o^2}} \tag{3.27}$$

We should not now make the assumption $\frac{vt}{R_o^2} << 1$ which led to (3.15) since it masks the asymmetry of the geometry caused by the squint angle. The Doppler frequency component associated with the changing distance $R(t)$, is given by the first time derivative of the associated two way change in phase:

$$f_{Doppler} = \frac{1}{2\pi}\frac{4\pi}{\lambda}\frac{dR(t)}{dt}$$

which, from (3.27) is

$$f_{Doppler} = \frac{2}{\lambda}\frac{v^2 t}{R_o}\left[1+\frac{(vt)^2}{R_o^2}\right]^{-1/2} = \frac{2}{\lambda}\frac{v^2 t}{R_o}\frac{R_o}{R(t)} = \frac{2v^2 t}{\lambda R_o}\cos\mu \qquad (3.28)$$

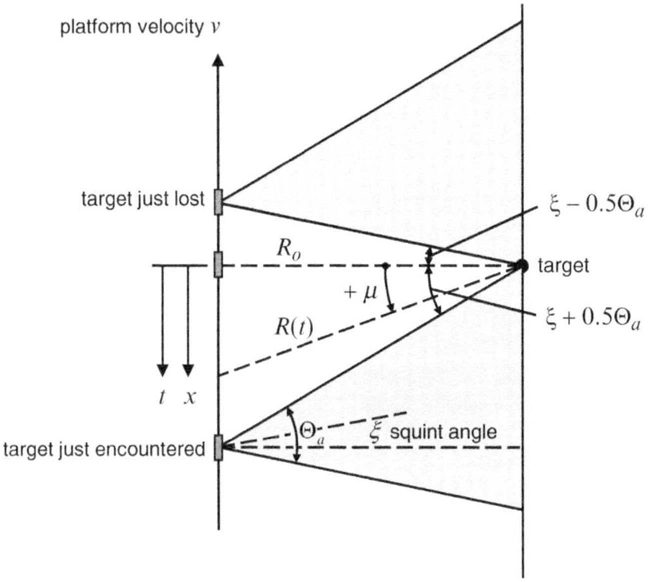

Fig. 3.14. Squint geometry in the slant plane; note that angles are measured in the positive sense anticlockwise from broadside

From Fig. 3.14 we can see that

$$t = \frac{R_o}{v}\tan\mu$$

so that (3.28) becomes

$$f_{Doppler} = \frac{2v}{\lambda}\sin\mu \qquad (3.29)$$

The Doppler frequency when the target is just encountered is given when $\mu = 0.5\Theta_a + \xi$, i.e.

$$f_{DopplerHigh} = \frac{2v}{\lambda}\sin(0.5\Theta_a + \xi)$$

The Doppler frequency when the target just disappears is given when $\mu = -(0.5\Theta_a - \xi)$, i.e.

$$f_{DopplerLow} = -\frac{2v}{\lambda}\sin(0.5\Theta_a - \xi)$$

Thus the Doppler bandwidth is

$$\mathcal{B}_c = f_{DopplerHigh} - f_{DopplerLow} = \frac{2v}{\lambda}[\sin(0.5\Theta_a + \xi) + \sin(0.5\Theta_a - \xi)]$$

i.e.

$$\mathcal{B}_c = \frac{4v}{\lambda}\sin(0.5\Theta_a)\cos\xi$$

The azimuth beamwidth of the antenna Θ_a is generally less than about 0.02, so that the last expression can be approximated

$$\mathcal{B}_c = \frac{2v}{\lambda}\Theta_a \cos\xi$$

Note $\Theta_a = \frac{\lambda}{l_a}$ where l_a is the azimuth length of the antenna, so that the chirp bandwidth becomes

$$\mathcal{B}_c = \frac{2v}{l_a}\cos\xi \text{ Hz} \tag{3.30}$$

Azimuth resolution is given in (3.15) in terms of the chirp bandwidth and the platform velocity which from (3.30) gives

$$r_a = \frac{l_a}{2\cos\xi} \text{ m} \tag{3.31}$$

which, by comparison with (3.8) shows that the effect of the squint is to lower the azimuth resolution. For 15° of squint the achievable resolution is 3% poorer than the theoretical value given when the radar points directly to broadside.

To achieve these results it is assumed that the azimuth chirp replica matches that induced in the squinted situation. That is not unreasonable since the chirp parameters can be assessed from the signal itself. Note also that while Doppler zero will still occur at broadside, that will no longer be the centre (the centroid) of the chirp. The Doppler centroid will be given as the arithmetic mean of the upper and lower Doppler frequencies, viz.

$$f_{DopplerCentroid} = \frac{2v}{\lambda}\cos 0.5\Theta_a \sin\xi \approx \frac{v}{\lambda}\sin\xi$$

If the antenna is squinted forward and then steered during platform motion such that it continues to illuminate the target as depicted in Fig. 3.15 then high resolution of that target region is possible at the expense of resolution and focussing in the remainder of the imaged domain. That is referred to as *spotlight mode* imaging, and is used when very high resolution of specific targets is desired.

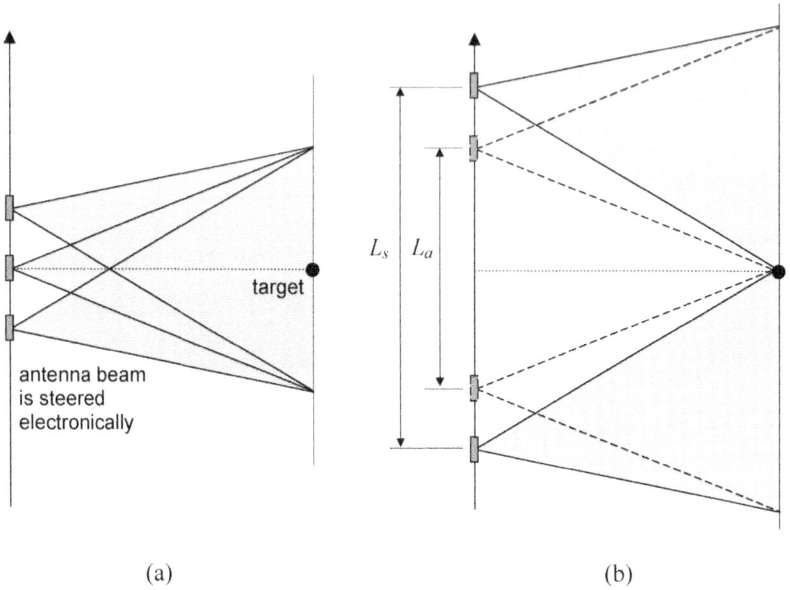

Fig. 3.15. (a) Spotlight mode imaging, in the slant plane and (b) the creation of a larger equivalent synthetic aperture L_s with steering than the synthetic aperture L_a without steering

Because the target is in view for a longer period of time with the steered antenna beam than it would have been if the antenna beam were fixed, the effect is equivalent to the creation of a larger synthetic aperture as depicted in Fig. 3.15b, thereby giving enhanced azimuth resolution.

Part B: The Target

3.11 The Radar Equation

The first part of this chapter has been devoted to understanding the operation of imaging radar, including how the landscape can be differentiated into resolution elements. We now turn our attention to the interaction of the incident radiation with the earth's surface. It is that interaction that determines the variations in brightness in a radar image and reveals properties of the earth's surface of interest. Here we set up the framework for describing the interaction; Chapt. 5 treats explicit earth surface cover types.

Before we look at scattering from the landscape consider the more traditional radar situation of the detection of a discrete target; the lessons we learn from this case readily transfer to understanding radar scattering in remote sensing. For the moment imagine the radar is an isotropic radiator as seen in Fig. 3.16. According to (2.1) it will produce a power density at the target, R metres away, of

$$p_i = \frac{P_t}{4\pi R^2}\ \text{Wm}^{-2}$$

The subscript i on the power density signifies that it is incident on the target.

If instead of an isotropic radiator the radar uses an antenna that concentrates the power in a preferred direction as shown in Fig. 3.16, the power density at the target will be

$$p_i = \frac{P_t G_t}{4\pi R^2}\ \text{Wm}^{-2}$$

where G_t is the gain of the transmitting antenna, defined as the ratio of power density it produces in the preferred direction compared with the power density produced by an isotropic radiator.

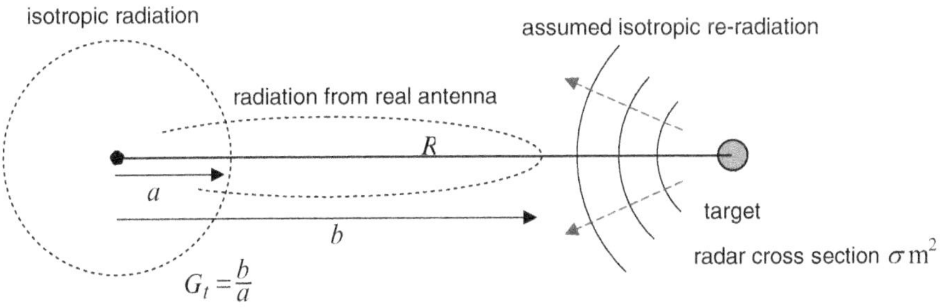

Fig. 3.16. Irradiation of a target with radar cross section $\sigma\,\text{m}^2$, and subsequent scattering

Suppose there is a target at position R. It could be an aircraft, a discrete element on the ground such as a tree, or a ship on the surface of the sea. The target will present an area or cross section to the incoming radiation. It may absorb some of the incident energy, but generally it will also reflect or scatter a significant portion of the energy. We now introduce the concept of the target's *radar cross section* (*RCS*). *RCS* has dimensions of area (orthogonal to the incident radiation); it describes how much power the target extracts from the power density of the incoming wave. Most of this intercepted power will be scattered. Irrespective of its shape, the target is assumed to scatter the intercepted power isotropically. While a real target will not behave isotropically this is nevertheless a very useful assumption that simplifies theoretical developments and leads to a measurable value for *RCS*[7]. Radar cross section – described by $\sigma\,\mathrm{m}^2$ – is usually not easily related to any physical cross sectional area of the target. If the target rotates with respect to the incoming radar beam then it will have a different *RCS*, defined by the implicit area needed at that orientation to account for the energy extracted from the wavefront and re-radiated back to the radar set isotropically.

The power "received" by the target and available for re-radiation is

$$P_\sigma = p_t\sigma = \frac{P_tG_t\sigma}{4\pi R^2} \quad \mathrm{W}$$

so that the power density produced back at the platform after scattering from the target is

$$p_r = \frac{P_tG_t\sigma}{(4\pi)^2R^4} \quad \mathrm{Wm^{-2}}$$

Note that there is an extra $4\pi R^2$ term in the denominator caused by the isotropic propagation back to the platform.

To find the actual power received the returned power density is multiplied by a property of the antenna referred to as its aperture A_r, which also has dimensions of area. Thus the received power is

$$P_r = \frac{P_tG_t\sigma A_r}{(4\pi)^2R^4} \quad \mathrm{W}$$

The aperture of an antenna can be written in terms of its gain according to[8]

$$G_r = \frac{4\pi}{\lambda^2}A_r \tag{3.32}$$

so that the power received by the radar system after scattering from the target is

$$P_r = \frac{P_tG_tG_r\lambda^2\sigma}{(4\pi)^3R^4} \quad \mathrm{W} \tag{3.33}$$

[7] It is significant to emphasise here that the property accorded to the target of a cross section, and the assumption of isotropic scattering (or re-radiation), are as observed in the received signal at the radar and not near the target itself. Not only do we not observe the scattering behaviour right at the target, but if we did we would have to account for so-called near field effects. Near field components complicate the situation but decay relatively quickly away from the scatterer as seen in Sect. 2.9. The equations in this section always assume we are in the far field of the transmitting antenna and the target.

[8] All antennas can receive and transmit and can thus be described by a gain or an aperture. Gain is often used to describe antenna behaviour in both transmission and reception whereas aperture is generally used only for reception.

This equation is called the *radar range equation* since it can be used to determine the maximum range of a radar if all the other terms are known and we know the limit of detection of received power. One of its celebrated features is the inverse fourth power dependence on the distance to the target. Targets at twice the range require sixteen times more power to detect! Because we will be working with existing radar remote sensing systems we will not encounter that problem explicitly. We can however easily see from (3.33) how the radar cross section of an object can be measured. If we choose a transmitter power and range, and measure the received power at the wavelength of interest then we can find σ. This assumes we know the antenna gains, which is always the case in practice. If we took several measurements of received power with different orientations of the target we would then be able to build up a picture of how the radar cross section of an object changes with the angle with which it is viewed.

3.12 Theoretical Expression for Radar Cross Section

The previous development can be used to derive an expression for radar cross section that we will employ when we come to describe target, and pixel, scattering properties. In words, the previous section says that the transmitted power creates a power density p_i incident on the target. The *RCS* of the target σ intercepts σp_i watts of power which it re-radiates isotropically, producing a power density at the receiver of

$$p_r = \frac{\sigma p_i}{4\pi R^2}$$

Using (2.7) average power density is related to electric field by $p = \eta |E|^2$ in which η is the impedance of free space – a constant that will soon cancel out of our expressions – and E is the rms value of the field. Using this in the above expression for received power density we have

$$|E^r|^2 = \frac{\sigma |E^i|^2}{4\pi R^2}$$

Re-arranging the last expression yields a definition for radar cross section

$$\sigma = \lim_{R\to\infty} 4\pi R^2 \frac{|E^r|^2}{|E^i|^2} \tag{3.34}$$

in which the limit on R reminds us that we need to be far enough away from the target so that near field effects can be ignored.

3.13 The Radar Cross Section in dB

Because its value can extend over an enormous range (less than 0.01m^2 for birds to more than 100m^2 for aircraft) it is usual to express radar cross section in decibels with respect to some reference level using the definition

$$\sigma = 10\log\frac{\sigma}{\sigma_{ref}} \quad \text{dB}$$

The most common reference is $\sigma_{ref} = 1\text{m}^2$; the unit of *RCS* is then dBm^2:

$$\sigma = 10\log\frac{\sigma}{1\text{m}^2} \quad \text{dBm}^2$$

3.14 Distributed Targets

Only some targets in radar remote sensing are of the nature of discrete scatterers as treated in the preceding section. More commonly scattering takes place from regions on the earth's surface that are distributed in nature, such as an area of soil or snow, an agricultural field or even the surface of the ocean. To accommodate those cover types the radar equation needs to be modified, commencing with a variation to the definition of radar cross section.

Radar cross section as a concept strictly refers only to discrete targets. To help formulate an alternative suited to distributed cover types consider a region composed of an infinite collection of infinitesimal elements of effective area ds as shown in Fig. 3.17, many of which make up an individual pixel. Further, suppose the radar cross section of each of those infinitesimal areas is $d\sigma$. On the average therefore the region exhibits a radar cross section per unit area of $d\sigma/ds$. This is denoted σ^0 and is referred to as the *scattering coefficient* of the region. From its definition its units are m^2m^{-2}. Colloquially, it is often called *sigma nought*.

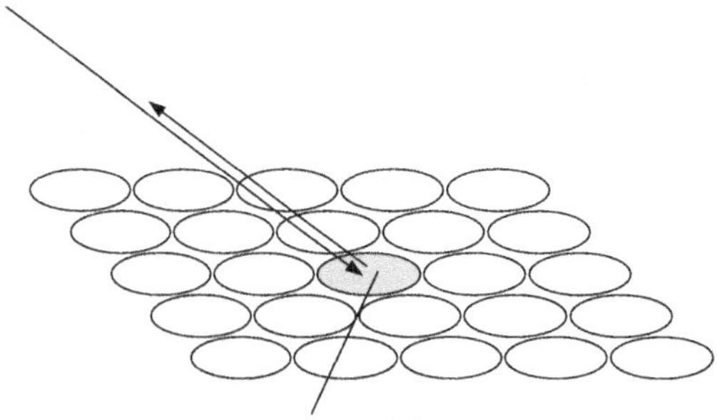

Fig. 3.17. Resolving a distributed region, such as an agricultural field, into a set of discrete incremental areas

From (3.33) the power received back at the platform after scattering from one of the incremental regions shown in Fig. 3.17 will be

$$dP_r = \frac{P_t G_t G_r \lambda^2 d\sigma}{(4\pi)^3 R^4} \quad \text{W}$$

or, in terms of the radar scattering coefficient for the region

$$dP_r = \frac{P_t G_t G_r \lambda^2 \sigma^o ds}{(4\pi)^3 R^4} \quad \text{W}$$

We can now find the total power returned to the platform from a particular resolution cell, or pixel, by integrating the last expression over the pixel area:

$$P_r = \iint_{pixel} \frac{P_t G_t G_r \lambda^2 \sigma^o ds}{(4\pi)^3 R^4} \quad \text{W}$$

If all the quantities inside the integral can be considered constant over pixel then the received power is

$$P_r = \frac{P_t G_t G_r \lambda^2 \sigma^o r_a r_g}{(4\pi)^3 R^4} \quad \text{W} \qquad (3.35)$$

in which r_a and r_g are the azimuth and ground range resolutions. This is the form of the radar equation most used in radar remote sensing since our interest centres mainly on the scattering properties of regions (forests, fields, ocean, etc) rather than discrete scatterers. If all other parameters are known through the design of the radar system σ^o can be determined by measuring P_r. σ^o describes the "tone" of the radar image and is analogous to the reflectance of earth surface materials at visible and infrared wavelengths used in optical remote sensing.

What is important now is to relate σ^o to the physical properties of the region being imaged – its composition, water content, physical properties and so on. This is an essential step in the interpretation of radar data and is the subject of Chapt. 5.

3.15 The Scattering Coefficient in dB

As with the radar cross section of discrete targets σ^o is commonly expressed in decibels. A reference of $1\text{m}^2\text{m}^{-2}$ is used so strictly absolute units (dBm^2) should be identified. In practice they are understood rather than written explicitly. Instead dB is just used:

$$\sigma^o = 10\log\frac{\sigma^o}{1\,\text{m}^2\text{m}^{-2}} \quad \text{dB}$$

Thus 0dB is a scattering coefficient of $1\text{m}^2\text{m}^{-2}$, 3dB means $2\text{m}^2\text{m}^{-2}$ and -20dB means $0.01\text{m}^2\text{m}^{-2}$. Table 3.1 shows a range of scattering coefficients expressed in both natural and dB form. Because the decibel is based on logarithms, and logarithms have the property that the log of a product is the sum of the individual logs, the table illustrates how easily dBs can be computed. For example a scattering coefficient of $20\text{m}^2\text{m}^{-2}$ is 2x10, which in dBs is 3+10=13dB. A scattering coefficient of -7dB is -10+3dB which in natural form will be $0.1\text{x}2=0.2\text{m}^2\text{m}^{-2}$.

Table 3.1 Converting Scattering Coefficients to dB Form

Scattering coefficient (m^2m^{-2})	dB	Scattering coefficient (m^2m^{-2})	dB
0.001	-30	10	10
0.005	-23	20	13
0.01	-20	50	17
0.02	-17	100	20
0.05	-13	200	23
0.1	-10	500	27
0.2	-7	1000	30
0.5	-3	2000	33
1	0	10,000	40
2	3	100,000	50
5	7		

3.16 Polarisation Dependence of the Scattering Coefficient

Section 2.8 tells us that the wavefronts we have described above in terms of power and power density are composed of electric and magnetic field vectors at right angles to the direction of propagation and to each other. That is the case both for the incident wave and for the wave after scattering from a discrete target or a distributed region. Sect. 2.10 describes the polarisation of the wave in terms of the orientation of the electric field vector. Although not strictly correct theoretically we describe polarisation as horizontal, if the field is horizontal to the earth's surface and vertical if it is in a plane that is vertical to the earth's surface.

Polarisation turns out to be a particularly important parameter in radar remote sensing because the scattering properties of earth surface materials can be different for different incident polarisations. The scattered wave can also have a different polarisation from that of the incident wave, a mechanism referred to as *polarisation rotation* or sometimes *depolarisation*. In the most general case the scattered wave can have both horizontal and vertical components even though the incident wave was simply horizontally or vertically polarised. This actually means that the polarisation of the scattered wave is in a plane different from vertical or horizontal which nonetheless can be resolved into horizontal and vertical components.

To account for the fact that the scattering coefficient is polarisation dependent we write it with subscripts σ^o_{PQ} which signify the polarisation of the incident wave and that of the wave scattered and received by the radar. The first subscript P indicates the received polarisation and the second Q the transmitted or incident polarisation. The subscripts are sometimes interpreted in the other order, so care is needed about which convention is being used when fully polarised data is employed. The convention used here is the most appropriate theoretically in the context of the matrix algebra we will use to describe multi-polarisation data.

Although many imaging radars in the past were single polarisation, in that the transmit and received polarisations were the same and fixed, more recent remote sensing radars can radiate both vertically and horizontally, and receive both the vertical and horizontal components of the scattered wave. In such a case there are, in principle, four relevant scattering coefficients, brought together in what is called the *sigma nought matrix*:

$$\begin{bmatrix} \sigma^o_{HH} & \sigma^o_{HV} \\ \sigma^o_{VH} & \sigma^o_{VV} \end{bmatrix} \tag{3.36}$$

Although not immediately obvious here it is assumed for monostatic radar systems that the two cross-polarised components σ_{HV} and σ_{VH} are the same, whereas the co-polarised components σ_{HH} and σ_{VV} can be quite different from each other. We will have more to say about that later.

We can define two measures at this point that find value in polarimetric radar remote sensing studies:

co-polarisation ratio $$p = \frac{\sigma^o_{HH}}{\sigma^o_{VV}} \tag{3.37}$$

cross-polarisation ratio $$q = \frac{\sigma^o_{HV}}{\sigma^o_{VV}} \quad \text{or} \quad \frac{\sigma^o_{VH}}{\sigma^o_{HH}} \tag{3.38}$$

As might be expected, the cross-polarisation ratio implicitly carries information about complex scattering events that may lead to a rotation of the polarisation state of the incident radiation.

3.17 The Scattering Matrix

For many applications it is sufficient to use the scattering coefficient defined above to describe the earth surface properties of interest. The full analytical power of imaging radar emerges, however, when we can perform *polarisation synthesis*. Although a radar would generally irradiate with vertically and horizontally polarised radiation, and detect both horizontally and vertically, some landscape features may be more evident, and more readily discriminated from other features, with different orientations of the field vectors. We need therefore to be able to synthesise the effect of other polarisation orientations from the ones available to us. To do so requires development via an electric field description of the scattering process as against the power density development we used to derive the concept of radar cross section and scattering coefficient. That leads to the concept of the scattering matrix which captures a description of a scatterer in terms of the relationship between incident and scattered electric fields. Just like the scattering coefficient, it is a property of the scatterer itself and embodies the landscape information of interest to us.

Fig. 3.18 shows coordinate systems[9] for the horizontal and vertical field components involved in backscattering from a discrete target, or from a pixel on the ground if the scattering coefficient is sufficiently uniform over the pixel that we can express the pixel's radar cross section as in (3.35) – i.e. $\sigma = \sigma^o r_a r_g$. R defines the direction of propagation of the transmitted (and thus incident) wave. Backscattering occurs in the $-R$ direction. This convention is referred to as back scatter alignment (BSA). It is possible to reverse the R coordinate for scattering; the convention is then called forward scattering alignment (FSA) which finds application in bistatic radar remote sensing. Appendix E discusses the

[9] We have chosen the horizontal and vertical orthogonal field components to use here since they are the ones encountered in imaging radar missions. We could have chosen any two components at right angles to each other and to the direction of propagation.

differences between the two systems. We adopt the BSA axes for most of the treatment in this book.

The only difference between the incident and transmitted fields is a result of propagation from the radar to the target. There will be a phase delay because of the travel of the wave over the distance R, and a drop in signal strength. Equation (2.1) shows that the power density falls in an inverse square fashion with distance. Equating (2.1) and (2.7c) shows that the rms field strength is

$$|E| = \sqrt{\frac{P_t}{4\pi\eta}}\frac{1}{R} = \frac{\text{constant}}{R} \tag{3.39}$$

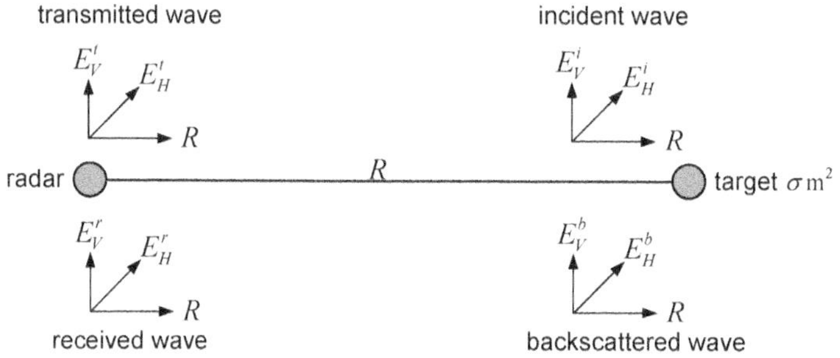

Fig. 3.18. Field components relevant to the scattering matrix, assuming that all components are transverse to the direction of propagation; this implies near field effects are ignored

Thus the field amplitudes fall in an inverse distance fashion with distance. Just as with the transmitted and incident waves, the only difference between the backscattered and received waves is a phase difference and an inverse distance drop in amplitude. It is the comparison of the incident and backscattered waves that is of most interest to us, because that is what contains information directly about the scattering properties of the target and, ultimately, the biophysical properties of the target itself.

We express the most general relationship between the incident and backscattered fields in the form of a matrix equation

$$\begin{bmatrix} E_H^b \\ E_V^b \end{bmatrix} = \begin{bmatrix} S_{HH} & S_{HV} \\ S_{VH} & S_{VV} \end{bmatrix} \begin{bmatrix} E_H^i \\ E_V^i \end{bmatrix}$$

or

$$\mathbf{E}^b = \mathbf{S}\mathbf{E}^i \tag{3.40}$$

where the field components are summarised in vector form and the matrix

$$\mathbf{S} = \begin{bmatrix} S_{HH} & S_{HV} \\ S_{VH} & S_{VV} \end{bmatrix} \tag{3.41}$$

is referred to as the *scattering matrix* or *Sinclair matrix* of the target. As with the sigma nought matrix of (3.36) note that the first subscript on each of the elements refers to the polarisation of the scattered wave while the second subscript refers to the polarisation of the incident wave. Equation (3.40) says that the horizontally polarised backscattered field can be viewed as the result of the target scattering a horizontally polarised component of the incident field and a depolarised vertically incident component:

$$E_H^b = S_{HH}E_H^i + S_{HV}E_V^i$$

If the incident field were just horizontally polarised – i.e. $E_V^i = 0$ – then $E_H^b = S_{HH}E_H^i$, so that the only target property of significance is S_{HH}. Likewise a vertically polarised backscattered field can be viewed as the result of the target scattering a vertically polarised component of the incident field and a depolarised horizontally incident component:

$$E_V^b = S_{VH}E_H^i + S_{VV}E_V^i$$

S_{VV} is the only property of importance for a vertically polarised radar.

The elements of the scattering matrix contain all the information we need about the target. Each is a complex quantity (having both an amplitude and phase angle) that is dependent on the frequency, or wavelength, of operation and the incidence angle at the earth's surface. In principle it is also dependent on the azimuth angle with which the target is viewed, although that is generally fixed by the broadside direction to the motion vector of the platform. Given that each element has an amplitude and phase, the scattering matrix contains eight pieces of information about the target, or region on the ground.

In practice we don't measure the backscattered components right at the target, nor are they theoretically available at the target itself. As noted earlier, that has to do with the difference between the near field of the target (which requires a detailed field theory description to understand fully) and the far field of the target (some distance away, beyond which the power density description and isotropic scattering representation we adopted in Sect. 3.10 can be used). Therefore (3.40) is usually written as though the scattering properties are observed back at the radar

$$\begin{bmatrix} E_H^r \\ E_V^r \end{bmatrix} = \frac{e^{j\beta R}}{R}\begin{bmatrix} S_{HH} & S_{HV} \\ S_{VH} & S_{VV} \end{bmatrix}\begin{bmatrix} E_H^i \\ E_V^i \end{bmatrix}$$

or

$$\mathbf{E}^r = \frac{e^{j\beta R}}{R}\mathbf{S}\mathbf{E}^i \qquad (3.42)$$

The exponential term accounts for the phase difference induced in transmission which can be ignored since it will affect all components equally; the distance term in the denominator comes from (3.39). Some authors include the $\sqrt{4\pi}$ in the denominator that is part of (3.39); that is not a problem, it is just taken up in the scaling of the field components.

Recall that it is often convenient to express the field components in complex exponential form, the real part of which is the sinusoidal form

$$E_0\cos(\omega t - \beta R) = \mathcal{Re}\{E_0 e^{j(\omega t - \beta R)}\} = E_0\mathcal{Re}\{e^{j\omega t}e^{-j\beta R}\}$$

Since all components have the same frequency the first exponential term is often omitted as is the real part operator, accepting that both are there implicitly should it be necessary to revert to the sinusoidal description. Therefore it is commonplace to write the field in the summary form $E_0 e^{-j\beta R}$ or even in phasor form $E_0\angle-\beta R$ which essentially just replaces the complex exponential by the angle sign. The exponential form is used in (3.42) but with the sign reversed (i.e. positive) since the backscattered wave travels in the negative R direction.

We now return to a consideration of the meaning of the scattering matrix elements and their use. The first question that comes to mind is their relationship to the radar cross section of the target. We will consider the simple case of *HH* polarisation to demonstrate this. From (3.34), for R large enough to be in the far field, we have

$$\sigma_{HH} = 4\pi R^2 \frac{\left|E_H^r\right|^2}{\left|E_H^i\right|^2}$$

in which E_H^r is the field observed at the receiver. From the same understanding of field propagation that led to (3.42) we can see this to be

$$\left|E_H^r\right| = \frac{e^{j\beta R}}{R}\left|E_H^b\right|$$

Ignoring the phase propagation term, which is irrelevant in power related quantities, gives the *HH* radar cross section as

$$\sigma_{HH} = 4\pi \frac{\left|E_H^b\right|^2}{\left|E_H^i\right|^2}$$

giving from (3.40)

$$\sigma_{HH} = 4\pi\left|S_{HH}\right|^2$$

In general we find

$$\sigma_{PQ} = 4\pi\left|S_{PQ}\right|^2 \qquad (3.43)$$

This shows the relationship between the scattering matrix element and the radar cross section of a discrete target rather than the backscattering coefficient of a distributed region of landscape. Under the assumption that the backscattering coefficient is constant across a pixel we can equate the scattering matrix element to the backscattering coefficient multiplied by the area of the pixel ($r_a x r_g$), i.e.

$$\sigma_{PQ}^o = \frac{4\pi\left|S_{PQ}\right|^2}{r_a r_g} \qquad (3.44)$$

It is important to note that some authors[10] define the sigma nought matrix as the relationship between the *incident* and *received* power densities rather than the *incident* and *backscattered* densities as done here in (3.36) thus avoiding problem with near field behaviour. If that approach is taken then there will be an additional R^2 multiplier in

[10] See G.T. Ruck, D.E. Barrick, W.D. Stuart and C.K. Krichbaum, *Radar Cross Section Handbook*, Plenum, N.Y., 1970.

(3.43). It is important when moving between scattering coefficients and the scattering matrix to be clear of the definition of scattering matrix being used.

As with the sigma nought matrix of (3.36) we can assume $S_{VH}=S_{HV}$ in the case of backscattering; this is called the *reciprocity condition.* There are some unusual circumstances when it doesn't apply, most notably as a result of Faraday rotation if the wave passes through the ionosphere, which it will do for spacecraft platforms. At the higher frequencies used in remote sensing radar imaging Faraday rotation is generally considered negligible. At longer wavelengths it can be significant and may need to be taken into account when studies based on the scattering matrix are of interest[11].

3.18 Target Vectors

The elements of the scattering matrix can be used to derive other pixel descriptors perhaps more suited to analysis by the classification techniques discussed in Chapt. 8. A *target vector* (a vector rather than a matrix that summarises the properties of the target) can be created by arranging the four elements of the scattering matrix in column form

$$\mathbf{k}=\begin{bmatrix} S_{HH} \\ S_{HV} \\ S_{VH} \\ S_{VV} \end{bmatrix}=[S_{HH} \quad S_{HV} \quad S_{VH} \quad S_{VV}]^{\mathrm{T}} \tag{3.45}$$

in which we have also used the vector transpose operation so the column vector can be written more compactly in row form[12]. Since for backscattering $S_{HV}=S_{VH}$ one of the elements of the vector is redundant and carries no additional information, so the vector is reduced to three dimensions

$$\mathbf{k}=[S_{HH} \quad S_{HV} \quad S_{VV}]^{\mathrm{T}} \tag{3.46}$$

Often this is written as

$$\mathbf{k}=[S_{HH} \quad \sqrt{2}S_{HV} \quad S_{VV}]^{\mathrm{T}} \tag{3.47}$$

so that the Euclidean norms (i.e. magnitudes) of the forms in (3.45) and (3.47) are the same. The norm is also called the *span* of the target vector.

Other target vectors can be formed using combinations of the scattering matrix elements. The most common alternative to (3.45), derived from the Pauli basis[12], is

$$\mathbf{k}_P=\frac{1}{\sqrt{2}}[S_{HH}+S_{VV} \quad S_{HH}-S_{VV} \quad S_{HV}+S_{VH} \quad j(S_{HV}-S_{VH})]^{\mathrm{T}} \tag{3.48}$$

For backscattering, in which the two cross-polar terms are equal, this reduces to

---

[11] See Sect. 3.24 for a discussion of this effect.

[12] When the elements are arranged as shown in (3.45) they are sometimes said to have lexicographical ordering. This is in contrast to the combinations of the elements of the scattering matrix in the form of the Pauli basis target vector of (3.48) which can be derived from the Pauli spin matrices used in quantum mechanics. See S.R. Cloude and E. Pottier, A review of target decomposition theorems in radar polarimetry, *IEEE Transaction on Geoscience and Remote Sensing*, vol. 34, no. 2, March 1996, pp. 498-518.

$$\mathbf{k}_P = \frac{1}{\sqrt{2}}[S_{HH}+S_{VV} \quad S_{HH}-S_{VV} \quad 2S_{HV}]^{\mathrm{T}} \tag{3.49}$$

3.19 The Covariance and Coherency Matrices

Another way of expressing target properties is through the *covariance matrix* defined as the expected value (i.e. an average over a number of measurements) of the product of the target vector and the transpose of its complex conjugate

$$\mathbf{C} = \mathcal{E}\left(\mathbf{k}\mathbf{k}^{*\mathrm{T}}\right) \tag{3.50}$$

Although involving complex elements, thus requiring the conjugation operation, this is not unlike the definition of the covariance matrix used in maximum likelihood classification of optical remote sensing data[13]. From (3.45) we can expand the covariance matrix as

$$\mathbf{C} = \begin{bmatrix} \langle S_{HH}S_{HH}^*\rangle & \langle S_{HH}S_{HV}^*\rangle & \langle S_{HH}S_{VH}^*\rangle & \langle S_{HH}S_{VV}^*\rangle \\ \langle S_{HV}S_{HH}^*\rangle & \langle S_{HV}S_{HV}^*\rangle & \langle S_{HV}S_{VH}^*\rangle & \langle S_{HV}S_{VV}^*\rangle \\ \langle S_{VH}S_{HH}^*\rangle & \langle S_{VH}S_{HV}^*\rangle & \langle S_{VH}S_{VH}^*\rangle & \langle S_{VH}S_{VV}^*\rangle \\ \langle S_{VV}S_{HH}^*\rangle & \langle S_{VV}S_{HV}^*\rangle & \langle S_{VV}S_{VH}^*\rangle & \langle S_{VV}S_{VV}^*\rangle \end{bmatrix}$$

in which we have used the angular brackets to indicate that the expected value can be obtained by averaging over the available samples (pixels). Remember that each of the scattering matrix elements is complex and can be written in the simple phasor form

$$S_{HH} = |S_{HH}|\angle S_{HH} \text{ so that } S_{HH}S_{HH}^* = |S_{HH}|\angle S_{HH}.|S_{HH}|\angle - S_{HH} = |S_{HH}|^2\angle 0$$

Thus the diagonal elements of the covariance matrix simplify to give

$$\mathbf{C} = \begin{bmatrix} \langle |S_{HH}|^2\rangle & \langle S_{HH}S_{HV}^*\rangle & \langle S_{HH}S_{VH}^*\rangle & \langle S_{HH}S_{VV}^*\rangle \\ \langle S_{HV}S_{HH}^*\rangle & \langle |S_{HV}|^2\rangle & \langle S_{HV}S_{VH}^*\rangle & \langle S_{HV}S_{VV}^*\rangle \\ \langle S_{VH}S_{HH}^*\rangle & \langle S_{VH}S_{HV}^*\rangle & \langle |S_{VH}|^2\rangle & \langle S_{VH}S_{VV}^*\rangle \\ \langle S_{VV}S_{HH}^*\rangle & \langle S_{VV}S_{HV}^*\rangle & \langle S_{VV}S_{VH}^*\rangle & \langle |S_{VV}|^2\rangle \end{bmatrix} \tag{3.51}$$

Comparing this last expression with (3.43) shows that the diagonal elements of the covariance matrix are, to within a multiplicative constant, the four scattering coefficients of the pixel: $c_{11} \propto \sigma_{HH}^o$, $c_{22} \propto \sigma_{HV}^o$, $c_{33} \propto \sigma_{VH}^o$ and $c_{44} \propto \sigma_{VV}^o$. The off-diagonal terms describe the interactions or correlations among the set of scattering mechanisms. From an

[13] Generally the computation of covariance in (3.50) entails the operation $\mathcal{E}\{[\mathbf{k}-\mathcal{E}(\mathbf{k})][\mathbf{k}-\mathcal{E}(\mathbf{k})]^{*T}\}$. It can be shown, though, that the expected value of the target vector itself is zero; see Sect. 8.4.4.1.

image perspective they tell us the degree of correlation of the two co-polarised (*HH* and *VV*) images and the degree of correlation of the like and cross-polarised (*HH* or *VV* and *HV* or *VH*) images.

We now look at three special cases of the covariance matrix.

Reciprocity

If the reciprocity relation holds (backscattering when Faraday rotation is not a problem) the covariance matrix simplifies to

$$\mathbf{C}=\begin{bmatrix}\langle S_{HH}S_{HH}^*\rangle & \langle\sqrt{2}S_{HH}S_{HV}^*\rangle & \langle S_{HH}S_{VV}^*\rangle\\ \langle\sqrt{2}S_{HV}S_{HH}^*\rangle & \langle 2S_{HV}S_{HV}^*\rangle & \langle\sqrt{2}S_{HV}S_{VV}^*\rangle\\ \langle S_{VV}S_{HH}^*\rangle & \langle\sqrt{2}S_{VV}S_{HV}^*\rangle & \langle S_{VV}S_{VV}^*\rangle\end{bmatrix}$$

$$=\begin{bmatrix}\langle|S_{HH}|^2\rangle & \langle\sqrt{2}S_{HH}S_{HV}^*\rangle & \langle S_{HH}S_{VV}^*\rangle\\ \langle\sqrt{2}S_{HV}S_{HH}^*\rangle & \langle 2|S_{HV}|^2\rangle & \langle\sqrt{2}S_{HV}S_{VV}^*\rangle\\ \langle S_{VV}S_{HH}^*\rangle & \langle\sqrt{2}S_{VV}S_{HV}^*\rangle & \langle|S_{VV}|^2\rangle\end{bmatrix} \tag{3.52}$$

which can be derived directly from (3.47), or from (3.51) by noting that the centre two rows are then identical, as are the centre two columns; span has been preserved by inserting √2. Note that the off diagonal elements, in pairs about the diagonal, are conjugates of each other.

Media with Reflection Symmetry

Suppose a scatterer exhibits symmetry in its scattering properties either side of the plane of incidence. That will be the case for a number of natural scatterers including many rough surfaces and foliage canopies. Essentially, if the medium looks geometrically to be symmetric either side of the plane of incidence then it will scatter that way. Such a medium is said to exhibit *reflection symmetry*. It is a feature of media with reflection symmetry that the like and cross polarised backscattering responses are not correlated; as a consequence the corresponding off diagonal terms in (3.52) are zero. Thus, the covariance matrix for backscattering is[14]

$$\mathbf{C}=\begin{bmatrix}\langle|S_{HH}|^2\rangle & 0 & \langle S_{HH}S_{VV}^*\rangle\\ 0 & \langle 2|S_{HV}|^2\rangle & 0\\ \langle S_{VV}S_{HH}^*\rangle & 0 & \langle|S_{VV}|^2\rangle\end{bmatrix} \tag{3.53}$$

Media with Azimuthal Symmetry

If a medium exhibits reflection symmetry not just in the plane of incidence but in any rotated plane that contains the indent ray, in the sense that the backscatter is insensitive to

[14] See S.V. Ngheim, S.H. Yueh, R Kwok and F.K. Lee, Symmetry properties in polarimetric remote sensing, *Radio Science*, vol. 27, no. 5, September-October 1992, pp. 693-711.

the orientation of the electric field vector, then the medium is said to have azimuthal symmetry and its covariance matrix is also[15]

$$\mathbf{C}=\begin{bmatrix}\left\langle |S_{HH}|^2\right\rangle & 0 & \left\langle S_{HH}S^*_{VV}\right\rangle\\ 0 & \left\langle 2|S_{HV}|^2\right\rangle & 0\\ \left\langle S^*_{HH}S_{VV}\right\rangle & 0 & \left\langle |S_{VV}|^2\right\rangle\end{bmatrix} \quad (3.54)$$

An alternative to the covariance matrix is the *coherency matrix*, developed from the Pauli basis target vector of (3.48)

$$\mathbf{T}=\mathcal{E}\left(\mathbf{k}_P\mathbf{k}_P{}^{*\mathrm{T}}\right) \quad (3.55)$$

which expands to

$$\mathbf{T}=\frac{1}{2}\begin{bmatrix}\left\langle k_ak^*_a\right\rangle & \left\langle k_ak^*_b\right\rangle & \left\langle k_ak^*_c\right\rangle & \left\langle k_ak^*_d\right\rangle\\ \left\langle k_bk^*_a\right\rangle & \left\langle k_bk^*_b\right\rangle & \left\langle k_bk^*_c\right\rangle & \left\langle k_bk^*_d\right\rangle\\ \left\langle k_ck^*_a\right\rangle & \left\langle k_ck^*_b\right\rangle & \left\langle k_ck^*_c\right\rangle & \left\langle k_ck^*_d\right\rangle\\ \left\langle k_dk^*_a\right\rangle & \left\langle k_dk^*_b\right\rangle & \left\langle k_dk^*_c\right\rangle & \left\langle k_dk^*_d\right\rangle\end{bmatrix} \quad (3.56)$$

in which

$$\begin{aligned}k_a&=S_{HH}+S_{VV}\\ k_b&=S_{HH}-S_{VV}\\ k_c&=S_{HV}+S_{VH}\\ k_d&=j(S_{HV}-S_{VH})\end{aligned}$$

Note that the off diagonal terms, in pairs about the diagonal, are conjugates of each other.
We now look at three special cases of the coherency matrix.

Reciprocity
If the reciprocity relation holds (again, backscattering when Faraday rotation is not a problem) the coherency matrix loses its last row and column, leaving

$$\mathbf{T}=\frac{1}{2}\begin{bmatrix}<(S_{HH}+S_{VV})(S^*_{HH}+S^*_{VV})> & <(S_{HH}+S_{VV})(S^*_{HH}-S^*_{VV})> & 2<(S_{HH}+S_{VV})S^*_{HV}>\\ <(S_{HH}-S_{VV})(S^*_{HH}+S^*_{VV})> & <(S_{HH}-S_{VV})(S^*_{HH}-S^*_{VV})> & 2<(S_{HH}-S_{VV})S^*_{HV}>\\ 2<S_{HV}(S^*_{HH}+S^*_{VV})> & 2<S_{HV}(S^*_{HH}-S^*_{VV})> & 4<S_{HV}S^*_{HV}>\end{bmatrix}$$

(3.57)

[15] See J.J. van Zyl, Application of Cloude's target decomposition theorem to polarimetric imaging radar data, *SPIE* Vol. 1748, *Radar Polarimetry*, 1992, pp. 184-191.

Media with Reflection Symmetry
If a scatterer exhibits symmetry in its scattering properties either side of the plane of incidence the co-polar and cross-polar terms are uncorrelated, as before, so that the coherency matrix reduces to

$$\mathbf{T} = \frac{1}{2}\begin{bmatrix} <(S_{HH}+S_{VV})(S_{HH}^*+S_{VV}^*)> & <(S_{HH}+S_{VV})(S_{HH}^*-S_{VV}^*)> & 0 \\ <(S_{HH}-S_{VV})(S_{HH}^*+S_{VV}^*)> & <(S_{HH}-S_{VV})(S_{HH}^*-S_{VV}^*)> & 0 \\ 0 & 0 & 4<S_{HV}S_{HV}^*> \end{bmatrix} \quad (3.58)$$

Media with Azimuthal Symmetry
If a medium exhibits azimuthal symmetry its coherency matrix takes on a diagonal form[16]

$$\mathbf{T} = \frac{1}{2}\begin{bmatrix} a & 0 & 0 \\ 0 & b & 0 \\ 0 & 0 & c \end{bmatrix} \quad (3.59)$$

3.20 Measuring the Scattering Matrix

The principal objective in radar remote sensing is to understand properties of the landscape by measuring either the scattering coefficient(s) of (3.36) or the scattering matrix of (3.41). Usually the scattering coefficient, *sigma nought*, is measured in a relative sense and external calibration devices are used to give it an absolute value. We have more to say about calibration in Chapt. 4. In this section we concentrate on measuring the scattering matrix, rather than the scattering coefficients. Not only does that give us a very concise and convenient summary of the properties of the region being imaged (or at least how those properties influence incident radiation) but it also allows the very powerful methodology of polarisation synthesis to be used, as seen in Sect. 3.22.

Measuring the scattering matrix requires an application of (3.42) along with knowing how the incident field is related to that transmitted. Applying (3.39) to find that relationship we have

$$\begin{bmatrix} E_H^r \\ E_V^r \end{bmatrix} = \text{constant}.\frac{e^{j2\beta R}}{R^2}\begin{bmatrix} S_{HH} & S_{HV} \\ S_{VH} & S_{VV} \end{bmatrix}\begin{bmatrix} E_H^t \\ E_V^t \end{bmatrix}$$

in which the phase delay accounts for the two way path and the pre-matrix denominator accounts for inverse distance propagation towards the target followed by inverse distance propagation of the scattered field back to the radar. This is the field equivalent to the inverse fourth power dependence on range in the power expression of (3.33).

An experiment described by this expression will, in principle, yield the four complex elements of the scattering matrix. There is a practical problem with obtaining accurate values of their phase angles because the transmission path between the radar and target,

[16] See S.R. Cloude and E. Pottier, A review of target decomposition theorems in radar polarimetry, *IEEE Transaction on Geoscience and Remote Sensing*, vol. 34, no. 2, March 1996, pp. 498-518, and S.R. Cloude, D.G. Corr and M.L. Williams. Target detection beneath foliage using polarimetric synthetic aperture radar interferometry, *Waves in Random and Complex Media*, vol.14, no. 2, April 2004, pp. S393–S414.

and return, induces the phase change indicated by the exponential term. To give that a sense of scale, note that if the radiation we are using is at say 1GHz the wavelength will be 30cm, which accounts for a full cycle (360°) of phase. Varying atmospheric conditions can change the wavelength of the radiation and thus induce changes in phase in transmission. Moreover, it is difficult to identify where the actual point of scattering lies for a discrete target or distributed region of ground, making precise specification of R (within better than 30 cm, for example) in these equations difficult. As a result we don't try to determine the actual (absolute) phase angles of the elements of the scattering matrix. Instead, we simply measure them with respect to the phase of the HH component, implicitly taking the phase of S_{HH} to be zero.

3.21 Relating the Scattering Matrix to the Stokes Vector

Recall from Sect. 2.13 that the Stokes vector, or its modified form, is a description of an electromagnetic wave in terms of power density quantities rather than field vectors, that nevertheless preserves information on the polarisation state of the radiation. The Stokes parameters are also able to account for any unpolarised component of a travelling wave so there is value in being able to describe the signal scattered from a target in terms of its Stokes vector.

Analogous to the development in Sect 3.17, which dealt with the field description of scattering, let $\mathbf{s}^r$ and $\mathbf{s}^i$ instead be the received and incident waves described in terms of their Stokes vectors. They will be related by some matrix equation of the form

$$\mathbf{s}^r = \frac{1}{R^2}\mathbf{H}\mathbf{s}^i \tag{3.60}$$

The 4x4 matrix $\mathbf{H}$ is called the *Kennaugh matrix*[17], or sometimes the *Stokes matrix*. The R^2 in the denominator accounts for the inverse square law of power density drop with distance between the target and the receiver. When dealing with power density an exponential phase term has no meaning. To find $\mathbf{H}$ we adopt (2.31) to give

$$\mathbf{R}\mathbf{g}^r = \frac{1}{R^2}\mathbf{H}\mathbf{R}\mathbf{g}^i$$

Pre-multiplying both sides by $\mathbf{R}^{-1}$ gives

$$\mathbf{g}^r = \frac{1}{R^2}\mathbf{W}\mathbf{g}^i \tag{3.61a}$$

in which

$$\mathbf{W} = \mathbf{R}^{-1}\mathbf{H}\mathbf{R} \tag{3.61b}$$

If we know $\mathbf{W}$ we can re-arrange the last expression to find the Kennaugh matrix $\mathbf{H}$. If we imagined (3.61) at the scatterer itself we could ignore the $1/R^2$ term provided we assume that we can work with far field quantities. We then have a relationship between the backscattered and incident vectors.

[17] Sometimes this is called the Mueller matrix, with the name Kennaugh matrix reserved for forward rather than backscattering situations. We will use Kennaugh matrix here, which is more common in radar.

$$\mathbf{g}^b = \mathbf{W}\mathbf{g}^i \tag{3.62}$$

in which from (2.33)

$$\mathbf{g}^b = \begin{bmatrix} E_H^b E_H^{b*} \\ E_V^b E_V^{b*} \\ E_H^b E_V^{b*} \\ E_V^b E_H^{b*} \end{bmatrix} = \begin{bmatrix} |E_H^b|^2 \\ |E_V^b|^2 \\ E_H^b E_V^{b*} \\ E_V^b E_H^{b*} \end{bmatrix} \text{ and } \mathbf{g}^i = \begin{bmatrix} E_H^i E_H^{i*} \\ E_V^i E_V^{i*} \\ E_H^i E_V^{i*} \\ E_V^i E_H^{i*} \end{bmatrix} = \begin{bmatrix} |E_H^i|^2 \\ |E_V^i|^2 \\ E_H^i E_V^{i*} \\ E_V^i E_H^{i*} \end{bmatrix}. \tag{3.63}$$

We can derive expressions for each of the elements of the vectors $\mathbf{g}^b$ and $\mathbf{g}^i$ by returning to (3.40) from which

$$E_H^b = S_{HH} E_H^i + S_{HV} E_V^i$$

so that

$$E_H^{b*} = S_{HH}^* E_H^{i*} + S_{HV}^* E_V^{i*}$$

and $$E_H^b E_H^{b*} = S_{HH} S_{HH}^* E_H^i E_H^{i*} + S_{HV} S_{HH}^* E_V^i E_H^{i*} + S_{HH} S_{HV}^* E_H^i E_V^{i*} + S_{HV} S_{HV}^* E_V^i E_V^{i*}$$

The last expression can be re-written as the product of two vectors, using a different order for the second and fourth terms to allow comparison with (3.63):

$$E_H^b E_H^{b*} = \begin{bmatrix} S_{HH} S_{HH}^* & S_{HV} S_{HV}^* & S_{HH} S_{HV}^* & S_{HV} S_{HH}^* \end{bmatrix} \begin{bmatrix} E_H^i E_H^{i*} \\ E_V^i E_V^{i*} \\ E_H^i E_V^{i*} \\ E_V^i E_H^{i*} \end{bmatrix}$$

This demonstrates that the first element in the backscattered vector $\mathbf{g}^b$ can be expressed in terms of the incident vector $\mathbf{g}^i$ and elements of the scattering matrix. We can do the same for the remaining three elements of the backscattered vector to show that the matrix $\mathbf{W}$ in (3.62) is given by

$$\mathbf{W} = \begin{bmatrix} S_{HH} S_{HH}^* & S_{HV} S_{HV}^* & S_{HH} S_{HV}^* & S_{HV} S_{HH}^* \\ S_{VH} S_{VH}^* & S_{VV} S_{VV}^* & S_{VH} S_{VV}^* & S_{VV} S_{VH}^* \\ S_{HH} S_{VH}^* & S_{HV} S_{VV}^* & S_{HH} S_{VV}^* & S_{HV} S_{VH}^* \\ S_{VH} S_{HH}^* & S_{VV} S_{HV}^* & S_{VH} S_{HV}^* & S_{VV} S_{HH}^* \end{bmatrix} \tag{3.64}$$

We can then get the Kennaugh matrix by inverting (3.61b)

$$\mathbf{H} = \mathbf{R}\mathbf{W}\mathbf{R}^{-1} \tag{3.65}$$

Note from (2.32) that

$$\mathbf{R} = \begin{bmatrix} 1 & 1 & 0 & 0 \\ 1 & -1 & 0 & 0 \\ 0 & 0 & 1 & 1 \\ 0 & 0 & -j & j \end{bmatrix} \text{ so that } \mathbf{R}^{-1} = \frac{1}{2}\begin{bmatrix} 1 & 1 & 0 & 0 \\ 1 & -1 & 0 & 0 \\ 0 & 0 & 1 & j \\ 0 & 0 & 1 & -j \end{bmatrix} \tag{3.66}$$

We now have all the material needed to use (3.60) to see how the Stokes vector is changed by scattering; the Kennaugh matrix that governs that transformation is specified entirely in terms of the scattering matrix through (3.64).

3.22 Polarisation Synthesis

A significant advantage of multi-polarisation radar becomes apparent when it is realised that having available the full scattering matrix for a target makes it possible to synthesise how the target would appear in other polarisation combinations, even though they were not recorded by the radar. It allows us to develop a very full description of the target's scattering properties both to assist in identifying it and to help discriminate it from other targets.

Essentially the signal received by a radar is the power density available at the receiving antenna after the transmitted signal has been scattered from the target. In Sects. 3.11-3.16 that has been described in terms of the target radar cross section or the scattering coefficient of the earth's surface being imaged. It is of value to revisit the radar cross section since it leads us to think about the measurements undertaken by radar polarimeters – devices that record the radar response in the four available polarisation combinations: *HH*, *VV*, *HV* and *VH*. In (3.34) the radar cross section is expressed

$$\sigma = \lim_{R\to\infty} 4\pi R^2 \frac{\left|E^r\right|^2}{\left|E^i\right|^2}$$

in which there is a subtle assumption. It assumes that the field illuminating the receiving antenna is efficiently converted to power in the radar receiver. That can only happen if the orientation of the receiving antenna matches the polarisation of the incoming electric field as illustrated in Fig. 2.23. Suppose that is not necessarily the case and that the orientation of the incoming electric field and the optimum antenna orientation is as shown in Fig. 3.19. Even though the electric field is not perfectly aligned to the antenna it will still induce a component of electric field on the antenna equal to its projection, as depicted in the figure. If we describe the orientation of the antenna in the plane at right angles to the incoming ray from the target by the spatial unit magnitude vector $\mathbf{p}^{ra}$ (which will, in general, be resolvable into horizontal and vertical components if the antenna is tilted) then the magnitude of the projected value of the received field that is detected by the antenna is given by the scalar (or dot) product as seen in (2.43a):

$$E^{r'} = \mathbf{p}^{ra}.\mathbf{E}^r = \mathbf{p}^{ra\mathrm{T}}\mathbf{E}^r \tag{3.67}$$

Even though the operations in (3.67) give rise to scalar quantities, the field component $E^{r'}$ is oriented along the antenna vector and, in principle, should be written

$$E^{r'}\mathbf{p}^{ra} \tag{3.68}$$

The field $\mathbf{E}^r$ incident on the receiving antenna can be expressed, ignoring changes in phase, as

$$\mathbf{E}^r = \frac{1}{R}\mathbf{E}^b = \frac{1}{R}\mathbf{S}\mathbf{E}^i$$

in which $\mathbf{E}^b$ is the (far field) backscattered field at the target and $\mathbf{E}^i$ is the field incident on the target; **S** is the scattering matrix of (3.41). Thus, from (3.67)

$$E^{r'} = \frac{1}{R}\mathbf{p}^{ra}.\mathbf{S}\mathbf{E}^i$$

which, when substituted into (3.34), and noting $|E_i| = |\mathbf{E}_i|$ gives the radar cross-section as

$$\sigma = \lim_{R\to\infty} 4\pi R^2 \frac{|E^{r'}|^2}{|E^i|^2} = 4\pi\left|\mathbf{p}^{ra}.\mathbf{S}\frac{\mathbf{E}^i}{|\mathbf{E}^i|}\right|^2$$

The construct $\frac{\mathbf{E}^i}{|\mathbf{E}^i|}$ is a vector of unit amplitude in the direction of polarisation of the electric field incident on the target[18]. That will be the same as the polarisation of the electric field actually transmitted from the radar, unless there are any unusual atmospheric properties in the path from the radar to the target. If we call this unit vector $\mathbf{p}^i \equiv \mathbf{p}^t$ then the last expression can be written

$$\sigma = 4\pi|\mathbf{p}^{ra}.\mathbf{S}\mathbf{p}^t|^2 \tag{3.69}$$

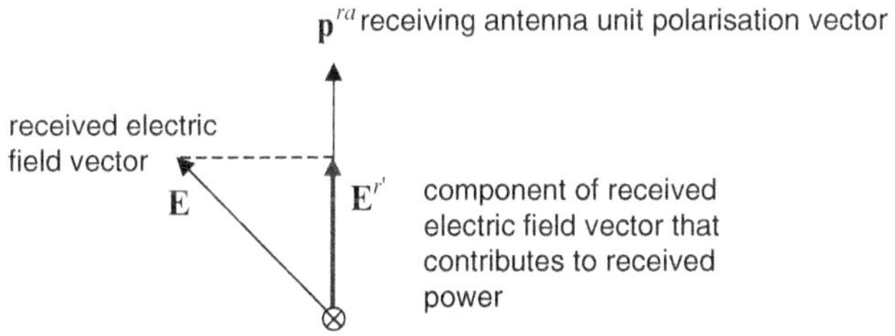

Fig. 3.19. Illustrating the effective component (projection) of the received electric field vector that is picked up by a linear receiving antenna, polarised differently from the field; note that the antenna vector has unit magnitude and the diagram is viewed towards the receiver as implied by the arrow tail (cross in circle) at the origin

This is our first equation for polarisation synthesis. It says that if we know the scattering matrix for the target then we can see how the target would appear if we used transmitted radiation with a polarisation described by the polarisation vector $\mathbf{p}^t$, and chose to receive the resulting scattered field in the direction of polarisation described by the polarisation vector $\mathbf{p}^{ra}$. Fig. 3.20 shows this diagrammatically, illustrating how the transmitted

[18] In vector algebra it is common to find a unit vector **a** that aligns with a given vector **A**, by normalising **A** by its magnitude, i.e. **a**=**A**/|**A**|. That concept is used frequently in electromagnetism and radar.

polarisation is modified after scattering by the target and how the received signal magnitude is affected by the respective polarisations of the received signal and the antenna. This assumes that the polarisation state of the antenna on reception is the same as on transmission.

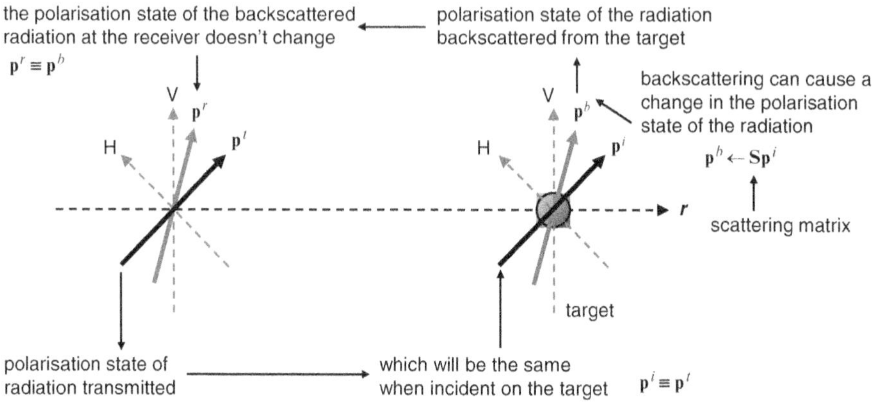

Fig. 3.20. The polarisation change sequence in radar scattering: the linearly polarised case

We now introduce an important generalisation. The derivation that led to (3.69) was based on the linearly polarised situation shown in Figs. 3.19 and 3.20. However (3.67) applies more generally, irrespective of the nature of the polarisation being considered; see Sect. 2.16. Rather than restrict ourselves to antennas that transmit and receive linearly polarised signals assume now we are dealing with antennas that use elliptical polarisation. We define the polarisation vector $\mathbf{p}^{ra}$ of an antenna as that unit amplitude vector that has the same relative components as the electric field that the antenna would transmit and can thus can represent elliptical as well as linear configurations. Likewise it represents the optimum polarisation of a received wave if the received signal at the antenna terminals were to be maximised. If the received electric field had a different polarisation from optimal then the component of the received electric field that leads to received power is the scalar product of the polarisation vector of the antenna and the electric field vector incident on the antenna – viz. (3.67). As a consequence (3.69), although derived by starting with a linearly polarised situation, actually applies for any general transmit and receive antenna polarisation vectors.

We now generalise further. For (3.69) to be used we must know the target scattering matrix. To handle more general situations it is better to derive a form of that equation in terms of Stokes vectors and the Kennaugh matrix. They can account for received signals that include unpolarised components. Also, suppliers of radar imagery often provide the data in the form of Kennaugh matrix elements or measures derived from them.

The following derivation is a little long but it results in an expression for radar cross section in terms of the properties of the polarisation ellipses that describes the transmitted and received fields (strictly the transmitting and receiving antennas), and the elements of the target's *Stokes scattering operator*, which derives from the Kennaugh matrix.

Using (3.67) the received power density will be

$$p^r = (\mathbf{p}^{ra\mathrm{T}}\mathbf{E}^r)(\mathbf{p}^{ra\mathrm{T}}\mathbf{E}^r)^*$$

This last expression may need a little explanation. Recall that power density is proportional to the square of the magnitude of the electric field and is a scalar quantity. The way that is written when the quantities are complex is to take the product of the field and its complex conjugate, which we have done here. Also, even though the bracketed entries are real as written they are the magnitudes of complex quantities "aligned" with the receiving antenna polarisation vector as described in (3.68) and as seen explicitly in Fig. 3.19 for the case of linear polarisation. Strictly that polarisation vector should also appear inside each bracket. We have left it out for simplicity since the product of the unit vector and its transpose will be a unity scalar, thus cancelling as expected. Finally, note that we are using the form of (3.67) based on the transpose rather than dot product operation (see Sect. 2.16). That also simplifies some of our subsequent notation.

Noting that we can write the vectors

$$\mathbf{p}^{ra} = \begin{bmatrix} p_H^{ra} \\ p_V^{ra} \end{bmatrix} \text{ and } \mathbf{E}^r = \begin{bmatrix} E_H^r \\ E_V^r \end{bmatrix}$$

then

$$\mathbf{p}^{ra\mathrm{T}}\mathbf{E}_r = p_H^{ra}E_H^r + p_V^{ra}E_V^r$$

$$(\mathbf{p}^{ra\mathrm{T}}\mathbf{E}_r)^* = p_H^{ra*}E_H^{r*} + p_V^{ra*}E_V^{r*}$$

since the complex conjugate of a product is the product of complex conjugates. Therefore the available power density on reception is

$$\begin{aligned} p^r = (\mathbf{p}^{ra\mathrm{T}}\mathbf{E}^r)(\mathbf{p}^{ra\mathrm{T}}\mathbf{E}^r)^* &= (p_H^{ra}E_H^r + p_V^{ra}E_V^r)(p_H^{ra*}E_H^{r*} + p_V^{ra*}E_V^{r*}) \\ &= p_H^{ra}p_H^{ra*}E_H^rE_H^{r*} + p_V^{ra}p_V^{ra*}E_V^rE_V^{r*} \\ &+ p_H^{ra}p_V^{ra*}E_H^rE_V^{r*} + p_V^{ra}p_H^{ra*}E_V^rE_H^{r*} \end{aligned}$$

which can be represented as the scalar product

$$\begin{bmatrix} p_H^{ra}p_H^{ra*} \\ p_V^{ra}p_V^{ra*} \\ p_H^{ra}p_V^{ra*} \\ p_V^{ra}p_H^{ra*} \end{bmatrix} \bullet \begin{bmatrix} E_H^rE_H^{r*} \\ E_V^rE_V^{r*} \\ E_H^rE_V^{r*} \\ E_V^rE_H^{r*} \end{bmatrix} = [p_H^{ra}p_H^{ra*} \quad p_V^{ra}p_V^{ra*} \quad p_H^{ra}p_V^{ra*} \quad p_V^{ra}p_H^{ra*}] \begin{bmatrix} E_H^rE_H^{r*} \\ E_V^rE_V^{r*} \\ E_H^rE_V^{r*} \\ E_V^rE_H^{r*} \end{bmatrix}$$

By reference to (2.33) we can write the first of these column vectors as $\mathbf{g}^{ra}$ which describes the polarisation state of the receiving antenna in terms of horizontal and vertical components. The second is a vector describing all products of the components of the electric field received at the antenna. It is related directly to the field backscattered from the target and can be described by the vector $\mathbf{g}^r$. Ignoring any phase effect we note that the magnitude of the electric field at the radar antenna is $1/R$ of that backscattered. Thus $\mathbf{g}^r$ will be $1/R^2$ of that backscattered, since it is proportional to the square of the field. We can therefore write the last expression which, recall, is the actual power density available for generating a signal in the receiver of the radar, as

$$p^r = \mathbf{g}^{ra}.\frac{1}{R^2}\mathbf{g}^b \tag{3.70}$$

How can this expression be related to the Stokes vectors of interest to us? Equation (2.31) shows that any Stokes-like vector can be written as

$$\mathbf{s} = \mathbf{Rg}$$

so that (3.70) becomes

$$p^r = \frac{1}{R^2}\mathbf{R}^{-1}\mathbf{s}^{ra}.\mathbf{R}^{-1}\mathbf{s}^b$$

where $\mathbf{R}$ is given in (2.32) and $\boldsymbol{s}^{ra}$ is a Stokes vector describing the polarisation state of the receiving antenna; effectively it is equivalent to the Stokes vector of the field the antenna would launch if used in transmission. We can move the left hand $\mathbf{R}^{-1}$ across the dot product sign by taking its transpose to give

$$p^r = \frac{1}{R^2}\mathbf{s}^{ra}.\left(\mathbf{R}^{-1}\right)^{\mathrm{T}}\mathbf{R}^{-1}\mathbf{s}^b$$

The backscattered Stokes vector in this last expression is related to the incident Stokes vector via the Kennaugh matrix of (3.60)

$$\mathbf{s}^b = \mathbf{Hs}^i \equiv \mathbf{Hs}^t$$

Here we have assumed that the incident Stokes vector is equivalent to that transmitted. That is a satisfactory assumption because we are seeking to do just two things: first, apply (3.67) which does not require knowledge of the transmitted power density, only that incident at the target. Secondly, we are only interested in the polarisation state incident at the target, which is the same as that transmitted. That is equivalent to normalising the transmitted and incident Stokes vectors (I_o=1 in (2.30)) and thus effectively the incident power density.

The received power density expression now becomes

$$p^r = \frac{1}{R^2}\mathbf{s}^{ra}.\left(\mathbf{R}^{-1}\right)^{\mathrm{T}}\mathbf{R}^{-1}\mathbf{Hs}^t$$

Applying (3.65) gives

$$p^r = \frac{1}{R^2}\mathbf{s}^{ra}.\left(\mathbf{R}^{-1}\right)^{\mathrm{T}}\mathbf{WR}^{-1}\mathbf{s}^t = \frac{1}{R^2}\mathbf{s}^{ra}.\mathbf{Ms}^t \tag{3.71}$$

in which $\mathbf{M}$ is called the *Stokes scattering operator*, defined in terms of $\mathbf{W}$:

$$\mathbf{M} = \left(\mathbf{R}^{-1}\right)^{\mathrm{T}}\mathbf{WR}^{-1} \tag{3.72}$$

which, in turn, can be completely specified by the elements of the scattering matrix of the target.

As an aside, note that from (3.65) and (3.72) the Kennaugh matrix and Stokes scattering operator are related by

$$\mathbf{H} = \mathbf{RR}^{\mathrm{T}}\mathbf{M} \tag{3.73}$$

Equation (3.34) defines radar cross section as

$$\sigma = \lim_{R\to\infty} 4\pi R^2 \frac{|E_r|^2}{|E_i|^2} \equiv \lim_{R\to\infty} 4\pi R^2 \frac{p^r}{p^i}$$

Since we have assumed that the incident power density at the target is unity this gives

$$\sigma = 4\pi \mathbf{s}^{ra}.\mathbf{M}\mathbf{s}^{t} \tag{3.74}$$

We know **R** from (2.32) and **W** from (3.64) and thus can compute the elements of the Stokes scattering operator. After lengthy manipulation we find

$$\begin{aligned} m_{11} &= 0.25\{S_{HH}S_{HH}^* + S_{HV}S_{HV}^* + S_{VH}S_{VH}^* + S_{VV}S_{VV}^*\} \\ &= 0.25\{S_{HH}S_{HH}^* + 2S_{HV}S_{HV}^* + S_{VV}S_{VV}^*\} \text{ for backscattering, with } S_{VH}=S_{HV}. \end{aligned}$$

$$\begin{aligned} m_{12} &= 0.25\{S_{HH}S_{HH}^* - S_{HV}S_{HV}^* + S_{VH}S_{VH}^* - S_{VV}S_{VV}^*\} \\ &= 0.25\{S_{HH}S_{HH}^* - S_{VV}S_{VV}^*\} \text{ for backscattering.} \end{aligned}$$

$$\begin{aligned} m_{13} &= 0.25\{S_{HH}S_{HV}^* + S_{HV}S_{HH}^* + S_{VH}S_{VV}^* + S_{VV}S_{VH}^*\} \\ &= 0.5\{\mathcal{Re}(S_{HH}S_{HV}^*) + \mathcal{Re}(S_{HV}S_{VV}^*)\} \text{ for backscattering} \end{aligned}$$

$$\begin{aligned} m_{14} &= 0.25j\{S_{HH}S_{HV}^* - S_{HV}S_{HH}^* + S_{VH}S_{VV}^* - S_{VV}S_{VH}^*\} \\ &= 0.5\{Im(S_{HV}S_{HH}^*) + Im(S_{VV}S_{HV}^*)\} \text{ for backscattering} \end{aligned}$$

$$m_{21} = 0.25\{S_{HH}S_{HH}^* + S_{HV}S_{HV}^* - S_{VH}S_{VH}^* - S_{VV}S_{VV}^*\} = m_{12} \text{ for backscattering}$$

$$\begin{aligned} m_{22} &= 0.25\{S_{HH}S_{HH}^* - S_{HV}S_{HV}^* - S_{VH}S_{VH}^* + S_{VV}S_{VV}^*\} \\ &= 0.25\{S_{HH}S_{HH}^* - 2S_{HV}S_{HV}^* + S_{VV}S_{VV}^*\} \text{ for backscattering} \end{aligned}$$

$$\begin{aligned} m_{23} &= 0.25\{S_{HH}S_{HV}^* + S_{HV}S_{HH}^* - S_{VH}S_{VV}^* - S_{VV}S_{VH}^*\} \\ &= 0.5\{\mathcal{Re}(S_{HH}S_{HV}^*) - \mathcal{Re}(S_{HV}S_{VV}^*)\} \text{ for backscattering} \end{aligned}$$

$$\begin{aligned} m_{24} &= 0.25j\{S_{HH}S_{HV}^* - S_{HV}S_{HH}^* - S_{VH}S_{VV}^* + S_{VV}S_{VH}^*\} \\ &= 0.5\{Im(S_{HV}S_{HH}^*) + \mathrm{Im}(S_{HV}S_{VV}^*)\} \text{ for backscattering} \end{aligned}$$

$$m_{31} = 0.25\{S_{HH}S_{VH}^* + S_{HV}S_{VV}^* + S_{VH}S_{HH}^* + S_{VV}S_{HV}^*\} = m_{13} \text{ for backscattering}$$

$$m_{32} = 0.25\{S_{HH}S_{VH}^* - S_{HV}S_{VV}^* + S_{VH}S_{HH}^* - S_{VV}S_{HV}^*\} = m_{23} \text{ for backscattering}$$

$$\begin{aligned} m_{33} &= 0.25\{S_{HH}S_{VV}^* + S_{HV}S_{VH}^* + S_{VH}S_{HV}^* + S_{VV}S_{HH}^*\} \\ &= 0.5\{\mathcal{Re}(S_{HH}S_{VV}^*) + S_{HV}S_{HV}^*\} \text{ for backscattering} \end{aligned}$$

$$m_{34} = 0.25j\{S_{HH}S_{VV}^{*} - S_{HV}S_{VH}^{*} + S_{VH}S_{HV}^{*} - S_{VV}S_{HH}^{*}\}$$
$$= 0.5Im(S_{VV}S_{HH}^{*}) \text{ for backscattering}$$

$$m_{41} = 0.25j\{S_{HH}S_{VH}^{*} + S_{HV}S_{VV}^{*} - S_{VH}S_{HH}^{*} - S_{VV}S_{HV}^{*}\} = m_{14} \text{ for backscattering}$$

$$m_{42} = 0.25j\{S_{HH}S_{VH}^{*} - S_{HV}S_{VV}^{*} - S_{VH}S_{HH}^{*} + S_{VV}S_{HV}^{*}\} = m_{24} \text{ for backscattering}$$

$$m_{43} = 0.25j\{S_{HH}S_{VV}^{*} + S_{HV}S_{VH}^{*} - S_{VH}S_{HV}^{*} - S_{VV}S_{HH}^{*}\} = m_{34} \text{ for backscattering}$$

$$m_{44} = 0.25\{-S_{HH}S_{VV}^{*} + S_{HV}S_{VH}^{*} + S_{VH}S_{HV}^{*} - S_{VV}S_{HH}^{*}\}$$
$$= 0.5\{S_{HV}S_{HV}^{*} - \mathcal{Re}(S_{HH}S_{VV}^{*})\} \text{ for backscattering} \quad (3.75)$$

With this expansion of **M**, (3.74) allows us to generate the radar response that would be observed if the transmitted wave was described by the normalised Stokes vector $\mathbf{s}^{t}$ and the scattered wave is received on an antenna with the normalised Stokes vector $\mathbf{s}^{ra}$. In summary, using the transpose instead of the dot product, and incorporating the symmetry of **M** for backscattering, this is

$$\sigma = 4\pi[1 \quad \cos 2\tau\cos 2\varepsilon \quad \sin 2\tau\cos 2\varepsilon \quad \sin 2\varepsilon]^{rec} \begin{bmatrix} m_{11} & m_{12} & m_{13} & m_{14} \\ m_{12} & m_{22} & m_{23} & m_{24} \\ m_{13} & m_{23} & m_{33} & m_{34} \\ m_{14} & m_{24} & m_{34} & m_{44} \end{bmatrix} \begin{bmatrix} 1 \\ \cos 2\tau\cos 2\varepsilon \\ \sin 2\tau\cos 2\varepsilon \\ \sin 2\varepsilon \end{bmatrix}^{trans}$$

(3.76)

To assist the readability of this last expression we have used expanded superscripts of *rec* and *trans* to signify received and transmitted. Fig. 3.21 shows the operation diagrammatically to emphasise the relationship between the elliptically transmitted polarisation, that scattered from the target and the dependence of the strength of the signal received resulting from the relative polarisation alignment of the scattered radiation and that of the receiving antenna. Again, this assumes that the polarisation state of the antenna on reception is the same as on transmission.

Clearly, there is a range of choices for the receiver and transmitter Stokes vectors. Conventionally, they are chosen to be the same, in which case the response generated is referred to as the *co-polarised* response. Alternatively, if the receiver polarisation is orthogonal to the transmitter polarisation the response is referred to as *cross-polarised.* Both are generally computed to give a description of target behaviour. To obtain the cross-polarised response the receiver normalised Stokes vector has the sign of its ellipticity angle ε reversed compared with its transmitted value to change the "handedness" on reception; it also has 90° added to its orientation angle τ to ensure received linear polarisation will be orthogonal to that transmitted.

We are now in the position to demonstrate the generation of polarisation response plots for a number of well defined discrete targets. That entails evaluating (3.76) for both co-polarised and cross-polarised reception over the full range of inclination and ellipticity angles, thereby showing how such a target appears for all polarisation combinations. We commence by looking at a large metallic plate aligned at right angles to the radar ray. A linearly polarised ray incident on such a plate will be totally reflected so we can write its scattering matrix as

$$S_{plate} = \begin{bmatrix} 1 & 0 \\ 0 & 1 \end{bmatrix}$$

This summarises the fact that a horizontally polarised field will be totally reflected (the "1" in the top left element) and a vertically polarised field will be totally reflected (the "1" in the bottom right element)[19]. There is no cross polarised reflection of the linearly polarised incident wave (signified by the zeros in the off-diagonal positions). From (3.64), (3.66) and (3.72), the corresponding Stokes scattering operator is

$$\mathbf{M} = \begin{bmatrix} 0.5 & 0 & 0 & 0 \\ 0 & 0.5 & 0 & 0 \\ 0 & 0 & 0.5 & 0 \\ 0 & 0 & 0 & -0.5 \end{bmatrix}$$

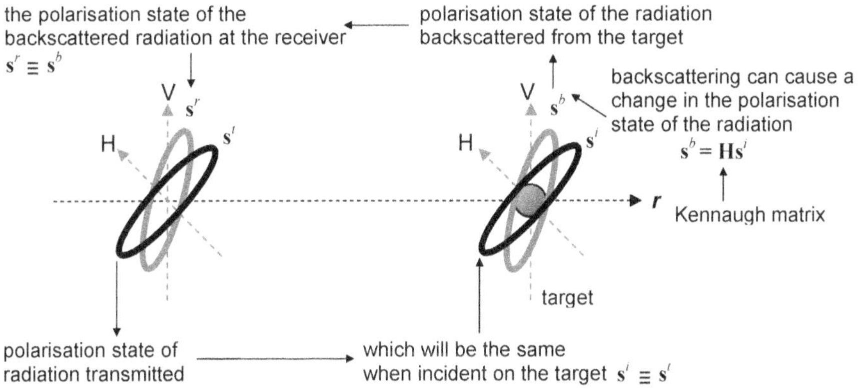

Fig. 3.21. The polarisation change sequence in radar scattering: the elliptically polarised case

When used in (3.76) the normalised polarisation responses in Fig. 3.22 are obtained. As expected, for any linear polarisation the plate gives a maximum co-polarised response (maximum reflection of the incident field). However, for elliptical polarisation the co-polarised response is less than maximum, reducing to zero for circular polarisation. To see why that is the case Fig. 3.23 demonstrates that the handedness of a circularly (elliptically) polarised wave is reversed, of necessity, on reflection. That also explains why the cross-polarised response is maximum for circular polarisation of either hand. Note that both the co- and cross-polarised behaviours are independent of orientation

[19] This is how it appears when the back scattering aligned coordinates are used. If a forward scattering alignment is adopted then one matrix element would be negative to preserve propagation conventions, as discussed in Appendix E.

angle, as would be expected since such a concept has no meaning when referring to a plate with no geometric boundaries or other aligned geometric features.

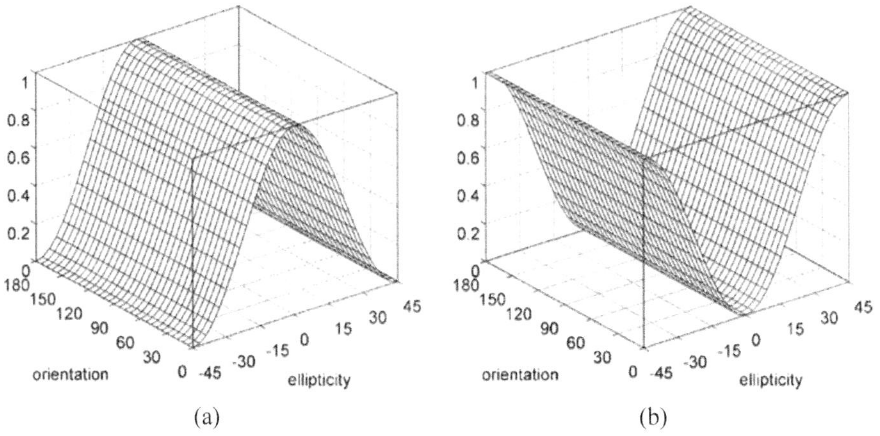

Fig. 3.22. (a) Co- and (b) cross-polarised responses of a flat metallic plate (and a trihedral corner reflector)

As a second example, Fig. 3.24 gives the polarisation response of the dihedral corner reflector shown in Fig. 3.25. That device is often used as a control point and a radiometric calibration target, as discussed in Sect. 4.2.2. As noted in Table 4.1 if it is constructed from square plates of side dimension a its maximum radar cross section, given when it is optimally aligned to the radar ray, is

$$\sigma = 8\pi\left(\frac{a^2}{\lambda}\right)^2$$

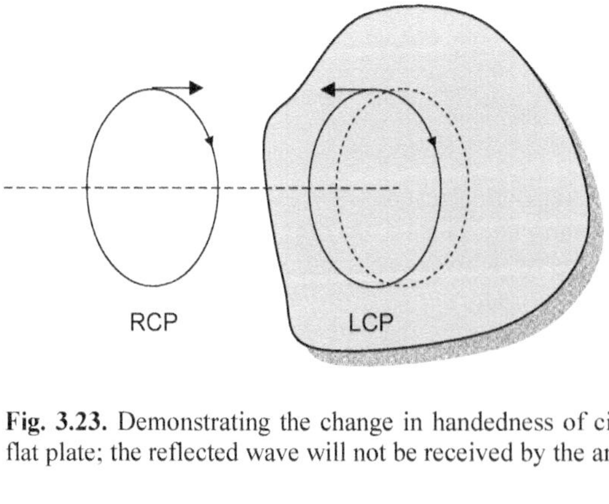

Fig. 3.23. Demonstrating the change in handedness of circular polarisation on reflection from a flat plate; the reflected wave will not be received by the antenna that launched the incident wave

This applies for both horizontal and vertical polarisation. It does not have a cross polarised response when aligned with its axis orthogonal to the incident ray as may be understood by looking at the reflected rays in Fig. 3.25. There is no opportunity over the two reflections for the polarisation to be rotated. Note also that the final polarity of a horizontally polarised wave is not affected by the two reflections, whereas that of a vertically polarised wave is reversed. Using these observations together with (3.43) we can see from the above expression for the radar cross section that the scattering matrix for the dihedral corner reflector is

$$S_{DCR} = \frac{\sqrt{2}a^2}{\lambda}\begin{bmatrix} 1 & 0 \\ 0 & -1 \end{bmatrix}$$

From (3.75) the corresponding Stokes scattering operator is

$$\mathbf{M} = \frac{a^4}{\lambda^2}\begin{bmatrix} 1 & 0 & 0 & 0 \\ 0 & 1 & 0 & 0 \\ 0 & 0 & -1 & 0 \\ 0 & 0 & 0 & 1 \end{bmatrix}$$

Using (3.76) the polarisation responses in Fig. 3.24 are obtained.

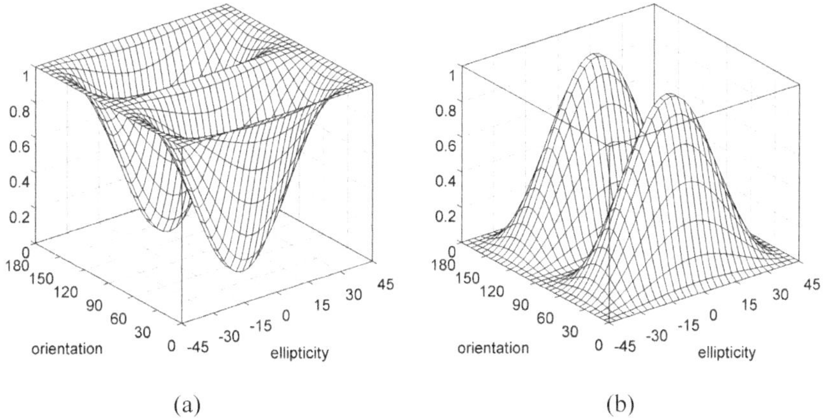

Fig. 3.24. (a) Co- and (b) cross polarisation responses of a dihedral corner reflector

The dihedral corner reflector is of limited value in calibration studies since it has to be aligned precisely along the flight direction of the radar platform; however, it is an important element in modelling the backscattering behaviours of landscape features that appear as vertical surfaces standing on horizontal surfaces. Buildings, ships at sea and even large tree trunks are examples that lend themselves to being described in that manner. A better calibration device is the trihedral corner reflector of Sect. 4.2.2, which doesn't suffer the alignment problem. Its *normalised* scattering matrix is the same as that above for an infinite flat plate, so that its normalised polarisation response is also given by Fig. 3.22.

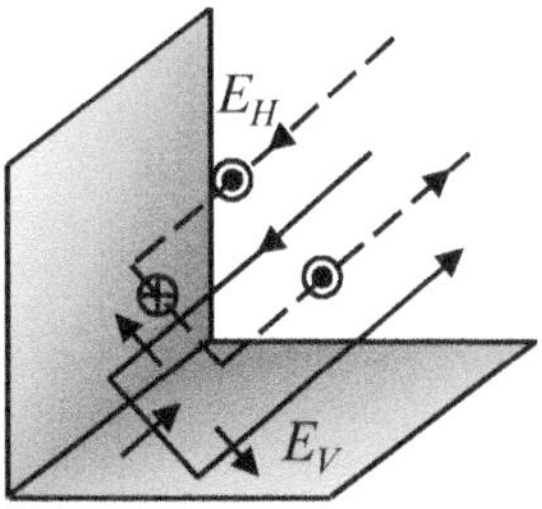

Fig. 3.25. Dihedral corner reflector in the back scatter alignment convention, showing how the polarity of a horizontal wave is not affected, whereas the polarity of a vertically polarised wave is reversed.

The responses represented by the radar cross sections in (3.74) and (3.76), and just illustrated, are for a single discrete target or for a dominant scatterer in a resolution cell (pixel). In most remote sensing applications each resolution element consists of a very large number of randomly distributed incremental scatterers, as depicted in Fig. 3.17. In that case we compute the response as the average over that ensemble. If the response is normalised by the size of the resolution element then the scattering coefficient for the pixel is

$$\sigma^o = \frac{4\pi}{r_a r_g}\mathbf{s}^{ra}.\langle\mathbf{M}\rangle\mathbf{s}^t \tag{3.77}$$

where, as before, the angular brackets signify the average. The average could also be taken over several pixels in which case the total area may need to be included in (3.77) rather than just that of the individual resolution element.

We can also average the cross section of (3.69) over a pixel or several pixels:

$$\sigma^o = \frac{4\pi}{r_a r}\left\langle\left|\mathbf{p}^{ra}.S\mathbf{p}^t\right|^2\right\rangle \tag{3.78}$$

However, whereas the average scattering coefficient in (3.77) is directly related to the average Stokes scattering operator **M**, in (3.78) the averaging cannot be taken conveniently inside the magnitude squared operation, requiring the scattering matrices of individual scatterers to be found and processed by the polarisation vectors before the average squared operation is applied. That renders (3.78) less convenient than (3.77). It can also be demonstrated that more multiplications are required to evaluate (3.78) compared with (3.77).

As a final illustration consider the scattering matrix for a slightly rough surface; in Chapt. 5 this will be seen to be of the form

$$\mathbf{S} = \begin{bmatrix} -0.41 & 0 \\ 0 & -0.57 \end{bmatrix}$$

which is based on the Bragg surface scattering model for a dry surface at an incidence angle of about 30°. The polarisation responses for this surface are shown in Fig. 3.26, which are very different again from the two already considered.

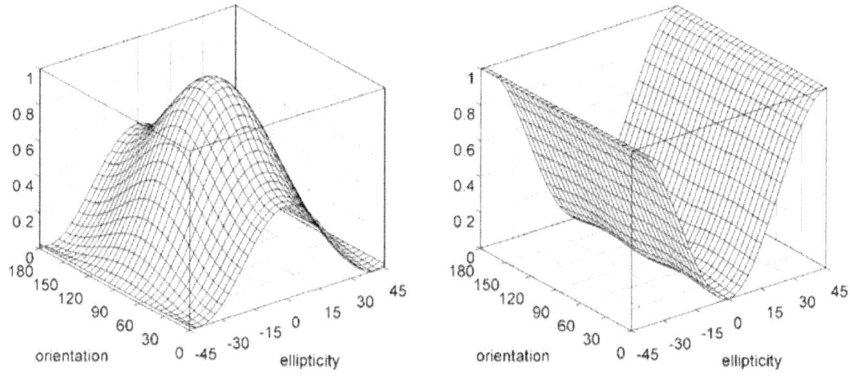

(a) (b)

Fig. 3.26. (a) Co- and (b) cross polarisation responses of a slightly rough dry surface

Fig. 3.27 shows the co-polar responses or three groups of pixels in an AirSAR scene at C, L and P bands. As a result of the specific examples just given we can see how the shapes of the responses might be used as analytical features when interpreting the likely cover types.

3.23 Compact Polarimetry

There are complexities associated with recording the full scattering matrix for a target that complicates the design and construction of fully quadrature polarised radar. The design solutions adopted impose limitations on parameters such as swath width, as demonstrated in Fig. 3.28. That figure shows the timing sequence of the transmitted ranging pulses and the received echoes for normal quad polarised radar. The ranging pulse is first transmitted on one polarisation. All the returns from that ranging pulse on the two orthogonal polarisations are received before the next ranging pulse is transmitted; this time the orthogonal polarisation is used in transmission. All the echoes are again received before the next ranging pulse is transmitted, but again with the polarisation reversed.

Such a sequence uses a single transmitter which has its output alternated between antennas that radiate on the orthogonal polarisations, and uses receiving antennas also sensitive to the orthogonal polarisations. While that has a number of hardware design advantages it means that all of the returns from a given ranging pulse have to be received in half the time interval between transmitted pulses of the same polarisation. As a consequence, from (3.18a), only half the swath width is possible compared with that if the full inter-pulse interval were available.

In order to achieve a better swath width, and coincidentally reduce average power requirements and simplify transmitting hardware, compact polarimetric systems have been proposed. Some are as simple as dual polarised radars in which just one polarisation is transmitted and both received. However, systems that are called partially polarimetric offer better prospects for understanding target behaviour by more closely approximating full quadrature polarisation.

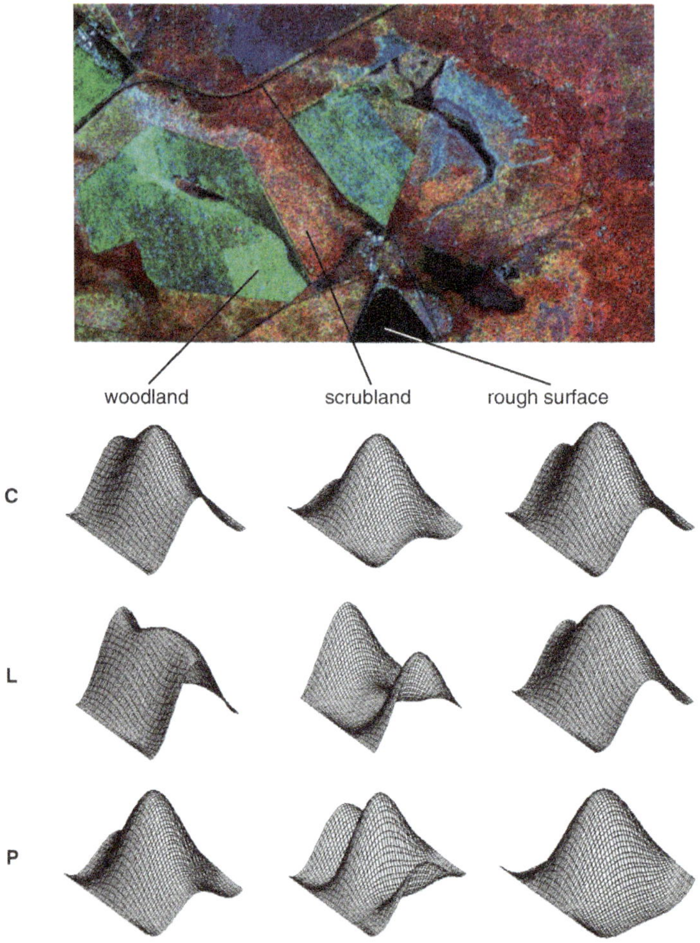

Fig. 3.27. The co-polarisation responses at C, L and P bands for three different regions in a AirSAR image (Carinda, NSW, Australia); the image has been displayed using VV polarisations with C band as red, L band as green and P band as blue, while the axes have been omitted for clarity; all processing was done using ENVI™ (ITT Visual Information Solutions)

The $\pi/4$ system[20] proposes transmission with a linear polarisation at the 45° orientation, mid-way between vertical and horizontal polarisation, as shown in Fig. 3.29. Reception uses both horizontal and vertical polarisations. If we refer to the incident (transmitted)

[20] See J-C Souyris, P.I.R. Fjortoft, S. Mingot and J-S Lee, Compact polarimetry based on symmetric properties of geophysical media: the $\pi/4$ mode, *IEEE Transactions on Geoscience and Remote Sensing*, vol. 43, no. 3, March 2005, pp. 634-646.

signal as $\mathbf{E}_D^i$ (for diagonal polarisation) then by resolving it into its horizontal and vertical components we can see that the backscattered (received) horizontal and vertical signals in terms of the symmetric scattering matrix of the target are given by

$$E_H^b = 0.707 S_{HH} \left|\mathbf{E}_D^i\right| + 0.707 S_{HV} \left|\mathbf{E}_D^i\right|$$
$$E_V^b = 0.707 S_{HV} \left|\mathbf{E}_D^i\right| + 0.707 S_{VV} \left|\mathbf{E}_D^i\right|$$

so that the effective scattering matrix elements recorded are

$$S_{HD} = S_{HH} + S_{HV} \quad (3.79a)$$
$$S_{VD} = S_{VV} + S_{HV} \quad (3.79b)$$

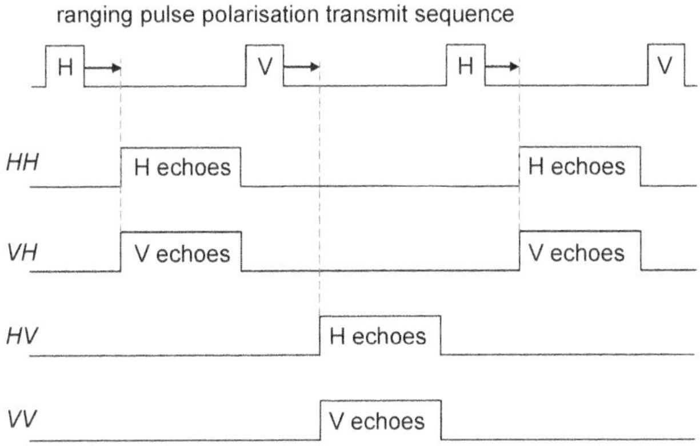

Fig. 3.28. The sequence of transmitted ranging pulses and received echoes in a fully quad-polarised imaging radar

The factors 0.707, which come from the trig functions of 45°, are generally ignored since they don't influence target properties, although they will be important for system level power considerations. As expected from the limited nature of the system (one transmit polarisation) (3.79) shows that we cannot recover the full scattering matrix for the target but only combinations of its elements. Provided the targets of interest are restricted to those with reflection symmetry the derived covariance information can be used as a target discriminator[21].

An alternative partially polarised system, referred to as having compact hybrid polarity, proposes circular transmission with linear horizontal and vertical reception as shown in Fig. 3.30; left and right circular reception would also be possible[22,23]. If right circular

[21] *Ibid.*
[22] See R.K. Raney, Hybrid-polarity SAR architecture, *IEEE Transactions on Geoscience and Remote Sensing*, vol. 45, no. 11, pt 1, November 2007, pp. 3397-3404.
[23] Note also that a fully quad-polarised system has been proposed based on alternate transmission of right and left circularly polarised pulses followed by vertical and horizontal linear reception; see R.K. Raney,

polarisation were transmitted then from (2.25) its equivalent horizontal and vertical components are $E_H = E_R/\sqrt{2}$ and $E_V = -jE_R/\sqrt{2}$ so that the backscattered horizontal and vertical components are

$$E_H^b = 0.707(S_{HH} - jS_{HV})E_R$$
$$E_V^b = 0.707(S_{HV} - jS_{VV})E_R$$

giving as the effective scattering elements

$$S_{HR} = S_{HH} - jS_{HV} \quad (3.80a)$$
$$S_{VR} = S_{HV} - jS_{VV} \quad (3.80b)$$

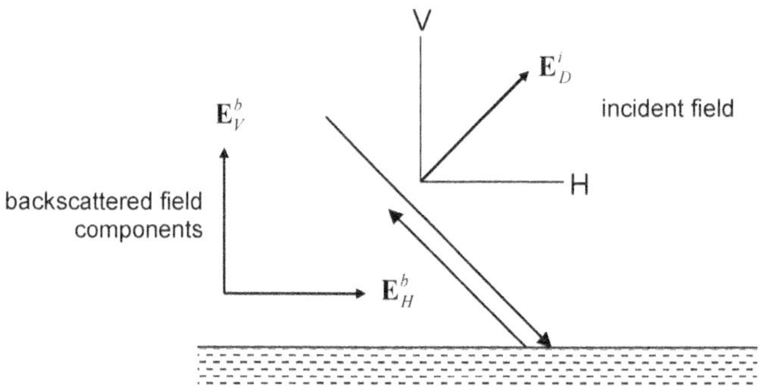

Fig. 3.29. The fields in π/4 compact polarimetry

3.24 Faraday Rotation

Having covered the concept of polarisation and scattering matrices we can now look at a peculiar influence of the atmosphere on radio wave propagation that has implications for radar remote sensing.

In the upper atmosphere there is a region of ionisation stretching from about 80km to 400km. Known as the ionosphere, it is formed by solar photons disassociating atmospheric molecules thereby creating free electrons that can interact with the passage of a radio wave. Because it is sunlight dependent the properties of the ionosphere vary continuously, and certainly with time of day, season and with the long and short term cycles of the sun, most notably the 11 year sunspot cycle. Because of the mix of atmospheric constituents and photon energies, the ionosphere breaks up into a number of layers of different electron densities[24]. Those layers are well known to the HF radio community since they are used to refract radio waves around the earth's curvature. In fact for frequencies at HF and lower the ionosphere will not permit the transmission of radio

Hybrid-quad-pol SAR, *Proceedings of the International Geoscience and Remote Sensing Symposium 2008* (*IGARSS08*), vol. 4, 7-11 July 2008 pp. 491- 493 Boston 2008.

[24] See J.A. Richards, Radio Wave Propagation: *An Introduction for the Non-Specialist*, Springer, Berlin, 2008.

waves; all energy transmitted upwards towards the ionosphere will be returned to the earth. As a consequence, any transmission to spacecraft has to happen at frequencies high enough that the ionosphere appears transparent. Similarly, any transmission from a space vehicle to the earth has to be at VHF and higher in order to pass through the ionosphere.

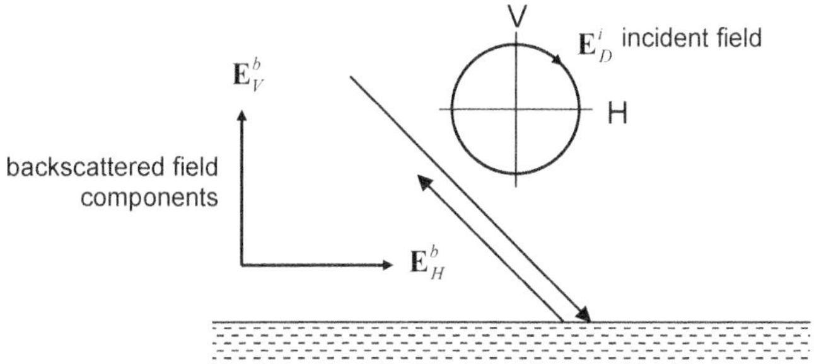

Fig. 3.30. The fields in hybrid compact polarimetry

Fortunately, the frequencies used in radar remote sensing are high enough that the ionosphere generally is not a problem and we can image from space. At the lower end of the frequency ranges of interest however, even though the signal passes though the ionospheric layers, there is an effect on the plane of polarisation of the wave. It will suffer *Faraday rotation*. The rotation can be quite severe at P and L bands but is much less of a problem at C and X bands.

Faraday rotation is the result of a wave propagating in a medium – such as the charged environment of the ionosphere – in the presence of a magnetic field (such as the earth's magnetic field) which has a component parallel to the direction of propagation. The degree of rotation can be expressed in several forms, but if the earth's magnetic field can be assumed not to change over the path travelled by the radar wave in the ionosphere then we can express the rotation angle as[25]

$$\Omega = K\lambda^2 \tag{3.81a}$$

where

$$K = 2.62x10^{-13} B\cos\chi \int_s N_e ds \tag{3.81b}$$

in which N_e is the electron density of the ionosphere (which varies with height and position) and the integration is over the path travelled by the ray through the ionosphere. $B\cos\chi$ is the component of the earth's magnetic field parallel to the direction of propagation in which B is the local value of the magnetic field and χ is the angle between a normal drawn to the propagation direction and the direction of the field. The integral of the electron density over the propagation path is referred to as the total electron count (TEC).

The angular sense of the rotation depends upon the direction of the parallel component of the magnetic field compared with the direction of propagation. If a ray passes through the ionosphere in both directions, such as to the ground on radar transmission and from

[25] See W.B. Gail, Effect of Faraday rotation on polarimetric SAR, *IEEE Transactions on Aerospace and Electronic Systems*, vol. 34, No. 1, January 1998, pp. 301-308, and A. Freeman and S. Saatchi, On the detection of Faraday rotation in linearly polarized L-band SAR backscatter signatures, *IEEE Transactions on Geoscience and Remote Sensing*, vol. 42, No. 8, August 2004, pp. 1607-1616.

the ground after scattering, the rotations will add. The sign of the rotation angle Ω is positive in the northern hemisphere and negative in the southern hemisphere based on a right handed coordinate system in which, for transmission, z is the propagation direction towards the earth, x is the direction of horizontal polarisation and y is the direction of vertical polarisation.

Consider now the impact of Faraday rotation on recorded radar imagery. First, for a simple single linearly polarised system the rotation will lead to a loss of signal at the receiver since the backscattered polarisation no longer aligns fully with the radar antenna, as seen in Sect. 2.16. How severe is the effect? Note from (3.81a) that it is strongly dependent on wavelength. Freeman and Saatchi[26] estimate that the worst case one-pass rotations are about 2.5° at C band, 40° at L band and 320° at P band. In general we can assume that the effect is negligible for C band and higher frequencies, important at L band and quite severe at P band, requiring correction. For a multi-polarised system Faraday rotation will cause significant cross-talk – i.e. coupling – among the polarisations as the following demonstrates.

Using (2.20), and noting that propagation will be out of the page in a right hand coordinate system with the axes shown in Fig. 2.20, the effect of Faraday rotation (on both transmission and reception) on the observed scattering matrix can be written

$$\mathbf{X} = \begin{bmatrix} X_{11} & X_{12} \\ X_{21} & X_{22} \end{bmatrix} = \begin{bmatrix} \cos\Omega & \sin\Omega \\ -\sin\Omega & \cos\Omega \end{bmatrix} \begin{bmatrix} S_{HH} & S_{HV} \\ S_{HV} & S_{VV} \end{bmatrix} \begin{bmatrix} \cos\Omega & \sin\Omega \\ -\sin\Omega & \cos\Omega \end{bmatrix} \tag{3.82a}$$

in which we have assumed symmetry (reciprocity) for the target scattering matrix. On expansion this shows that the observed scattering matrix is

$$\mathbf{X} = \begin{bmatrix} S_{HH}\cos^2\Omega - S_{VV}\sin^2\Omega & S_{HV} + 0.5\sin 2\Omega(S_{HH} + S_{VV}) \\ S_{HV} - 0.5\sin 2\Omega(S_{HH} + S_{VV}) & -S_{HH}\sin^2\Omega + S_{VV}\cos^2\Omega \end{bmatrix} \tag{3.82b}$$

which demonstrates explicitly the coupling between polarisations and the loss of symmetry (since $X_{12} \neq X_{21}$). If Ω=0 $\mathbf{X}$ reduces to $\mathbf{S}$.

In (3.82) we have assumed that the radar system is properly calibrated and that any noise is negligible; otherwise the observed matrix $\mathbf{X}$ will contain additive noise terms and other matrices that multiply (distort) the observations resulting from mis-calibration. If we assume that calibration is good and noise is not a problem we can, in principle, recover the actual scattering matrix by inverting (3.82a) provided we know the rotation angle Ω:

$$\mathbf{S} = \begin{bmatrix} \cos\Omega & -\sin\Omega \\ \sin\Omega & \cos\Omega \end{bmatrix} \begin{bmatrix} X_{11} & X_{12} \\ X_{21} & X_{22} \end{bmatrix} \begin{bmatrix} \cos\Omega & -\sin\Omega \\ \sin\Omega & \cos\Omega \end{bmatrix} \tag{3.83}$$

[26] Freeman and Saatchi, *loc cit.*

CHAPTER 4

CORRECTING AND CALIBRATING RADAR IMAGERY

As with any imagery acquired by airborne and spacecraft sensors the data recorded by imaging radar can be distorted in both brightness and geometry as the result of a number of environmental and system factors. In this Chapter we explore the most significant sources of radiometric and geometric distortions and treat methods for removing, or at least minimising, them. We commence with errors in geometry.

Closely related to distortions in brightness is the need to calibrate data, including the need to ensure that polarimetric responses are cross-calibrated. Calibration methods are also treated in this Chapter.

4.1 Sources of Geometric Distortion

4.1.1 Near Range Compressional Distortion

Equation (3.2) shows that the ground range resolution of an imaging radar depends on the reciprocal of the sine of the local angle of incidence. That means that the minimum resolvable region on the ground in the across track direction is larger for smaller angles of incidence than it is for larger angles, as seen in Fig. 4.1. In the azimuth (along track) direction the minimum resolvable distance is the same at near and far range, and we need not consider it further.

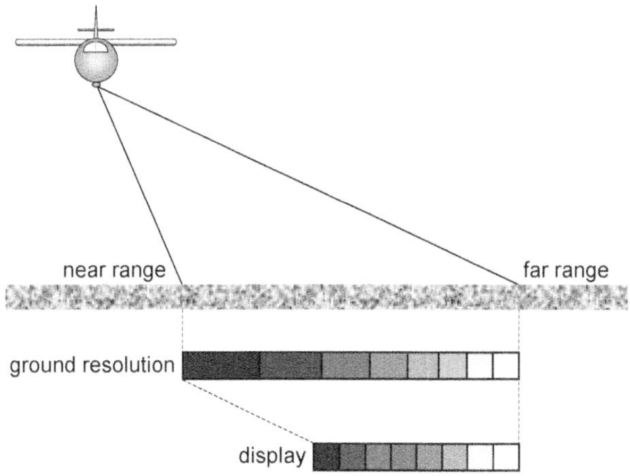

Fig. 4.1. Resolution of the image into range cells defined by the ground range resolution of the radar system, and the compression of near range cells when displayed or printed

J.A. Richards, *Remote Sensing with Imaging Radar*, Signals and Communication Technology,
DOI: 10.1007/978-3-642-02020-9_4,

Even though range resolution varies across the swath that is not the way the data is generally represented. Imagery is most often displayed using uniformly-sized pixels on a uniform grid of pixel centres, either on a computer screen or in hard copy form. Consequently, the larger regions of ground covered by a resolution cell at near range are displayed with the same dimensions as smaller regions on the ground resolved at far range. As a result the ground detail at near range is compressed into a smaller cell in comparison to the detail at far range, as illustrated in Fig. 4.1.

For imaging radars with a large change in local incidence angle across the swath (most notably aircraft systems) the compression of detail at near range has a dramatic effect on the range appearance of the image. That is perhaps best illustrated by boundaries or roads at angles to the flight line. Consider a region on the ground as shown in Fig. 4.2a in which there is a square of grid-like feature such as field boundaries. Within each of the square cells there could be many pixels. Imagine also that there are some diagonal lines as shown – they could be roads connecting across field corners. Fig. 4.2b shows how that region on the ground will appear in recorded and displayed radar imagery. Not only do the near range features appear compressed but linear features at angles to the flight line appear curved. Indeed the combined effect is as if the image were rolled backwards on the near swath side. If the variation of incidence angle is not great across the swath then the effect will be small, particularly if the radar operates at higher incidence angles. An actual example of near range compressional distortion is seen in the aircraft radar image shown in Fig. 4.3a.

This type of distortion also occurs for wide field of view optical systems; however it then happens at far range where the spatial resolution is poorest. Because optical scanners work to both sides of the flight line the distortion appears on both edges, leading to what is known as "S-bend" distortion in optical scanner imagery[1].

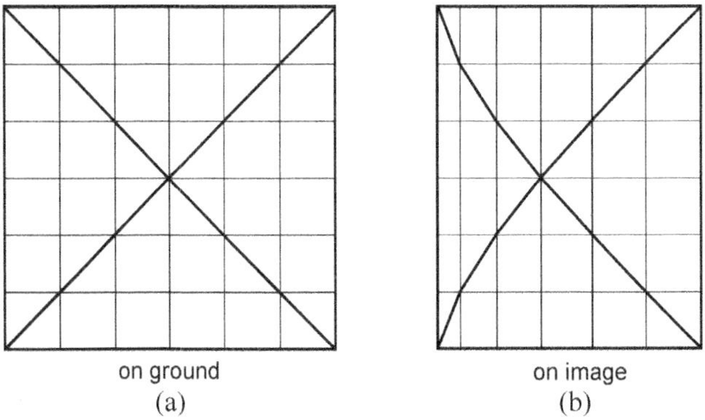

Fig. 4.2. (a) Region on the ground consisting of rectangular fields and diagonal features and (b) how it would appear in radar imagery subject to near range compressional distortion

Since near range compressional distortion is the result of the sine of the local angle of incidence in (3.2) it can easily be corrected as seen in Fig 4.3b, either by compensating for the mathematical dependence on θ in (3.2) or by resampling the image in the range direction on to a regular grid. Regrettably the low level of detail at near range resulting

[1] See J.A. Richards and X. Jia, *Remote Sensing Digital Image Analysis*, 4th ed., Springer, Berlin, 2006, Chapt. 2.

from the large resolution cells cannot be improved and, notwithstanding the good geometry, the detail at near range is usually still poor in a geometrically correct product.

(a)

(b)

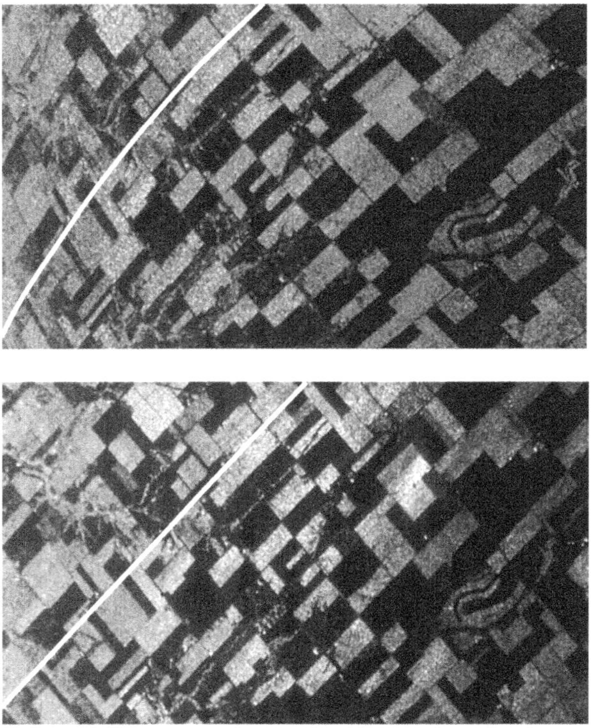

Fig. 4.3. (a) Aircraft radar image data showing near range compressional distortion and (b) corrected version; the white lines emphasise the degree of distortion and correction (*imagery courtesy of NASA JPL*)

4.1.2 Layover, Relief Displacement, Foreshortening and Shadowing

Consider how a tall tower would appear in the range direction in a radar image with sufficient resolution, as illustrated in Fig. 4.4. Because the radar echo from the top of the tower arrives back at the radar before that from the base (it travels a shorter two way path) the tower appears to lie over towards the radar. To see that effect most clearly it is of value to draw concentric circles from the radar set as indicated. Any points lying on one of those circles will be at the same slant range from the radar and thus will create echoes with the same time delay. By projecting the circle which just touches the top of the tower onto the ground it is seen that the tower is superimposed on ground features closer to the radar set than the base of the tower. For obvious reasons this effect is referred to as *layover*. It is interesting to recall that in optical imagery vertical objects appear to lie away from the imaging device since they are superimposed on features further from the device than the base (just as we see them in everyday life).

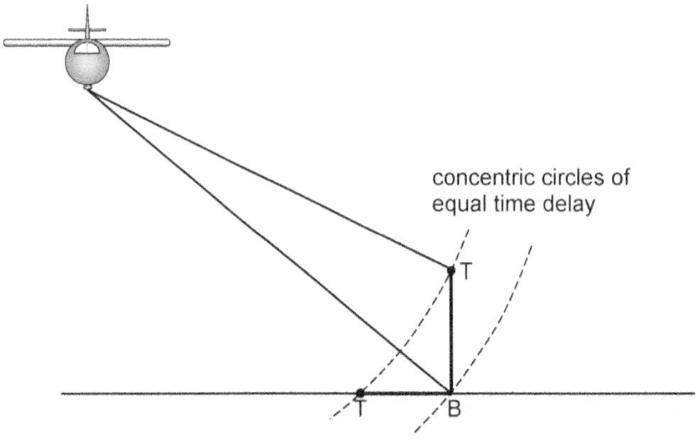

Fig. 4.4. Illustration of why tall objects lay over towards the radar

Now consider how a vertical feature with some horizontal dimension, such as the model mountain depicted in Fig. 4.5, will appear. Using the same principle of concentric circles to project the vertical relief onto the horizontal ground plane, several effects are evident. The front slope is foreshortened and the back slope is lengthened. In combination these two effects suggest that the top of the mountain is displaced towards the radar set. When displayed in ground range format, as shown, they again give the effect that the mountain is lying over towards the radar. If we know the local height then the amount of relief displacement can be calculated. In principle, therefore, the availability of a digital terrain map for the region should allow relief displacement distortion to be corrected

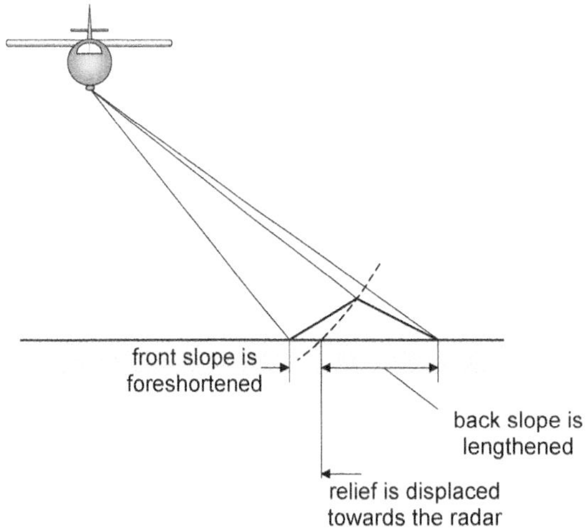

Fig. 4.5. Illustration of relief displacement: foreshortening of front slopes and lengthening of back slopes

As well as causing range displacement effects, topographic relief also manifests itself in a modification of the brightness of the image. On front slopes the local angle of incidence will be smaller than expected and thus the slopes will appear bright. On back slopes the angle of incidence will be larger than expected making them darker than would otherwise be the case. Fig. 4.6 shows why that happens, using typical scattering characteristics of a surface. Fig. 4.7 demonstrates the effect using a Seasat radar image for which the local incidence angle is 20°. At such small angles relief distortion can be quite severe in mountainous terrain.

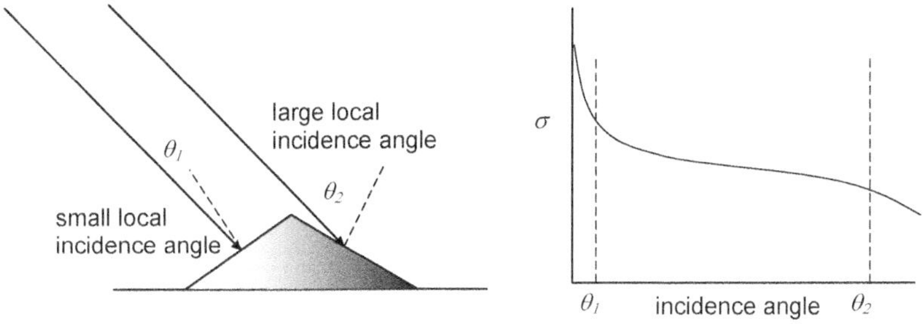

Fig. 4.6. Demonstration of brightness modulation caused by terrain relief and the angular dependence of surface backscattering coefficient

Now consider the potential for shadowing, as seen in Fig. 4.8. Shadowing is absolute in radar imaging and cannot be corrected; by contrast, for nadir viewing optical sensors it is possible sometimes to detect measurable signals in shadow zones because of atmospheric scattering of incident radiation into the shadow regions at (short) optical wavelengths. Radar shadowing is likely to be most severe in the far range and for larger angles of incidence, whereas it is often non-existent for smaller incidence angles.

We can now draw some conclusions from our observations so far that are of value in choosing look (incidence) angles suited to particular purposes:

- For low relief regions larger look angles will emphasise topographic features through shadowing, making interpretation easier.
- For regions of high relief larger look angles will minimise layover and relief distortion, but will exaggerate shadowing.
- Relief distortion is worse for smaller look angles.
- From spacecraft altitudes reasonable swath widths are obtained with mid range look angles (35° – 50°) for which there is generally little layover and little shadowing. Look angles in this range are good for surface roughness interpretation.

4.1.3 Slant Range Imagery

Recall from (3.1) that the radar system fundamentally resolves detail in the *slant* range direction. But we, as users, are interested in imagery that lies along the ground plane, leading to (3.2) as the range resolution expression most often used. As a result we create radar images that are projections onto the ground plane, as we must of course if we want them to be as close, planimetrically, to the actual detail on the ground, or if we want to join images side-by-side to form mosaics.

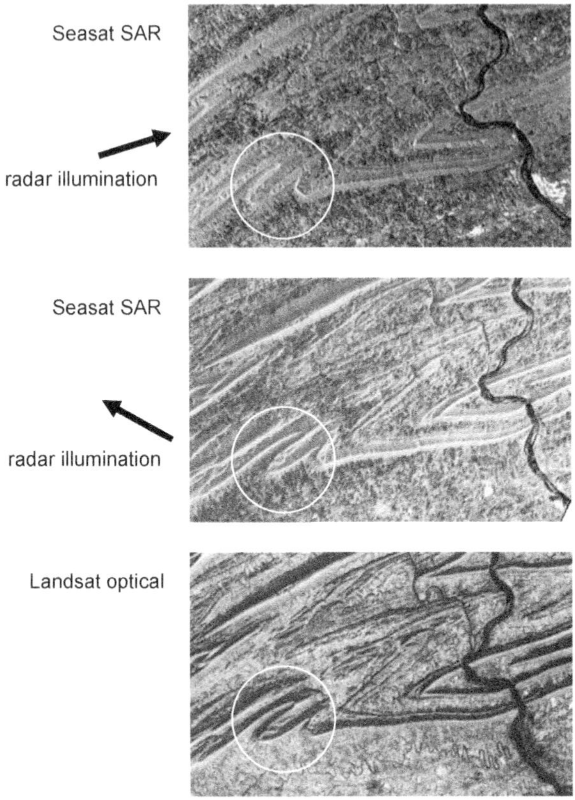

Fig. 4.7. Example of significant relief distortion of mountainous regions at low incidence angles; the image was acquired by the Seasat SAR in 1978 of the Appalachian mountains in Pennsylvania; a Landsat optical image is shown for comparison; note that the slopes facing the radar illumination direction appear bright, whereas those away from the illumination appear darker; note also the rather severe terrain distortion evident within the small circle[2] (from J.P. Ford et al., *Seasat Views North America, the Caribbean, and Western Europe with Imaging Radar*, JPL Publication 80-67, 1 November, 1980)

It is possible, nevertheless, to create an image product that represents detail on the slant plane, rather than the ground plane. This is illustrated in Fig. 4.9. In such a view the image has range coordinates measured along the slant direction rather than along the ground. A simple way to envisage the slant plane is to project it out to the side of the platform as shown. An advantage of slant range imagery is that it doesn't suffer the near range compressional distortion encountered when the ground range form is used; this can be appreciated by looking at the series of full concentric rings in the figure, the distance between pairs of which represent the slant range resolution of the system. The dotted curves illustrate that relief distortion occurs in both forms of imagery.

[2] The distortions can be assessed qualitatively by viewing the region in Google Earth™ or Google Maps™.

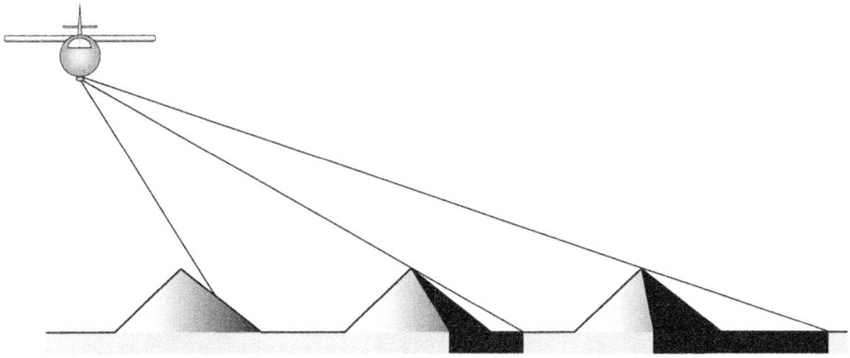

Fig. 4.8. Shadows in radar imagery

4.2 Geometric Correction of Radar Imagery

4.2.1 Regions of Low Relief

When the influence of relief is small the severe geometric distortions of layover, foreshortening and lengthening of back slopes are not significant. The remaining geometric errors are near range compressional distortion and the spatial errors associated with platform motion and attitude variations, and earth rotation, much the same as with optical remote sensing data.

Those errors can be corrected, first, by removing compressional distortion via a knowledge of the local angle of incidence, and its variation across the swath, followed by the use of control points and mapping polynomials[3]. Control points include natural and cultural features that can be identified both on the image data and a map of the region covered by the image.

Because of the presence of speckle in radar imagery (see Sect. 4.3.1) it is often difficult to locate naturally occurring control points to the required degree of accuracy. As a result, artificial control points are often created, prior to recording the image data, by deploying devices that will give recognisable returns in the received imagery. If the positions of those devices are accurately known, usually through having been determined using GPS/GNSS techniques, then rectification is assisted.

Two types of device can be used for that purpose, both of which return incident radar energy to the platform. One is passive, similar to optical retro-reflectors used to reflect laser beams. The other is active, working on the same principle, but incorporating electronics to amplify and re-transmit a calibrated level of power back to the platform.

[3] See J.A. Richards and X. Jia, *loc cit.*

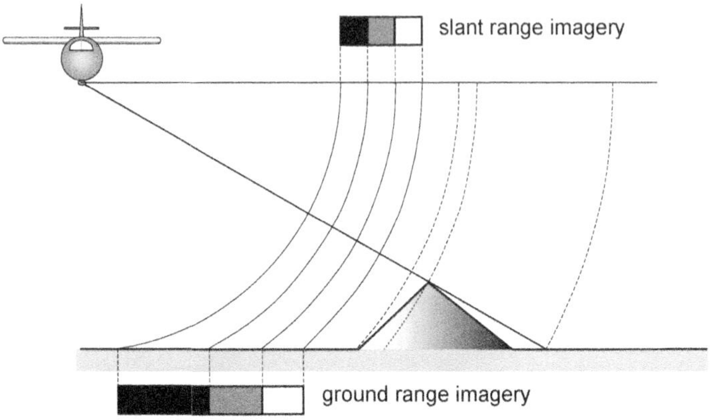

Fig. 4.9. Slant plane and ground plane views in the range direction

4.2.2 Passive Radar Calibrators

Any passive device that is capable of reflecting the incident microwave energy back to the radar could be used as a point of spatial reference in an image. A flat metal plate is a simple example. However, it must be aligned very accurately at right angles to the incoming beam for it to be of value. Instead a metallic corner reflector is generally preferred. There are four types in common use: the dihedral, the triangular trihedral, the square trihedral and the circular trihedral reflectors shown in Fig. 4.10. As well as providing a spatial reference, the radar cross sections of those devices are well known so that, in principle, they could also be used to calibrate the received power level in the radar response. Table 4.1 summarises the properties of the four devices. Their radar cross sections indicate the strength of the response they will provide to the radar set if properly aligned; their beamwidths indicate the range of angles over which their responses remain above half that at bore sight (i.e. less than 3dB down on maximum[4]).

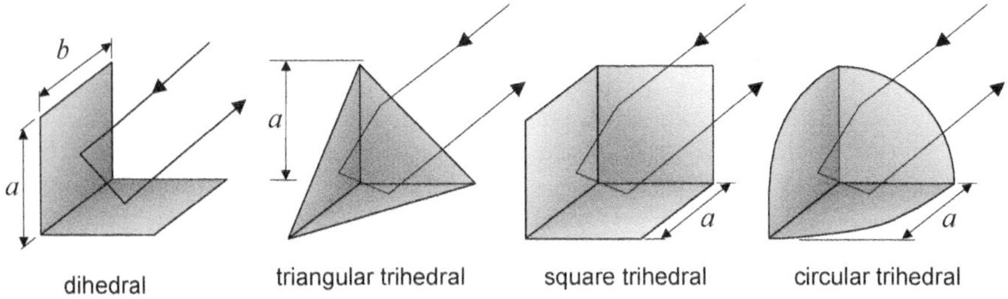

Fig. 4.10. Passive corner reflectors used for geometric correction and calibration of radar imagery

The dihedral corner reflector must be aligned so that the boundary between its horizontal and vertical planes is parallel to the flight line of the platform; only then does it

[4] "3dB down" means the angle away from bore sight at which the response is -3db compared with the maximum. From the definition of the decibel we can see that -3dB is equivalent to a ratio of 0.5.

provide good reflection over a range of angles about bore sight (the angle of view for which it appears symmetrical). Trihedral corner reflectors are more forgiving in their alignment and will give a moderately good return off-bore sight in both directions.

Table 4.1. Properties of corner reflectors

Device	Maximum radar cross section	3dB beamwidth
dihedral	$\sigma = 8\pi\left(\frac{ab}{\lambda}\right)^2$	$\pm 15^{\circ}$
triangular trihedral	$\sigma = \frac{4\pi}{3}\frac{a^4}{\lambda^2}$	40° cone angle about bore sight
square trihedral	$\sigma = 12\pi\frac{a^4}{\lambda^2}$	23° cone angle about bore sight
circular trihedral	$\sigma = 15.6\frac{a^4}{\lambda^2}$	32° cone angle about bore sight

4.2.3 Active Radar Calibrators (ARCs)

Instead of relying on passive reflection of the incident radar beam for localisation and calibration, it is possible to build a radio receiver which detects the energy and then transmits back to the radar an amplified signal at a known level and thus equivalent radar cross section. Such a device is generically called a *transponder*, and is shown in Fig. 4.11. Its main benefits are that the signal transmitted can be much larger than that scattered by passive devices and alignment problems are not so severe since simple communications antennas, with moderately broad beams, can be used both for reception and transmission. In radar remote sensing the device is generally referred to as an *active radar calibrator* (ARC).

One matter that is sometimes important with radar transponders is that the electronics between the receiving and transmitting antennas introduces a small time delay into the returning signal, additional to that resulting from the distance between the radar set and the target. Measurement of the combined time delay would therefore suggest that the transponder is further from the radar set in range, by an unknown amount, than it really is. To overcome that uncertainty, a deliberate delay element is sometimes incorporated into the transponder, as seen in the figure, so that the overall time delay from reception at the transponder to transmission of its response is known exactly and corresponds to a fixed distance in slant range. Any range measurements from the radar to the ground can then have that known delay subtracted to give the correct slant range. Interestingly, if the transponder radiates on a slightly different carrier frequency the ARC will appear in a different position in azimuth. That effect can be employed to shift the ARC response to an area of an image where it can more easily be seen[5].

Transponders are designed to have a specified radar cross section, σ. From (3.33) it can be seen that radar cross section, normally a passive quantity, is related directly to the ratio of the power received at the radar set to that transmitted by:

$$\sigma = \frac{(4\pi)^3 R^4}{G_t G_r \lambda^2} \frac{P_r}{P_t}$$

4.2.4 Polarimetric Active Radar Calibrators (PARCs)

The polarisation of the receiving and transmitting antennas in an active radar calibrator can be different. For example, the device might receive horizontally and transpond vertically. In such a case the transponder, known as a *polarimetric active radar calibrator* (PARC), can be used to calibrate cross-polarised (*HV*, *VH*) imagery.

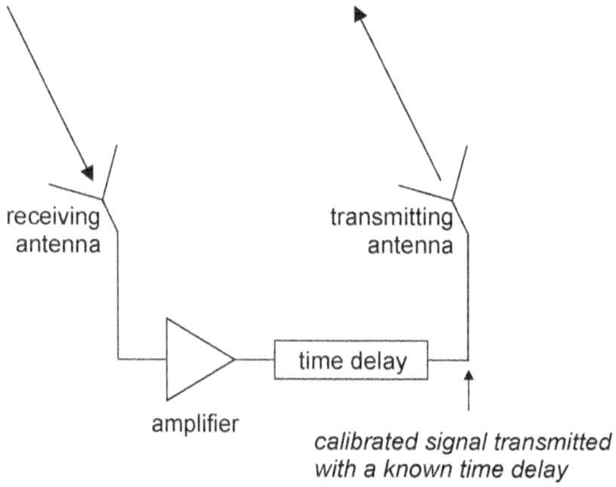

Fig. 4.11. Schematic of an active radar calibrator

4.2.5 Regions of High Relief

Figs. 4.5 and 4.9 show that topographic features are distorted planimetrically in both ground and slant range imagery because distance in the range direction is derived from the measurement of time delay. More particularly, we see from the manner in which the apex of a feature is shifted that there is a bunching towards the radar in the vicinity of regions of localised relief which can be so severe on occasions that fore slopes could compress to a single range position. Fortunately the shift towards the radar is easily modelled. As seen from Fig. 4.12 the distortion (shift) in range towards the radar is

$$\Delta r = h \cot\theta \tag{4.1}$$

[5] See M. Shimada, H. Oaku and M. Nakai, SAR calibration using frequency-tunable active radar calibrators, *IEEE Transactions on Geoscience and Remote Sensing*, vol. 37, no. 1, pt. 2, January 1999, pp. 564-573.

in which θ is the angle of incidence measured with respect to the assumed horizontal surface; to avoid confusion with the local angle of incidence formed with sloping terrain, we sometimes refer to θ as the system angle of incidence. Note that there is no shift distortion in the azimuth direction.

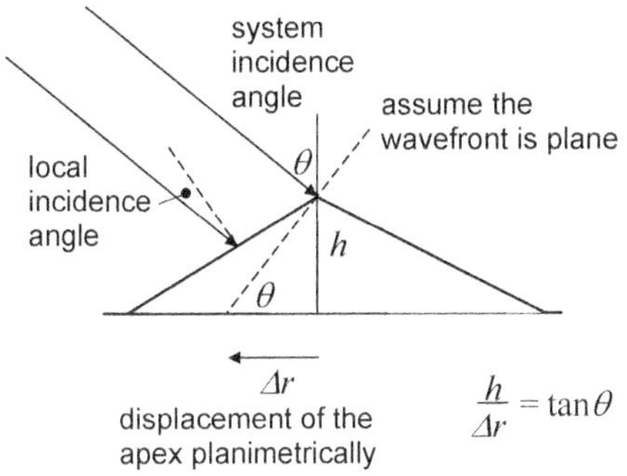

Fig. 4.12. Extent of distortion in the range position of relief above a datum

If we had available a digital terrain map (DTM) of the region covered by the radar at about the same spatial resolution, so that we were able to determine the local height of each pixel with reference to an appropriate datum, then we would be able to correct the local distortion resulting from relief. The difficulty of course is associating the pixel with the respective position on the terrain map to establish its elevation in the first place. Fortunately, though, we can actually simulate a radar "image" using the DTM data and our knowledge of terrain distortion in the range direction described by (4.1). That entails relocating the individual cells (or points) in the DTM according to (4.1). The artificial image can be shaded using a model of how the scattering coefficient of a surface varies with the local incidence angle at each cell given by the local slope calculated from the DTM (see Fig. 4.6). An alternative shading strategy is to assign brightness proportional to the cosine of the local angle of incidence. This will give maximum brightness to a facet facing the radar, with brightness gradually diminishing as the angle increases.

With an artificial image generated in this manner it is usually possible to recognise mutual features in that image (which is really a distorted DTM) and the recorded ground range real radar data. Using those features the two can be registered, after which the distortions in range can be removed.

While the foregoing is the principle of correcting imagery in the range direction in regions of high relief, the approach generally employed uses slant range imagery in the following manner. An artificial slant range image is created by computing the slant range to each cell in the DTM, using its height information and its position referred to the platform as shown in Fig. 4.13, according to

$$R_{ij} = \sqrt{(H - h_{ij})^2 + (X + x_{ij})^2} \tag{4.2}$$

Again, the artificial image is shaded using the cosine of the local angle of incidence. It is then possible to identify common control points in the artificial and real slant range images. Those control points only need to allow a relationship between the range positions of features in the two images to be established, since terrain effects do not distort the azimuth direction. If the range coordinate for the jth resolution cell in the ith scan line in the artificial image is called R_{ij}, and the range coordinate in the actual radar image is called ρ_{ij} then, using the control points, we can estimate the constants a and b that relate the two coordinates:

$$\rho_{ij} = aR_{ij} + b \tag{4.3}$$

We then proceed in the following manner. For each address in the DTM we compute the equivalent slant range as in Fig 4.13. Substituting that value in (4.3) allows the corresponding radar pixel to be identified. The brightness of that pixel is then placed at the DTM address. By doing that over all cells in the DTM a radar image is built up using the recorded radar data but with the DTM coordinates as the map base.

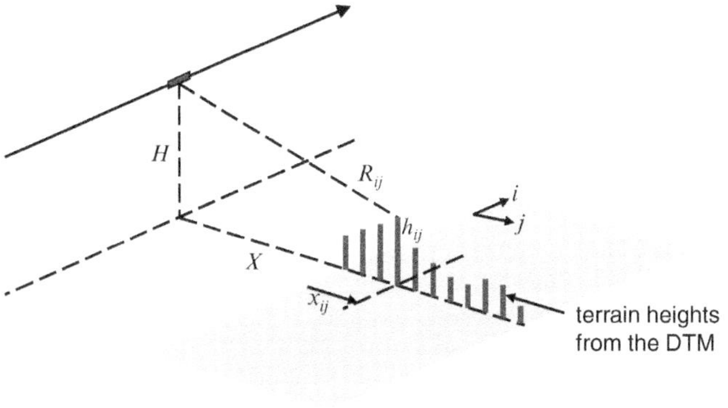

Fig. 4.13. Geometry for using a DTM to generate an artificial radar image

4.3 Radiometric Correction of Radar Imagery

4.3.1 Speckle

One of the most striking differences in the appearance of radar imagery compared to optical image data is its poor radiometric quality, caused by the overlaid speckled nature of the radar data. Fig. 4.14 shows a portion from an image of a fairly homogeneous region in which the speckle clearly affects the ability to interpret the data.

Speckle is a direct result of the fact that the incident energy is coherent – that is, it can be assumed to have a single frequency and the wavefront arrives at a pixel with a single phase. If there were a single large dominant scatterer in the pixel, such as a corner reflector or building, then the returned signal would be largely determined by the response of that dominant element, and any scattering from the background would be negligible. More often, though, the pixel will be a sample of a very large number of incremental scatterers; their returns combine to give the resultant received signal for that pixel. Such a situation is illustrated in Fig. 4.15.

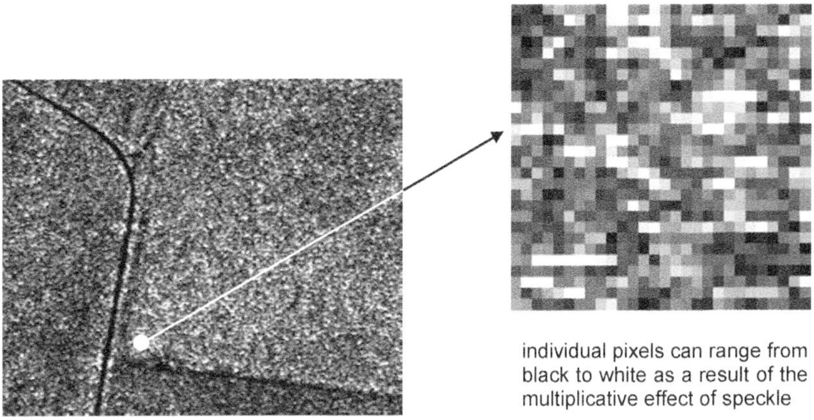

Fig. 4.14. Radar image showing the effect of speckle caused by the coherent interaction of the incident radiation with many incremental scatterers within a resolution cell.

Each of the individual return signals from within the pixel, received back at the radar, can be expressed in the convenient exponential form (shown here as the kth signal):

$$e_k = E_k \exp j(\omega t + \Phi_0 + \phi_k)$$

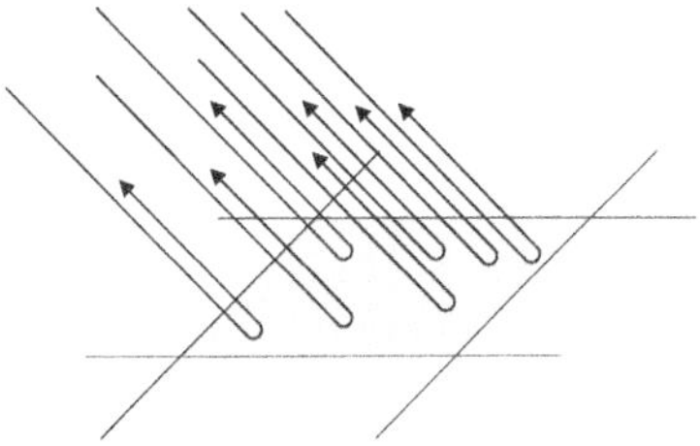

Fig. 4.15. Simulating the generation of speckle through the interference of a very large number of rays scattered from within a pixel

in which the amplitude E_k is directly related to the scattering properties of the pixel (strictly to the square root of the scattering coefficient) and the phase angle $\Phi_0+\phi_k$ is the result of the path travelled by that particular ray on its journey from the transmitter to the receiver. The common phase angle, Φ_0, corresponds to the average path of length R. The combined signal at the radar receiver is the sum of all the rays shown in Fig. 4.15:

$$E_{rec} = \sum_k E_k \exp j(\omega t + \Phi_0 + \phi_k) = \exp j(\omega t + \Phi_0) \sum_k E_k \exp j\phi_k \qquad (4.4)$$

The term outside the sum is common to all rays and thus does not affect how they combine. Only their individual amplitudes and relative phases are important in that respect. For simplicity suppose all the individual scatterers in Fig. 4.15 are identical so that the amplitudes in the last expression can all be considered the same and equal to E. After ignoring the common phase term (4.4) becomes

$$E_{rec} = E\sum_k \exp j\phi_k = E\sum_k (\cos\phi_k + j\sin\phi_k) = E\sum_k \cos\phi_k + jE\sum_k \sin\phi_k \tag{4.5}$$

which can be written

$$E_{rec} = E(I + jQ) = E\sqrt{I^2 + Q^2}e^{j\psi} \text{ with } \psi = \tan^{-1}\frac{Q}{I} \tag{4.6}$$

The resultant phase angle ψ is not important, since again it refers to the pixel as a whole, but the magnitude is. We can write the power density received at the radar as

$$p_{rec} = |E_{rec}|^2 = |E|^2(I^2 + Q^2) \tag{4.7}$$

(ignoring the impedance of free space term, as is often done in these sorts of calculation). What we need to do now is analyse $I^2 + Q^2$ because that is the source of speckle.

It is reasonable to assume that the incremental scatterers in Fig. 4.15 are randomly distributed over the pixel. As a result the phase angles ϕ_k in (4.5) can be assumed to be uniformly distributed over the allowable range of 0 to 2π. Under that assumption we can simulate the brightnesses (power density) of a group of adjacent image pixels, within each of which (4.7) applies. For this exercise we assume that there are 50 scatterers in each of 20x20 pixels, with the phases uniformly distributed within each pixel. Fig. 4.16 shows four different results for the same set of pixels.

One way of reducing the effect of speckle is to average over several supposedly independent images of the same region. We can demonstrate the benefit of doing so by using the four images of Fig. 4.16. Their average is shown in Fig. 4.17 in which the variation of the speckle is reduced. While this is not easily discerned visually from the image itself it is readily apparent in the computed standard deviation of the speckle.

Table 4.2 shows the mean and standard deviation of the aggregate of the four sets of pixels in Fig. 4.16 and the mean and standard deviation of the averaged image of Fig. 4.17. Several points are noteworthy. First, the standard deviation of the averaged image is half that of the original speckled images. Secondly, the mean and standard deviation of the original speckled data are the same.

Table 4.2. Means and standard deviations of speckle

	Raw speckle images Fig. 4.16	**Averaged speckle image Fig. 4.17**
Mean	50.70	50.70
Standard Deviation	50.05	25.42

Fig. 4.16. Speckle images generated from (4.7), the magnitude squared of (4.5)

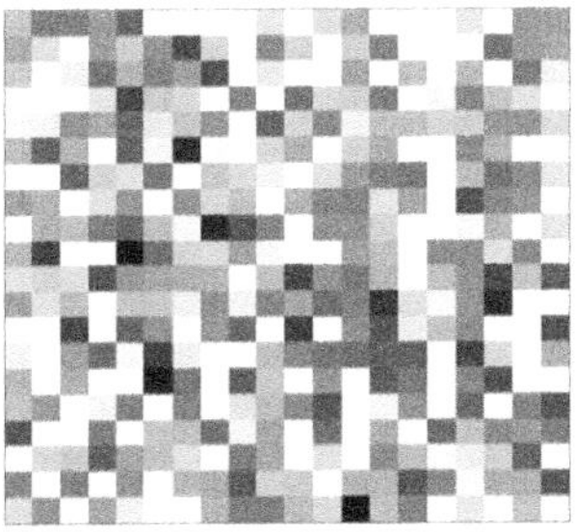

Fig. 4.17. Average of the four images shown in Fig. 4.16

These observations are readily explained by the statistics of the speckle itself. Fig. 4.18 shows histograms for the distributions of pixel brightness for the aggregated four sets of speckle images and the histogram for the averaged image of Fig. 4.17. As observed, the raw speckle data has an exponential-like distribution. The density function for the exponential probability distribution is

$$f(x) = \frac{1}{\gamma} e^{-x/\gamma} \quad 0 \le x < \infty \tag{4.8}$$

where γ is called the scale parameter of the distribution; it governs the rate at which the function falls off, and is numerically equal to the number of scatterers per pixel used to generate a sample of speckle (in this case 50). It is a property of the exponential distribution that the standard deviation equals the mean and is given by γ. Note that the distribution doesn't exist, or is zero, for negative x.

The histogram of the averaged data is the distribution of a random variable that is, essentially, one quarter of the sum of four exponentially distributed random variables. The distribution function of the sum of N exponential samples taken from the same distribution (with scale parameter γ) is a gamma distribution[6] with density function

$$f(x) = \frac{x^{N-1}}{\gamma^N \Gamma(N)} e^{-x/\gamma} \tag{4.9}$$

$\Gamma(N)$ is the gamma function which, for N real, is $(N-1)!$. If $N=1$, and noting that $0!=1$, the gamma distribution reduces to the exponential, as required. The mean and standard deviations of the gamma distribution are $N\gamma$ and $\sqrt{N}\gamma$, so that

$$\mu_{gamma} = N\mu_{exponential}$$
$$STDEV_{gamma} = \sqrt{N} STDEV_{exponential}$$

Since we have averaged rather than summed the N raw images, these values need to be divided by the number of terms so that

$$\mu_{gamma} = \mu_{exponential} \text{ for averaging} \tag{4.10a}$$

$$STDEV_{gamma} = \frac{STDEV_{exponential}}{\sqrt{N}} \text{ for averaging.} \tag{4.10b}$$

Thus the means are the same and the standard deviation of the averaged image is that of the raw speckled image divided by the square root of the number of terms that have been averaged. In the example here the standard deviation has been halved while the mean is the same, demonstrating that the averaging process reduces the variability in brightness of the image resulting from speckle.

In imaging radar systems a number of simultaneously recorded raw images of the same region are summed in the above manner to reduce speckle. In the terminology of radar image processing these simultaneous (or sub-) images are called "*looks*". From (4.10b) the standard deviation of the speckle will be reduced by the square root of the number of looks. Clearly, more looks will reduce the speckle further. We can see how the individual looks are created when we examine the formation of SAR images in Appendix D in which it will also be seen that look averaging leads to loss of spatial resolution.

[6] The gamma distribution is identical to the chi squared distribution in radar studies.

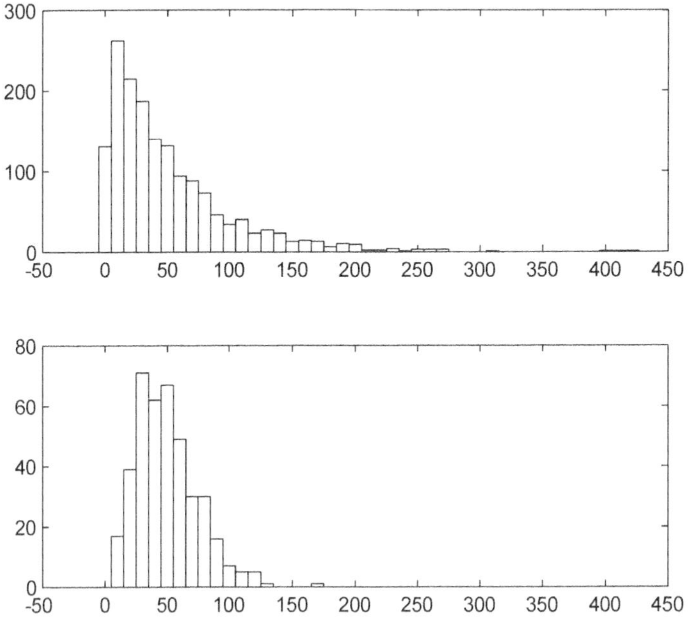

Fig. 4.18. Exponential histogram of the raw speckle data at top, and the four look average gamma histogram at bottom

Speckle is sometimes referred to as a multiplicative noise: in other words every pixel in the image has its implicit brightness *multiplied* by the computed speckle for that pixel[7]. To see that simply return to (4.7) and recognise that I^2+Q^2 is a variate drawn from an exponential distribution with mean γ, as just demonstrated, where γ is the number of samples within the pixel used to generate the speckle outcomes in Fig. 4.16. Let this variate be called s_γ and note that it can be written as

$$s_\gamma = \gamma s$$

in which s is an exponential variate drawn from a distribution with unity mean (and thus unity standard deviation). Using this expression for the samples of I^2+Q^2 we can write (4.7) as

$$p_{rec} = \gamma |E|^2 s \tag{4.11}$$

Since γ is numerically equal to the number of incremental scatterers within the pixel and $|E|^2$ is the power density received from one incremental scatterer then $\gamma|E|^2$ is the power

[7] Speckle is erroneously referred to as noise but it is in reality just the way the reflections appear because of irradiation with coherent energy, as outlined in this section. It is not imposed noise in the sense used in signal transmission.

density received from the full pixel, which will be a function of the scattering properties of the surface being imaged. We could write (4.11) as

$$p_{rec} = p_{pixel}s \tag{4.12}$$

which demonstrates explicitly that the received power density is that from the pixel (which is really what we want) *multiplied* by an exponential variate from a distribution with unity mean and standard deviation and which exists over the range $[0,\infty)$.

Although derived on the basis of a single pixel, within which the scattering properties do not vary, (4.12) applies in general to each of the pixels in a homogeneous region of image data. If we assume that the pixels within that region are by and large composed of the same scattering material then the only substantive differences among them result from the different values of s, drawn randomly from the exponential distribution.

It is the standard deviation of the speckle term that causes the noisy appearance of a radar image. As seen in the example above, averaging the four separate images reduces the standard deviation. We can express this also in terms of the signal to "noise" ratio of the image. From (4.12) the signal level is the average value (mean) of the received power density. Since the speckle has unity mean, the mean signal level is the same as the average pixel power density. The "noise" in the received signal is the standard deviation of the signal, which is the speckle standard deviation multiplied by the average pixel power. Given that the speckle standard deviation in (4.12) is also unity, then the signal to noise ratio of a radar image that has not been processed to reduce speckle is

$$SNR_{single\ look\ image} = \frac{\text{signal mean}}{\text{signal standard deviation}} = \frac{p_{pixel} \times \text{speckle mean}}{p_{pixel} \times \text{speckle standard deviation}} = 1 \quad (0\text{dB})$$

On the other hand, from (4.10) we can see that the signal to noise ratio is improved by look averaging:

$$SNR_{N\ look\ average\ image} = \sqrt{N} \times SNR_{single\ look\ image}$$

In the example in Figs 4.16, 4.17 we have examined speckle properties using (4.7); that assumes that the image product of interest is expressible in terms of power density (or power at the terminals of the receiving antenna) and thus is directly related to scattering coefficient. Some image products are in amplitude rather than intensity form, such as the type used, in principle, for polarisation synthesis. They are described by (4.6). It is of value to know the statistics of the speckle in this field (or received voltage) version. Fig. 4.19a shows the histogram of $\sqrt{I^2 + Q^2}$ using the data of Fig. 4.16. It has a Rayleigh distribution, with density function

$$f(x) = \frac{x}{\tau^2} e^{-x^2/2\tau^2} \tag{4.13}$$

where τ is a shape parameter that specifies the mode (the most likely value or maximum of the distribution). Since in our case this distribution has arisen as the description of a random variable $\sqrt{I^2 + Q^2}$ that is the square root of a variable that has an exponential distribution, then it can be shown that $\tau^2 = \gamma/2$. The mean and standard deviation of the Rayleigh distribution are given by[8]

$$\text{mean}=\sqrt{\frac{\pi}{2}}\tau = 1.253\tau\text{, standard deviation}=\sqrt{\frac{4-\pi}{2}}\tau = 0.655\tau$$

which, for this example, have the values of 6.30 and 3.21 respectively.

Amplitude images can also be averaged, or look summed, to reduce speckle; again the standard deviation of the speckle diminishes with the square root of the number of looks used. Fig. 4.19b shows the speckle histogram that results from averaging the four images of Fig. 4.16 in amplitude format. The mean in this case is still 6.30 while the standard deviation has been reduced to 1.62.

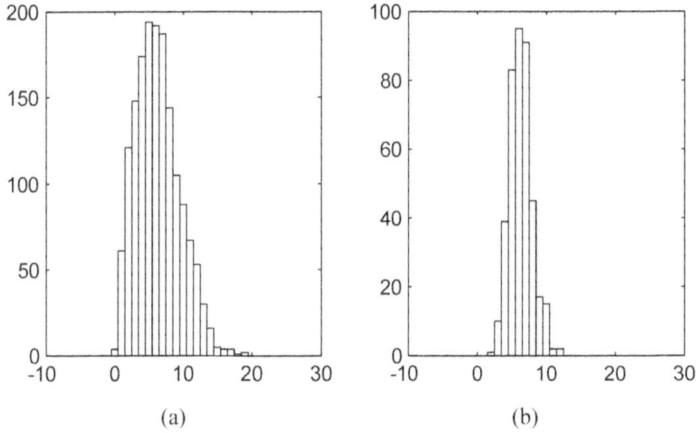

Fig. 4.19. Histograms of the speckle amplitude (a) single look (b) four look

4.3.2 Radar Image Products

We can now describe the products likely to be available from radar remote sensing missions. We commence by assuming a fully polarimetric radar so that the most general product is likely to be the scattering or Sinclair matrix

$$\mathbf{S} = \begin{bmatrix} S_{HH} & S_{HV} \\ S_{VH} & S_{VV} \end{bmatrix}$$

for which $S_{VH}=S_{HV}$ in backscattering. Each of the elements of the scattering matrix is complex, and can thus be written

$$S_{PQ} = A_{PQ} + jB_{PQ} = |s_{PQ}|e^{j\phi_{PQ}} \tag{4.14}$$

[8] Later we will be interested in the standard deviation when the mean is normalised to unity. That value is 0.655/1.253=0.523.

If imagery is provided in the form of a Sinclair matrix or, for a single polarisation radar, in terms of the complex scattering coefficient for that polarisation, then it is called *single look complex*, because it is complex and because it has not had speckle reduced through look averaging. It is also usually in slant range format and is full resolution, in the sense that there has been no trade off of resolution to provide speckle reduction.

More often than not the recorded multi-polarisation data is used to produce imagery in the form of the Stokes scattering operator of (3.72), which is readily suited to polarisation synthesis and which can accommodate unpolarised data.

Single polarisation imagery is often available in the form of the scattering coefficient of Sects. 3.14 and 3.15.

If data has been look averaged in order to reduce speckle then generally it will be provided in the form of the scattering coefficient σ^o_{PQ} that results from look summing during SAR image formation, as outlined in Appendix D. If N looks have been used, to reduce the speckle by $\sqrt{N}$, and degrade the spatial resolution in azimuth, the available data is then said to be *N-look imagery*.

Slant to ground range conversion can then be applied to generate products that are able to be registered subsequently to a planimetric grid.

4.3.3 Speckle Filtering

Even though multi-look radar image products have speckle variance reduced through look summing, it is sometime desirable to reduce the speckle further to improve the potential interpretability of the data. Data that has been processed as single look (complex) will almost certainly require speckle filtering at some stage since the 0dB signal to noise ratio is generally too poor in most applications to be usable.

As seen in the example of Figs. 4.16 and 4.17 averaging is an effective measure to use. It is feasible therefore to use simple mean value (box car) smoothing for speckle reduction by running a moving template or box over the image, centred on each pixel in turn, and then replacing the brightness value of that pixel by the mean value of all the image pixels covered by the template[9]. While effective within homogeneous regions, the problem with mean value smoothing is that it blurs edges and generally distorts high spatial frequency detail. What is required is a speckle filter that reduces speckle variance in the relatively homogeneous regions of an image while preserving edges and boundaries. In other words it needs to be adaptive, in that the amount of smoothing it applies should vary with position in the image.

Before proceeding we note that (4.12) is written in terms of the power density received at the radar. After calibration that would be expressed in terms either of the elements of the scattering matrix (amplitude image) or as a scattering coefficient (intensity or power image), noting that the speckle statistics will be either Rayleigh or exponential as appropriate . To accommodate both possibilities, we re-write (4.12) as

$$z = xs \tag{4.15}$$

where x is the pixel property in the absence of speckle, which is what we are trying to estimate by speckle filtering; z is the measured property of the pixel and, as before, s is the speckle variate. It has a unity mean, but we now describe its standard deviation by the

[9] See J.A. Richards and X. Jia, *Remote Sensing Digital Image Analysis*, 4th ed, Springer, Berlin, 2006, Chapt. 5.

symbol η_s, to distinguish it from the special case of an exponential distribution treated above.

Many adaptive smoothing filters seek an estimate of x using the expression

$$\hat{x} = \langle z \rangle + b(z - \langle z \rangle) \tag{4.16}$$

in which b is an adaptive weighting coefficient and $\langle z \rangle$ is the average of the radar measurements over a neighbourhood about the pixel whose smoothed value $\hat{x}$ is sought. Equation (4.16) is applied by moving over the image pixel by pixel and examining the neighbours in a window centred on the pixel of interest. The window can be any size although clearly if it is too large too much averaging will occur whereas if it is too small not enough speckle reduction will result. The weight b can be chosen in several ways, although the best performance is often obtained when it is calculated as[10]

$$b = \frac{\text{var}(x)}{\text{var}(z)} \tag{4.17a}$$

in which var(z) is the variance of measured radar values within the chosen window about the pixel of interest and var(x) is the real underlying variance of the image region in the absence of speckle. If the region in which we are interested is very uniform with little natural variation then $b \to 0$ and the pixel reflectance gets replaced by the average over the neighbourhood. If, on the other hand, the region possessed significant natural variance, as in the vicinity of rapidly changing reflectance with position, then var(z)≈var(x) giving b=1 and thus the pixel value will be left unmodified – i.e. unfiltered. However, we don't know var(x) in general so it has to be estimated from the available measurements.

By minimising the mean of the squared error between the estimate and the actual signal in the absence of speckle – i.e. $E\{(\hat{x} - x)^2\}$ – it can be shown[11] that the reflectance variance in the absence of speckle can be estimated by

$$\text{var}(x) = \frac{\text{var}(z) - \langle z \rangle^2 \eta_s^2}{1 + \eta_s^2} \tag{4.17b}$$

As with $\langle z \rangle$, var(z) is computed over the window about the pixel of interest, while the speckle standard deviation η_s is known from the distribution function for s in (4.15). Table 4.3 summarises the range of values that are relevant.

A slightly simpler speckle filter is the Lee Sigma Filter[12]. It also runs a sliding (usually square) window over the image and replaces the central pixel under the window by the average of the most likely pixels in the window. The pixels chosen to form the average are those lying within two standard deviations ("sigmas" and hence the name of the filter)

---

[10] See J.M. Durand, B.J. Gimonet, and J.R. Perbos, SAR Data filtering for classification, *IEEE Transactions on Geoscience and Remote Sensing*, vol. 25, no. 5, September 1987, pp. 629-637, and J-S. Lee, M.R. Grunes and G. de Grandi, Polarimetric SAR speckle filtering and its implication for classification, *IEEE Transactions on Geoscience and Remote Sensing*, Vol. 37, no. 5, pt 2, September 1999, pp. 2363-2372.

[11] See Lee et al, 1999, *loc cit.*

[12] J. S. Lee, A simple speckle smoothing algorithm for synthetic aperture radar images, *IEEE Transactions on Systems, Man and Cybernetics*, vol. 13, no. 1, January 1983, pp. 85–89.

of the central pixel's value. Clearly, for a heterogeneous region fewer window pixels will lie within the two sigma range and less averaging will occur, whereas for homogeneous regions there will be substantial averaging and thus speckle reduction. Typically window sizes of 7x7 to 11x11 are used.

Table 4.3 Speckle standard deviation when the speckle mean is unity

	single look	N look
amplitude image	0.523	$0.523/\sqrt{N}$
intensity image	1	$1/\sqrt{N}$

While simple in principle, the sigma filter introduces a bias into the estimate used for the central pixel[13] because the two sigma range about the mean, as a method for capturing the 96% most likely pixels, assumes a symmetric, Gaussian distribution from which the samples are to be taken. As we have seen earlier, though, speckle statistics can be as skewed as exponential, so that an equal two-sided range about the mean will not capture the right set of pixels and will lead to a bias in the mean estimate used for the pixel at the centre of the window. That is illustrated in Fig. 4.20.

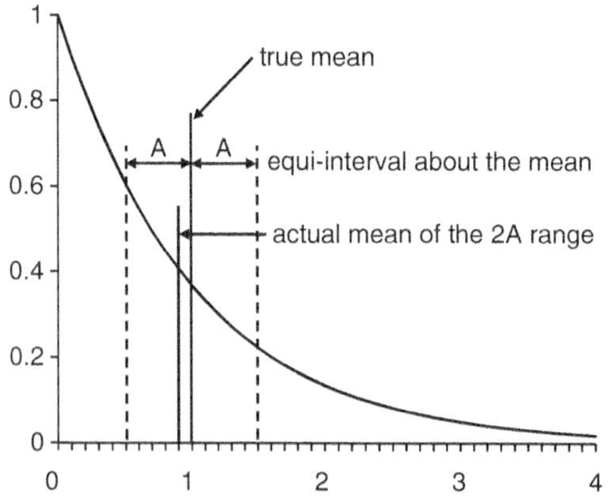

Fig. 4.20. Demonstrating the bias in mean estimate of the exponential distribution (with unity mean) when computed over a range centred on the true mean

To remove the bias the bounds either side of the mean for including a given percentage of the population need to be asymmetric. If the centre pixel (or an estimate of the mean) is $\hat{x} \approx z$ then let the range of pixel brightnesses to use within the search window to compute a new mean for the central pixel be bounded by $(T_1\hat{x}, T_2\hat{x})$. The values of the

[13] J-S. Lee, J-H. Wen, T.L. Ainsworth, K-S. Chen, and A.J. Chen, Improved sigma filter for speckle filtering of SAR imagery, *IEEE Transactions on Geoscience and Remote Sensing*, vol. 47, no. 1, pt 2, January 2009, pp. 202-213.

limit multipliers T_1 and T_2, which take the place of twice the standard deviation in the original sigma filter, are now usually based on 90% of the possible population. They depend upon the actual distribution and thus on whether we are dealing with amplitude or intensity images and how many looks have been averaged in producing those images. Table 4.4 gives values for those limits for a range of image types. Using those values leads to better filtering performance overall than when a simple two sigma range is used.

The Lee Sigma filter can be improved further if, instead of taking the simple mean of the pixels selected using the Table 4.4 limits, the estimate of (4.16) is used. However, since the distribution function has been truncated using the limits in Table 4.4 the standard deviation has to be re-computed so that the estimate in (4.17) remains an optimum minimum squared error measure. Those revised standard deviations are shown in the last column of Table 4.4.

Table 4.4. Upper and lower limit multipliers on the mean that will enclose 90% of the population of the distribution functions relevant to each of the image types listed. Also shown are the revised population standard deviation for use with the minimum mean square error estimate of the window mean in (4.16) (from J-S. Lee, J-H. Wen, T.L. Ainsworth, K-S. Chen, and A.J. Chen, Improved sigma filter for speckle filtering of SAR imagery, *IEEE Transactions on Geoscience and Remote Sensing*, vol. 47, no. 1, pt 2, January 2009, pp. 202-213, ©2009 IEEE)

Image type	T_1	T_2	revised η_s
Intensity 1 look	0.084	3.941	0.8191
Intensity 2 look	0.221	2.744	0.5699
Intensity 3 look	0.313	2.320	0.4624
Intensity 4 look	0.378	2.094	0.3991
Amplitude 1 look	0.286	2.043	0.4264
Amplitude 2 look	0.467	1.673	0.2911
Amplitude 3 look	0.557	1.531	0.2342
Amplitude 4 look	0.613	1.452	0.2010

A further improvement to the filter results if a better estimate for $\hat{x}$ could be found to use with the limits in Table 4.4. Recall above that the standard estimate used is just the value of the recorded brightness of the central pixel in the window – the one for which a speckle reduced version is sought. A better estimate is to average the pixels in a smaller window around the central pixel, or better still to apply (4.16) in that smaller window but with the original standard deviation rather than the adjusted one from Table 4.4. Once that estimate is available it can be used with the limits of Table 4.4 and the estimator of (4.16) to get an optimal estimate for the central pixel using a larger, say 11x11, window for speckle reduction.

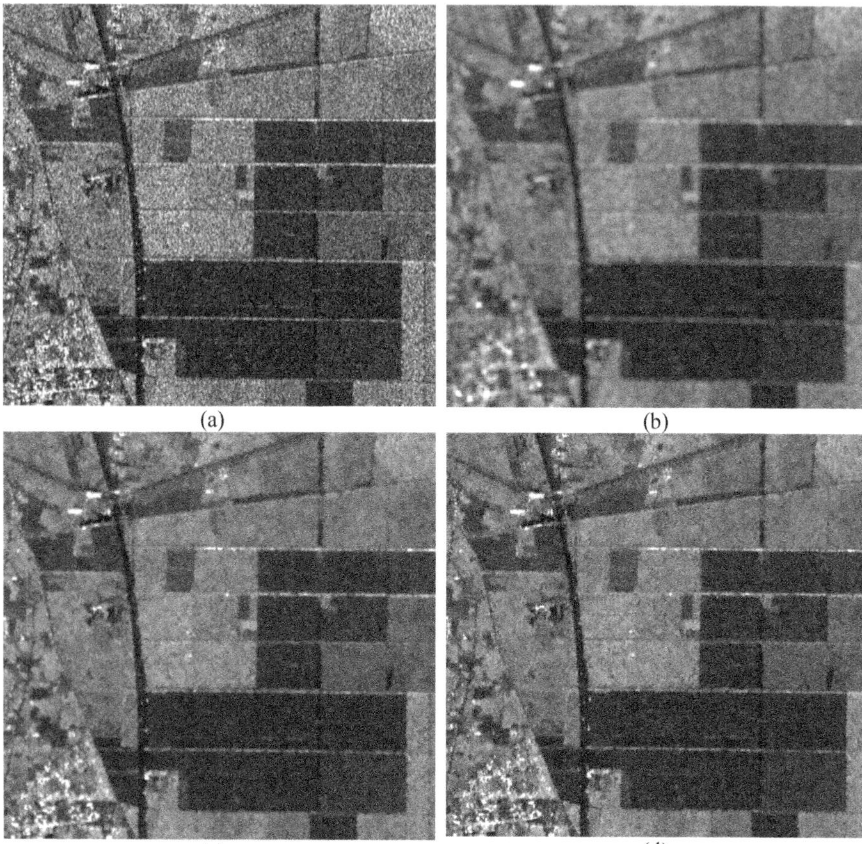

Fig. 4.21. Demonstration of the improved Lee Sigma Filter for speckle reduction using a four look amplitude image: (a) original 4 look image, (b) result of applying a simple 5x5 smoothing filter, (c) result with the improved Lee filter but with simple 3x3 smoothing filter used to extract the estimate of the mean, and with a 6 object threshold used for identifying and isolating bright objects and (d) the same as c but with (4.16) used to estimate the mean within the a smaller window, and to estimate the new value for the central pixel in the larger window (from J-S. Lee, J-H. Wen, T.L. Ainsworth, K-S. Chen, and A.J. Chen, *loc cit.* ©2009 IEEE)

A problem that occurs with all speckle reducing filters is how to avoid averaging out bright target responses, such as those from individual trees or buildings. It is not as simple as considering individual bright pixels as point targets since they might be the result of a mid range background multiplied by a speckle variate from the tail of the speckle distribution. A better identifier of likely point targets is to see if there is a set of adjacent bright pixels[14], since the point spread function of the radar will almost certainly smear the energy from a point target over several resolution cells as discussed in Appendix D. Fig.

[14] *ibid*

4.21 shows the effect of this modification to the Lee Sigma Filter along with the other improvements just outlined.

Finally, in applying any form of filter for speckle reduction it is important not to disturb the brightness relativities in multi-polarisation imagery lest subsequent analysis is prejudiced, in the same way that caution is exercised when any form of spatial processing is applied to optical imagery – in that case changes in band to band relativities can seriously impact on any algorithms to be applied for thematic mapping.

4.3.4 Antenna Induced Radiometric Distortion

The ideal pattern projected on the earth's surface by the real antenna carried on a SAR platform would be a rectangle, equal to the swath width in the across track direction and, in the along track direction, equal to the length of the synthetic aperture. Of course, real antennas cannot generate such precise patterns. Nor can they ensure that the level of power density created at the surface is the same over the full extent of the actual projected pattern.

The power density generated by an antenna depends on angular direction. It is summarised in a three dimensional polar pattern similar to that depicted in Fig. 4.22. The antenna is designed so that the power density generated in the main lobe is optimised while any power radiated in the directions of the side lobes is minimised. The relative sizes of the side lobes indicate the relative levels of power density created in those directions. Thus if the sides lobes are of any significant magnitude the radar is likely to receive measurable echoes from targets different from those intended (in the main beam).

We will return to that shortly, but it is also important to note that the main lobe profile in the so-called elevation plane will lead to non-uniform illumination across the swath. Antennas are designed to make that illumination as uniform as possible, often by shaping their elevation pattern in a cosecant squared fashion. Any residual variation in the pattern across the swath is inverted when the received signal is converted to image form; in other words the antenna pattern is used to calibrate the across track radiation to be as uniform as possible. That requires an accurate knowledge of the elevation beam pattern, which is can be obtained by measurements of the power density at the earth's surface once the platform is in orbit.

Careful antenna design can minimise side lobes, especially with the array antennas that are a feature of SAR systems. However, any residual side lobes can cause distortions in brightness of the recorded imagery if scattered power is received on the side lobes in the time window within which valid ranging pulses can occur – i.e. in time delays between the near and far swaths. Any side lobes forward or aft of the real antenna beam in azimuth and any elevation pattern side lobes that might lead to energy reflected from the parts of the platform are problematic in this regard. In some radar systems, particularly using aircraft, the effect of the side lobes is to create striping in brightness in the near range.

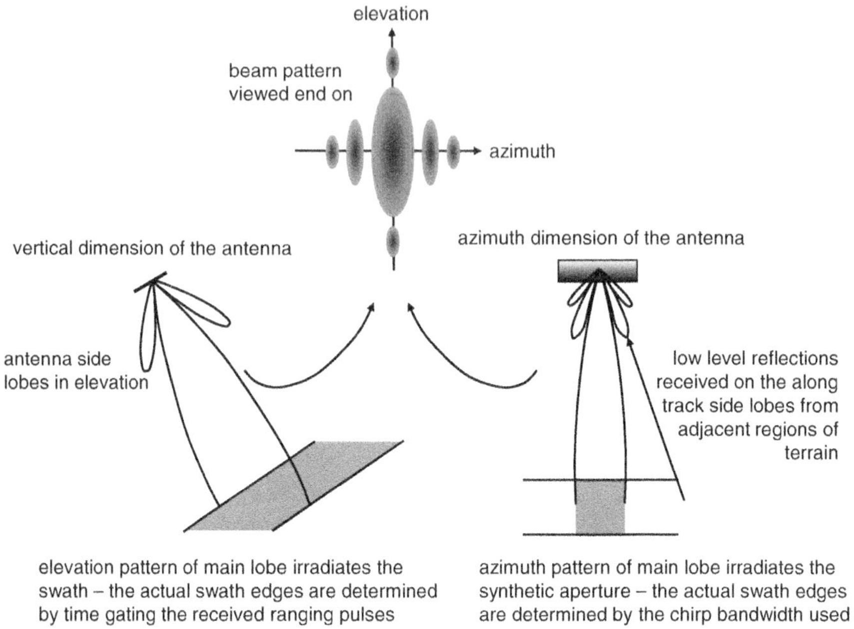

Fig. 4.22. Demonstrating how the side lobes of the antenna pattern appear and the effect they may have on recorded signal data

CHAPTER 5

SCATTERING FROM EARTH SURFACE FEATURES

5.1 Introduction

Remote sensing depends upon measuring the reflection or scattering of incident energy from earth surface features; emission from a surface is also possible, but that is beyond the scope of the discussion in this chapter.

If the incident energy is in the optical range of wavelengths – i.e. in the visible or near infrared – it is scattered largely by the surface of the material being imaged. Sometimes there is penetration into a medium, such as short wavelengths into water, but by and large the energy received by an optical sensor reflects from surfaces.

Because the wavelength of the microwave energy used in radar remote sensing is so long by comparison to that used in optical sensors[1], the energy incident on earth surface materials can often penetrate so that scattering can occur from within the medium itself as well as from the surface. Indeed, there are several mechanisms by which energy can scatter to the sensor, and they can be quite complex. In order to be able to interpret radar imagery it is necessary to have an understanding of the principal mechanisms so that received energy can be related to the underlying biophysical characteristics of the medium.

It is the purpose of this chapter to provide an introduction to the complex field of electromagnetic scattering as an aid to the interpretation of radar image data. A semi-quantitative treatment is given of a field usually based on electromagnetic theory and scattering concepts that are well beyond the level of presentation of this book. Nevertheless, our coverage is sufficient to permit the interpretation of radar imagery and to allow the development of backscatter models to be understood.

5.2 Common Scattering Mechanisms

Figure 5.1 shows the three most common scattering mechanisms that occur in radar remote sensing of the land surface. The first is *surface scattering* (analogous to that in optical imaging) in which the energy can be seen to scatter or reflect from a well-defined interface. The second is *volume scattering*, for which there is no identifiable single or countable number of scattering sites; instead, the reflections are seen to come from a myriad of scattering elements, such as the components of a tree canopy. The third is called *strong* or *hard target scattering* and can come in a variety of forms. Two types are shown in Fig. 5.1: corner reflector behaviour and facet scattering, both of which give particularly strong responses in radar imagery. If a surface is very dry the incident energy can penetrate, refract and scatter from sub-surface features, as depicted.

---

[1] Radar wavelengths are of the order of 10cm while optical wavelengths are of the order of 1μm – about five orders of magnitude different.

J.A. Richards, *Remote Sensing with Imaging Radar*, Signals and Communication Technology, DOI: 10.1007/978-3-642-02020-9_5,

It is now useful to examine each type of scattering behaviour in a more detail, although in a real situation several of the scattering pathways might occur together in a given pixel. We will have more to say about that later. We will also comment separately on scattering from the sea surface because it can involve a particularly interesting form of coupling of the radar energy with the surface. Scattering from sea ice will also be looked at separately because it is an interesting composite situation with a long term time variation.

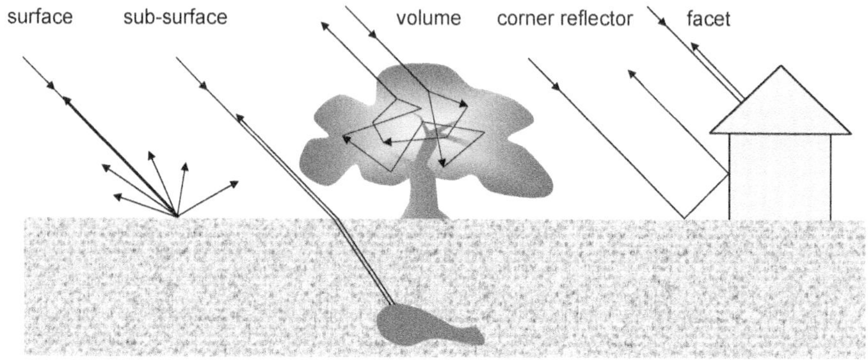

Fig. 5.1. Common scattering mechanisms

5.3 Surface Scattering

5.3.1 Smooth Surfaces

Consider a smooth surface between the air and a medium with *dielectric constant* ε_r. The dielectric constant[2] of a medium is one of its three electromagnetic properties. The others are *permeability*, which describes its magnetic behaviour but which is less important in our studies, and *conductivity*, which describes its lossiness or tendency to absorb energy from the wave as it propagates. We will have more to say about conductivity and losses later. For now we will concentrate on dielectric constant. All media have a dielectric constant $\varepsilon_r \geq 1$, including a vacuum for which it is unity. Unless air is very moist we assume it also has a unity dielectric constant.

The strength of surface scattering depends on the roughness of the surface and the dielectric constant of the material from which scattering occurs. In order to distinguish its behaviour better from volume scattering we say that scattering from a surface occurs when there is an identifiable discontinuity in dielectric constant (such as from air to water, air to soil, etc). In the case of volume scattering such a single abrupt change in dielectric constant cannot be distinguished although the individual scattering events within the volume occur at many dielectric discontinuities (air-leaves, air-twigs, etc).

The simplest form of surface scattering is reflection from a smooth surface. Understanding how energy interacts with such a surface provides significant insight into scattering from natural surfaces.

Imagine a ray of radar energy normally incident from the air onto the surface, as shown in Fig. 5.2a. Not all the incident energy will be reflected. Some will be transmitted into

[2] Dielectric constant is also called *relative permittivity*. A medium's *refractive index* is the square root of its dielectric constant.

the medium. If the transmitted ray does not encounter any subsequent variations in the dielectric properties of the medium it will continue to travel forward, gradually being absorbed by losses. If the medium contains embedded dielectric discontinuities the wave will be scattered, including being backscattered, about which we will have more to say in the context of volume scattering. Alternatively, if it encounters another abrupt dielectric constant discontinuity it will be reflected back up through the medium. We will assume for the present that the medium below the interface in Fig. 5.2a is homogenous and continues to infinity. It is the component reflected from the interface that is of interest here since by measuring it we hope to determine the properties of the surface material. The reflected power density relative to the incident power density is described by the *power reflection coefficient*

$$R = |\rho^2| \tag{5.1}$$

where ρ is called the *Fresnel reflection coefficient* of the air-surface interface. It relates the reflected and incident field phasors (each of which has amplitude and phase):

$$\rho = \frac{E^r}{E^i}$$

The field that crosses the interface is described by a *transmission coefficient*. That is examined in Sect. 5.3.3.

For the case of normal incidence where the medium beyond the interface is lossless, the Fresnel reflection coefficient is given by[3]

$$\rho_{normal} = \frac{1-\sqrt{\varepsilon_r}}{1+\sqrt{\varepsilon_r}} \tag{5.2}$$

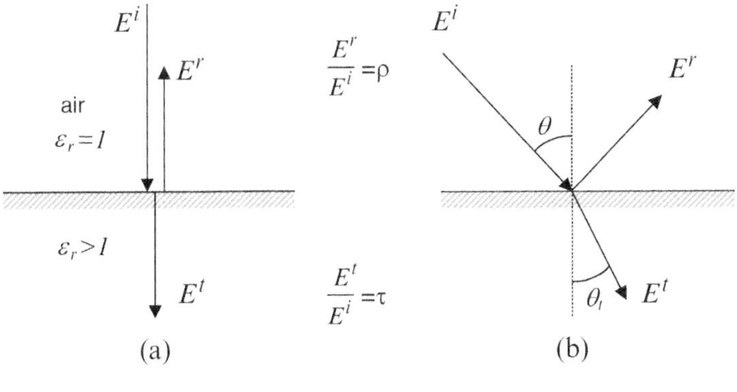

Fig. 5.2. Definition of reflection and transmission coefficients (a) vertical incidence and (b) oblique incidence

The dielectric constant of dry soil is about 4, so that the power reflection coefficient of (5.1) is 0.11 – thus only about 11% of the incident power is reflected. On the other hand

[3] See J.D. Kraus and D.A. Fleisch, *Electromagnetics with Applications*, 5th Ed., McGraw-Hill, 2000.

the dielectric constant of water is about 81 so that the power reflection coefficient is 0.64 – 64% of the incident power density is now reflected. Therefore, if we were to view a surface vertically, a water body would appear considerably brighter than adjacent regions of dry soil. Because the dielectric constant of water is so much larger than that of dry soil, the dielectric constant, and thus radar reflectivity, of soil is a strong function of moisture content, as shown in Fig. 5.3 for sand. Two components of dielectric constant are shown in the figure: the real part corresponds to the dielectric constant discussed here; the imaginary component is related to the lossiness of the medium as discussed in Sect. 5.3.3. Most dry natural media, not just soils, have low dielectric constants; it is the presence of moisture that leads to much greater values.

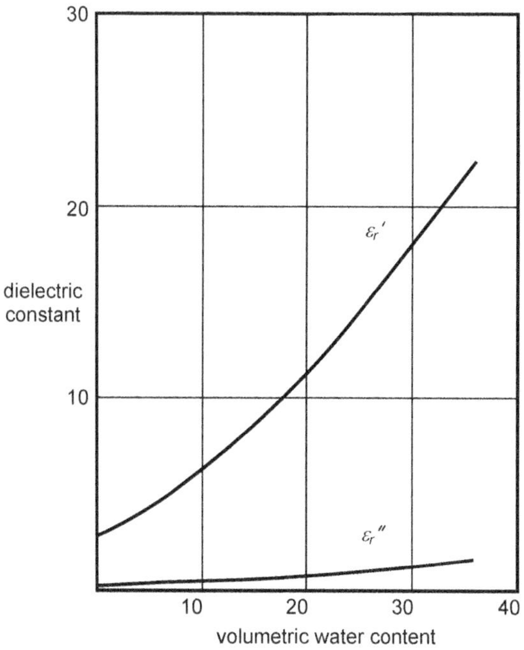

Fig. 5.3. Complex dielectric constant $\varepsilon_r' - j\varepsilon_r''$ of sand as a function of moisture content; the real part is the same as the dielectric constant used in (5.2) while the imaginary component is related to the absorption of energy by the moist sand (from J.A. Richards, *Radio Wave Propagation: An Introduction for the Non-Specialist*, Springer, Berlin, 2008)

Note that since $\varepsilon_r \geq 1$ the sign of the reflection coefficient of (5.2) is negative, which indicates that the reflected electric field is 180° out of phase with the incident field.

The case of vertical incidence shown in Fig. 5.2a is of little practical interest since, as shown in Chapt. 2, no range resolution is then available. Instead, the surface must be viewed at an angle out to the side of the platform. If the incoming ray has an angle of incidence of θ with the interface as shown in Fig. 5.2b then the reflection coefficient becomes polarisation dependent, and is given by[4]

[4] Kraus and Fleisch, *loc cit.*

$$\rho_H = \frac{\cos\theta - \sqrt{\varepsilon_r - \sin^2\theta}}{\cos\theta + \sqrt{\varepsilon_r - \sin^2\theta}} \quad \text{for horizontal (perpendicular) polarisation} \qquad (5.3a)$$

$$\rho_V = \frac{-\varepsilon_r\cos\theta + \sqrt{\varepsilon_r - \sin^2\theta}}{\varepsilon_r\cos\theta + \sqrt{\varepsilon_r - \sin^2\theta}} \quad \text{for vertical (parallel) polarisation} \qquad (5.3b)$$

Even though range resolution is now available, the reflected ray is away from the radar for both polarisations so that there is no energy backscattered. The surface will therefore appear black in recorded imagery. Nevertheless, these expression are of value when considering strong scattering and composite scattering situations.

A smooth surface such as that depicted in Fig. 5.2b is called *specular* since it acts like a mirror. Calm water bodies and very smooth soil surfaces are typical specular reflectors at radar wavelengths. How do we assess a surface as being specular? A little thought will suggest that it will be related to the wavelength of the radiation; less obviously perhaps, it is also related to the angle of incidence. If there is a vertical height variation of h on the surface then the surface is regarded as specular if

$$h < \frac{\lambda}{8\cos\theta} \qquad (5.4)$$

which is called the *Rayleigh criterion.*

5.3.2 Rough Surfaces

It is to be expected that as the roughness of a surface increases there will be more scattering back to the radar, and that the rougher the surface the lighter it will appear in radar imagery. Fig. 5.4 depicts qualitatively how the level of roughness affects backscatter and the existence or otherwise of a specular component in the scattered signal. If the surface is only slightly rough there will be a sizable specular component, with only a small component of backscatter, whereas for a very rough surface significant scattering will occur in all directions, including back to the sensor.

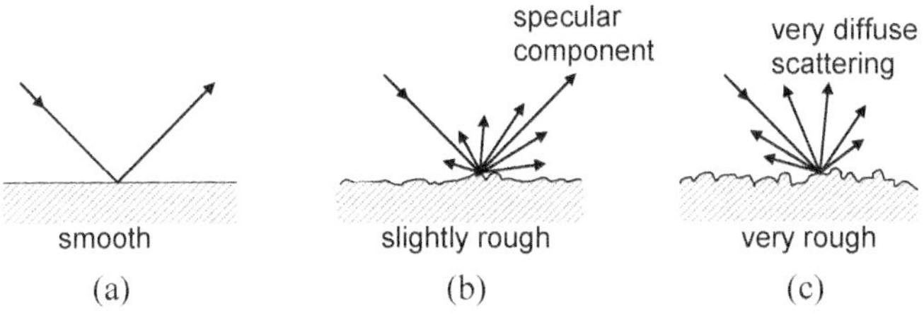

Fig. 5.4. Depicting the trend to diffuse surface scattering as roughness increases

It is relatively easy to understand scattering from the two extremes of roughness. In Sect. 5.3.1 we have already looked at the ideally smooth case. At the other extreme a "totally" rough surface is called *Lambertian*, well known in the theory of the scattering of light[5]. If the incident ray makes an angle θ with the surface normal then the bistatic scattering coefficient in the scattering direction θ_s is given by

$$\sigma^o(\theta,\theta_s) = \sigma_o^o \cos\theta \cos\theta_s \tag{5.5a}$$

which, for backscattering, is
$$\sigma^o(\theta) = \sigma_o^o \cos^2\theta \tag{5.5b}$$

σ_o^o is the backscattering coefficient for vertical incidence, which is polarisation independent. Expressed in decibels with respect to a reference of $1\text{m}^2\text{m}^{-2}$ (5.5b) is

$$\sigma^o(\theta_i)\, dB = \sigma_o^o\, dB + 20\log_{10}\cos\theta \tag{5.5c}$$

Fig. 5.5. shows a plot of the Lambertian surface scattering model (with $\sigma_o^o = 0.02$) as a function of incidence angle, along with curves computed from two other models that show typical scattering from (i) a very smooth surface and (ii) a surface of moderate roughness. As observed, when surface roughness increases the dependence on incidence angle is weaker, while for smoother surfaces there is a strong dependence. Allied with this observation is that smooth surfaces will appear considerably darker in radar imagery than rougher surfaces, particularly at moderate to large angles of incidence. Indeed, if one were interested in discriminating surface roughness, imaging with larger angles of incidence is preferred.

Along with the Lambertian model, two other models have been used in the construction of Fig. 5.5. They are just two of a number of approaches that are employed to describe surface scattering behaviour. Modelling backscattering is not simple and, even those models that are available suffer limitations.

As its name implies, the *small perturbation model* (SPM) is a reasonable descriptor of surface scattering when roughness is slight[6]. It is also referred to as the Bragg model, and is written as the sum of two components, one that describes coherent (specular) behaviour and the other that describes non-coherent (non-specular) behaviour:

$$\sigma^o(\theta) = \sigma_c^o(\theta) + \sigma_n^o(\theta)$$

The coherent component is not polarisation sensitive, whereas the non-coherent component is polarisation dependent. Those properties arise from the behaviour of the reflection coefficients of (5.2) and (5.3) which feature in the expressions for each component. The coherent term is given by

$$\sigma_c^o(\theta) = 4|\rho(0)|^2 \exp\{-4(k^2 s^2 + \theta^2/\Theta^2)\}/\Theta^2 \tag{5.6}$$

in which Θ is the beamwidth of the antenna that irradiates the surface, s is the rms variation in surface height and $k=2\pi/\lambda$ is the wave number, which is also sometimes

[5] See P.N. Slater, *Remote Sensing: Optics and Optical Systems*, Addison-Wesley, Reading Mass., 1980.
[6] For other candidate models see F.T. Ulaby and C. Elachi, *Radar Polarimetry for Geoscience Applications*, Artech House, Massachusetts, 1990, and M.C. Dobson and F.T. Ulaby, Mapping soil moisture distribution with radar, Chapt. 8 in F.M. Henderson and A.J. Lewis (eds) *Principles and Applications of Imaging Radar*, Vol. 2, *Manual of Remote Sensing, 3rd ed.*, Wiley, N.Y., 1998.

written as the phase constant β. Although we mainly use β throughout this book we have kept k here to make comparison with source material easier. $\rho(0)$ is the Fresnel reflection coefficient at vertical incidence.

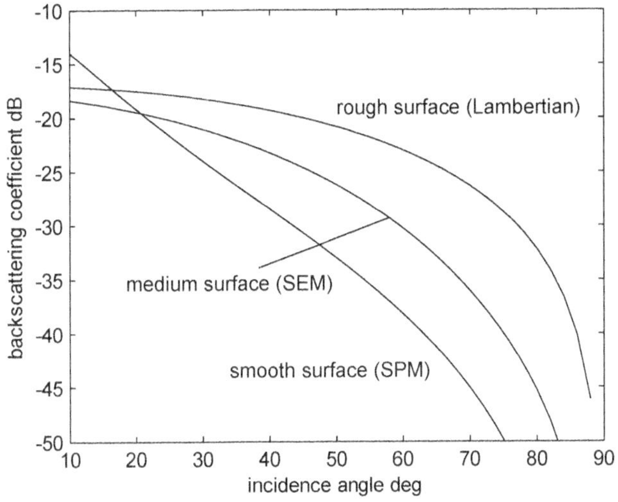

Fig. 5.5. Using three different scattering models to illustrate the effect of roughness on HH surface backscattering; the Lambertian model illustrates typical scattering from very rough surfaces; the small perturbation model depicts relatively smooth surfaces and the semi-empirical model indicates moderately rough surface scattering behaviour

The non-coherent component depends on what is called the correlation length of the surface roughness as well as the rms variation in surface height. The correlation length characterises the longitudinal variation in surface height variation. A surface which varies rapidly in height with position has a short correlation length whereas a more undulating, slowly varying surface has a larger correlation length. It is computed from the autocorrelation function of the surface roughness variation[7] which measures how correlated two points are along the surface with increasing separations between them. Adjacent points are highly correlated whereas correlation decreases as the spacing increases. How fast the correlation drops is determined by the nature of the surface variation. The separation at which it drops to 1/e of its maximum is the correlation length. If we denote the correlation length by $\boldsymbol{l}$ then the non-coherent component in the SPM is given by

$$\sigma_n^o(\theta) = 4\beta^4 \mathbf{s}^2 \boldsymbol{l}^2 \cos^4\theta[1 + 2(k\boldsymbol{l}\sin\theta)^2]^{-3/2}|\alpha_{xx}|^2 \tag{5.7}$$

in which the reflectivity parameters α_{xx} are

$$\alpha_{HH} \equiv \rho_{HH} = \frac{\cos\theta - \sqrt{\varepsilon_r - \sin^2\theta}}{\cos\theta + \sqrt{\varepsilon_r - \sin^2\theta}} \tag{5.8a}$$

[7] Dobson and Ulaby, *loc cit.*

$$\alpha_{VV} = (\varepsilon_r - 1)\frac{\sin^2\theta - \varepsilon_r(1+\sin^2\theta)}{\left[\varepsilon_r\cos\theta + \sqrt{\varepsilon_r - \sin^2\theta}\right]^2} \tag{5.8b}$$

$$\alpha_{HV} = \alpha_{VH} = 0 \tag{5.8c}$$

Equation (5.7) is based on modelling the autocorrelation function of the surface by an exponential expression which is a good representation of the surfaces most often found in practice. Other models of the surface autocorrelation will lead to slightly different versions of (5.7). Note that there is no cross polarised component in the SPM. The smooth surface curve in Fig. 5.5 was computed using this expression based on an rms height variation of 0.04cm and a correlation length of 1.5cm at a wavelength of 3.2cm. The antenna beamwidth was 0.1rad.

An empirically based model applicable to a wider range of surface roughness measures and which shows good agreement with measured data is the *semi-empirical model* (SEM) derived at the University of Michigan[8]. By drawing on the general forms of theoretical models, but choosing specific terms to allow fitting to experimental measurements, the SEM sets up an expression for co-polarised vertical backscattering and then finds the horizontal and cross-polarised components via co-polar p and cross-polar q ratios, as in the following:

$$\sigma^o_{VV}(\theta) = \frac{g\cos^3\theta}{\sqrt{p}}\{|\rho_V|^2 + |\rho_H|^2\} \tag{5.9a}$$

in which

$$g = 0.7\{1 - \exp[-0.65(k\mathbf{s})^{1.8}]\} \tag{5.9b}$$

Further

$$\sigma^o_{HH}(\theta) = p\sigma^o_{VV}(\theta) \tag{5.9c}$$

and

$$\sigma^o_{HV}(\theta) = q\sigma^o_{VV}(\theta) \tag{5.9d}$$

with

$$p = \left[1 - \left(\frac{2\theta}{\pi}\right)^{0.33/|\rho(0)|^2} \exp(-k\mathbf{s})\right]^2 \tag{5.9e}$$

and

$$q = 0.23|\rho(0)|[1 - \exp(-k\mathbf{s})] \tag{5.9f}$$

This model has been shown to work well provided the incidence angle is not too small; it will not predict specular behaviour near vertical incidence but is generally seen to be acceptable for angles in excess of about 20-30°, which is the range most appropriate to radar remote sensing. It also ignores the effect of the horizontal scale of surface roughness on scattering. The medium roughness curve of Fig. 5.5 was computed using this model, with an rms height variation of 0.1cm. Measurements of the effect of surface roughness on backscattering coefficient will be found in Ulaby et al (1978)[9].

[8] See Dobson and Ulaby, *loc cit.*

[9] F.T. Ulaby, P.P. Bratlivala and M.C. Dobson, Microwave backscatter dependence on surface roughness, soil moisture, and soil texture: Part 1-bare soil, *IEEE Transactions on Geoscience Electronics*, vol. GE-

Notwithstanding its empirical derivation a number of useful general observations can be made about surface scattering behaviour using the SEM of (5.9). Fig. 5.6 shows the modelled co-polar and cross-polar backscatter responses for a soil surface with a dielectric constant of 15 and an rms height variation of 0.1cm when irradiated at 9.5GHz. As seen, the cross-polarised response is always below the co-polar responses and VV is slightly higher than HH; all are weak functions of incidence angle until about 50° and larger.

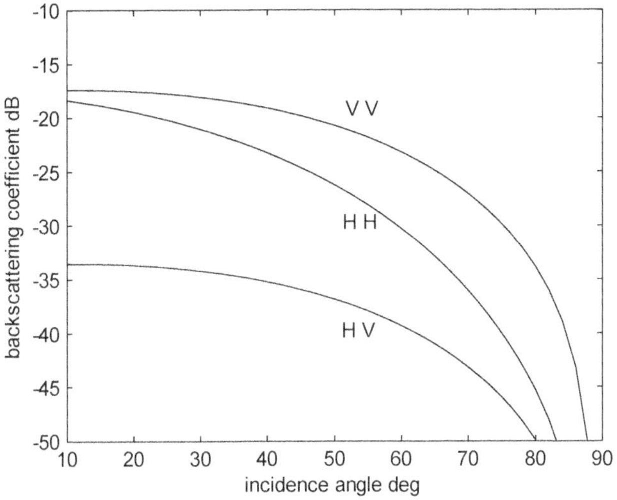

Fig. 5.6. Demonstration of the differences in horizontal and vertical polarisation responses for a rough surface, and an illustration of typical cross-polarised scattering; generated using the semi-empirical model

If the surface is almost smooth, so that s approaches zero, (5.9f) shows q approaches zero, indicating that there is little cross polarisation in the scattering from a smooth surface. At the other extreme, when s is very large – or more particularly when ks is large since wavelength is an important consideration in describing roughness as we saw in (5.4) – then q approaches $0.23|\rho(0)|$. With a dielectric constant of 15, $|\rho(0)| = 0.59$ giving q=0.136. From (5.9d) that would place the cross-polar response always about 8.7dB below the VV response for an extremely rough surface. The curves of Fig. 5.6 were computed with a smaller ks of 0.2, making the *HV* response 16dB lower than the *VV* response, which is about the difference observed in the diagram.

Figure 5.7 shows how the cross-polarisation ratio depends on surface roughness, measured as a fraction of a wavelength of the incident radiation, again generated using the semi-empirical model. The ratio is also shown as a function of the moisture content of the surface, which comes into the expressions of (5.9) through the dependence of the surface material's dielectric constant on moisture content. As expected the greatest cross polarised response occurs for very rough, highly reflective (moist) surfaces, whereas very smooth surfaces generate little depolarisation.

16, no. 4, October 1978, pp. 286-295. See also F.T. Ulaby, R.K. Moore and A.K. Fung, *Microwave Remote Sensing: Active and Passive*, Vol 2, Addison-Wesley, Reading, Mass., 1982.

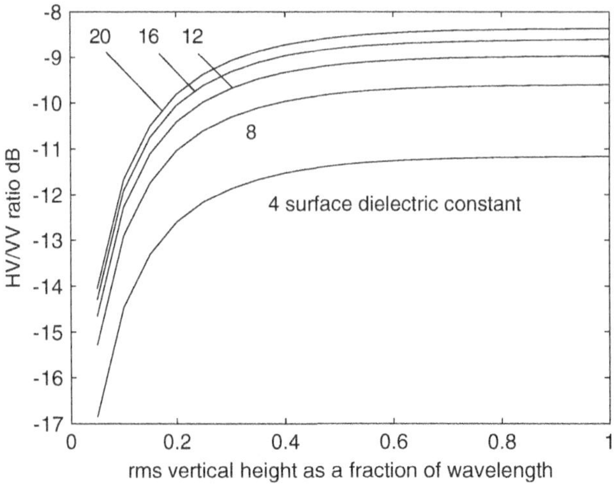

Fig. 5.7. The cross polar ratio as a function of surface roughness and dielectric constant; the dependence on surface moisture content can be determined by referring to relationships such as that in Fig. 5.3 for sand

If we now examine the co-polar ratio in (5.9e) we see that it approaches unity for extremely rough surfaces (as assessed against wavelength) meaning that there is then little difference between *HH* and *VV* behaviour. On the other hand, for smooth to moderately rough surfaces, (5.9e) demonstrates that the *HH* response will be lower than the *VV* response, although they converge at smaller incidence angles. Fig. 5.8 shows the co-polar ratio for the same values of dielectric constant and the same range of roughness used in constructing Fig. 5.7. Unlike the cross-polar ratio of (5.9f) the co-polar ratio is also a function of incidence angle, as seen in (5.9e). For the purpose of illustration two angles are used in Fig. 5.8: 20°, corresponding to those angles adopted in radars principally designed for oceanographic applications, and 40°, typical of those used for land-based applications. Note that the two like-polarised responses are approximately the same for very rough surfaces and only diverge significantly for smooth dry surfaces, and for larger angles of incidence.

Finally we can examine the expression in (5.9a) to gain an impression of the dependence of the surface scattering coefficient itself on factors, again such as surface dielectric constant and roughness. Fig. 5.9 shows the *VV* backscattering coefficient for incidence angles of 20° and 40°. As might be expected the surface is brighter at the smaller angles, consistent with Fig. 5.6; it is also brighter with increasing roughness and increasing dielectric constant (and thus moisture content).

Fig. 5.10 shows vertically polarised backscattering at the wavelengths commonly used in imaging radar systems, as a function of dielectric constant and roughness. A mid range incidence angle of 30° has been used. As seen, all wavelengths show about the same sensitivity to soil moisture (as captured in variations of the surface dielectric constant) but longer wavelengths provide better discrimination of surface roughness variations even though they have a lower absolute value of backscatter.

As a practical illustration of the enhanced backscatter resulting from increased soil moisture, and thus surface dielectric constant, Fig. 5.11 shows an image of an agricultural region in the vicinity of Ames, Iowa acquired by Seasat on 16 August 1978. Late on the previous day a large storm moved in from the west, which then separated into a number of isolated storm cells that moved to the north east. The lighter tone on the west of the image is the result of the storm and the light stripes show the paths of the storm cells.

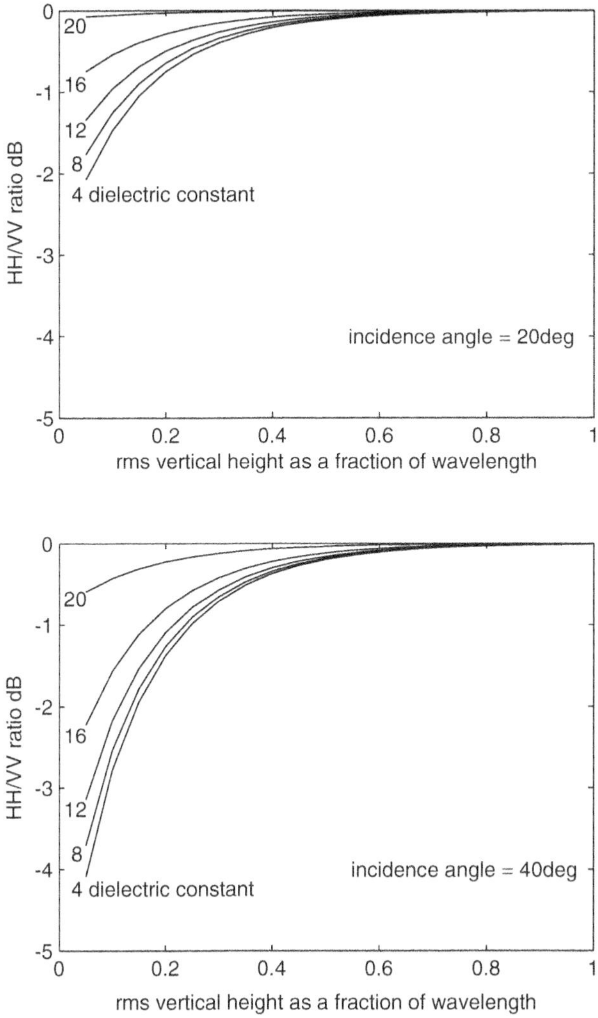

Fig.5.8. The co-polarised ratio as a function of surface roughness and dielectric constant for two values of incidence angle

In the development so far we have concentrated on expressions for the scattering coefficient itself. It is possible also to derive the scattering matrix of (3.41) for surface scattering behaviour, from which polarisation synthesis plots can be constructed. If a surface has only small scale height variations we can use the small perturbation/Bragg model of (5.7) for this purpose. The only element in (5.7) that is polarisation dependent is the reflectivity factor α_{xx}. Following Cloude and Pottier[10], and using (3.44), it is possible to induce that the scattering matrix for a slightly rough surface is

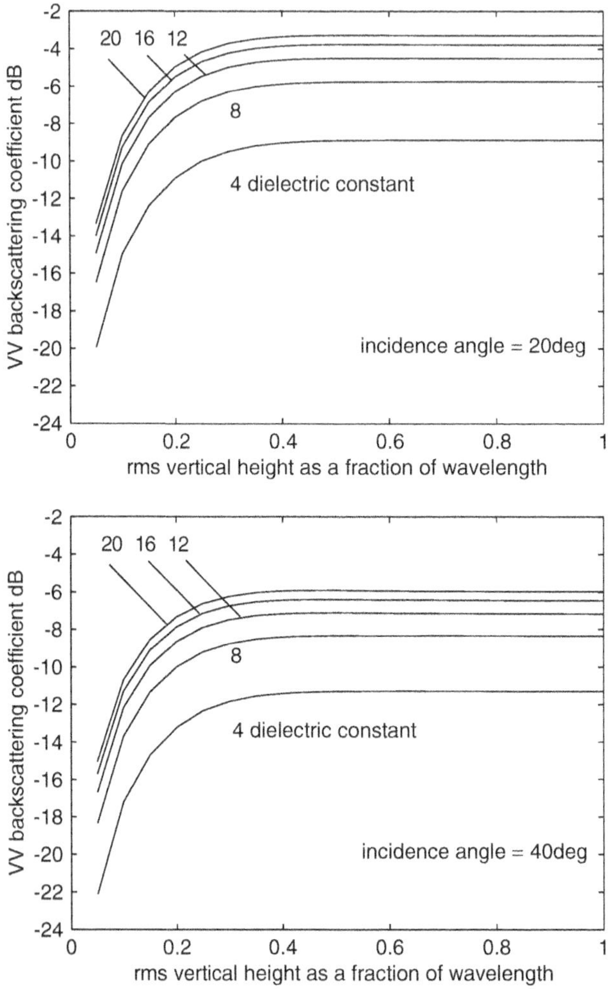

Fig. 5.9. The vertically polarised surface backscattering coefficient as a function of surface roughness and surface dielectric constant for two angles of incidence

[10] S.R. Cloude and E. Pottier, A Review of target decomposition theorems in radar polarimetry, *IEEE Transactions on Geoscience and Remote Sensing*, vol. 34, no. 2, March 1996, pp.498-518.

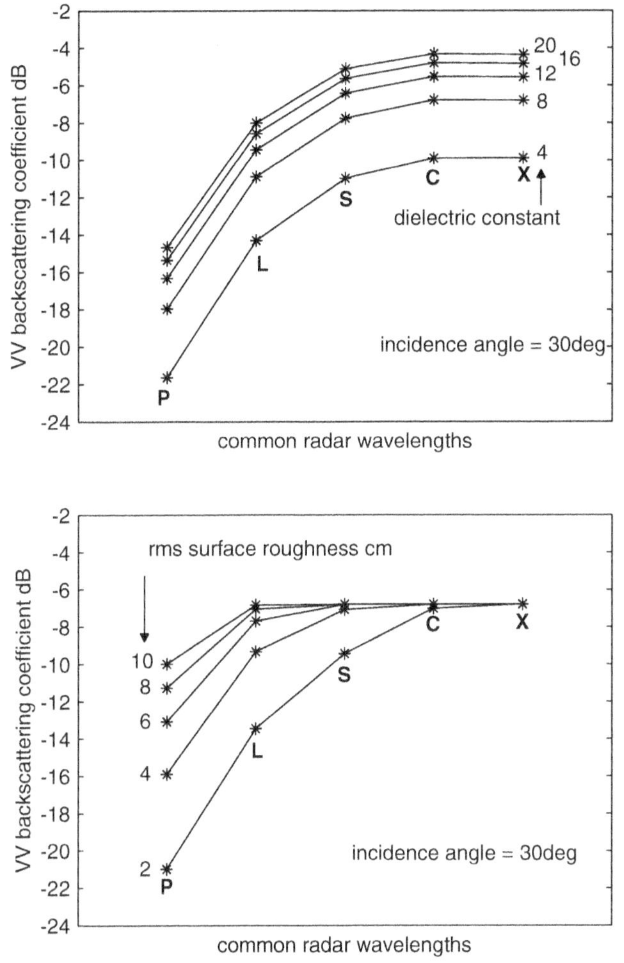

Fig. 5.10. The dependence of *VV* backscattering on dielectric constant (top, with an rms surface roughness of 3cm) and surface roughness (bottom, with a dielectric constant of 8) at the commonly used remote sensing radar wavelengths

$$\mathbf{S} = A\begin{bmatrix} \alpha_{HH} & 0 \\ 0 & \alpha_{VV} \end{bmatrix} \qquad (5.10a)$$

in which α_{HH} and α_{VV} are given by (5.8a,b) and $A = k^2 \mathbf{s}\,l \cos^2\theta[1+2(k\,l\sin\theta)^2]^{-3/4}/\sqrt{\pi}$. If we are interested in constructing normalised polarisation plots to examine the response of a pixel as a function of polarisation configuration the amplitude term is not important. Instead we can concentrate of the relative values of α_{HH} and α_{VV}, which depend only on angle of incidence and dielectric constant. The level of surface roughness does not affect the relative polarisation response when using the Bragg model. We can also normalise (5.10a) by the value of α_{VV} and use the normalised scattering matrix

$$\mathbf{S} = \begin{bmatrix} \alpha_{HH}/\alpha_{VV} & 0 \\ 0 & 1 \end{bmatrix} \qquad (5.10b)$$

A table of the ratio α_{HH}/α_{VV} is shown for range of dielectric constants and incidence angles in Fig. 5.12. Also shown are polarisation plots for the two extremes. As observed, as the ratio departs further from unity the plots depart from that of Fig. 3.22, which applies to a smooth surface (flat plate). The shape corresponding to the smaller ratios is typical of rough surface scattering.

If a surface has a periodic or near periodic structure enhanced returns can often be observed resulting from Bragg resonance, a condition treated in Sect. 5.5.4.

Sometimes it is important to understand bistatic scattering from rough surfaces. Bistatic radar is one reason; another is to be able to examine more complex scattering situations involving surfaces, such as the strong corner reflector behaviours considered in Sect. 5.5.2. If a surface has a standard deviation of roughness s, then the reflection coefficients of (5.3) can be modified according to

$$\rho_{effective} = \rho \exp[(-s\beta\cos\theta)^2] \qquad (5.11)$$

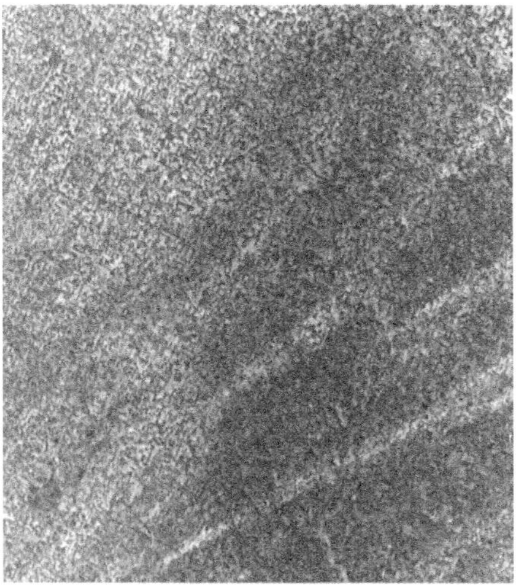

Fig. 5.11. Seasat image of Ames, Iowa showing the enhanced backscatter resulting from increased soil moisture owing to the effect of a storm to the west and subsequent storm cells that travelled to the north east late on the day prior to image acquisition (from J.P. Ford et al., *Seasat Views North America, the Caribbean, and Western Europe With Imaging Radar*, JPL Publication 80-67, NASA, 1 November 1980)

5.3.3 Penetration into Surface Materials

In the previous sections we have concentrated only on the component of energy that is reflected or backscattered from surfaces. In many instances energy can also cross the boundary and travel within the medium; that is the basis for detecting the sub-surface features depicted in Fig. 5.1. Most often there is substantial energy loss associated with transmission in the medium. We need now to understand the degree of loss that is likely, and the conditions under which sub-surface features might be imaged.

The transmitted components of the fields shown in Fig. 5.2 are related to the incident fields by the transmission coefficient τ as indicated in the diagram. We can obtain values for the transmission coefficients from the following equations via expressions for reflection coefficients[11]:

$$\tau_H = 1 + \rho_H \tag{5.12a}$$

$$\tau_V = \frac{\cos\theta}{\cos\theta_t}(1 + \rho_V) \tag{5.12b}$$

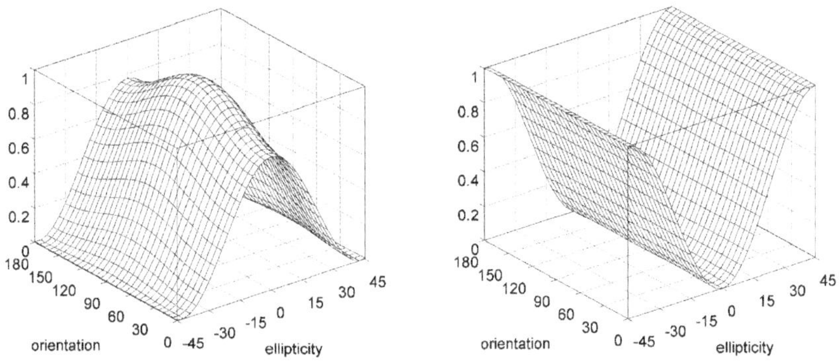

θ \ ε_r	5	10	15	20
20	0.879	0.853	0.842	0.835
30	0.762	0.712	0.691	0.679
40	0.634	0.565	0.536	0.519
50	0.511	0.429	0.395	0.375

$\alpha_{HH} / \alpha_{VV}$

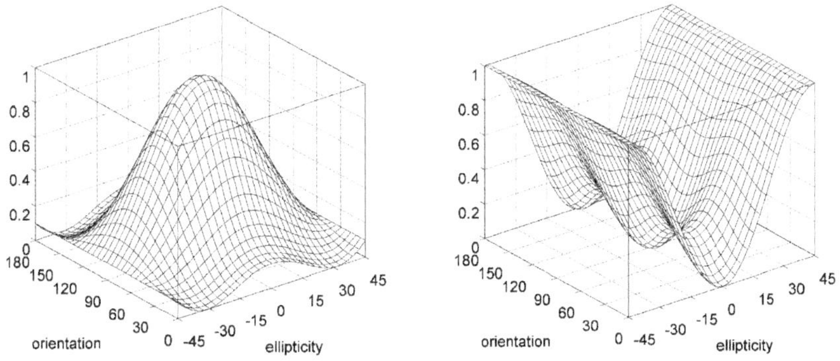

Fig. 5.12. Polarisation plots for surface scattering corresponding to different incidence angles and dielectric constants.

[11] See Kraus and Fleisch, *loc cit.*

θ_t is the transmission or refraction angle. Once the wave has crossed the boundary it propagates as an electric field according to

$$E(r) = E_o \exp(-\gamma R) \tag{5.13}$$

in which R is the direction of travel, E_o is the value of the field just under the surface and γ is the propagation constant, which determines how the field strength is modified with transmission. It is a complex number, the imaginary part of which simply describes the changing phase of the field as it propagates. That is of no interest here. Instead, it is the real part of the propagation constant that is important, since it describes the reduction in signal resulting from energy loss in the medium.

A wave's propagation constant is determined by its frequency ω and the properties of the medium in which it is travelling. We met those material properties briefly in Sect 5.3.1. We now need to be a bit more precise: in the most general terms they are conductivity σ, permittivity ε and permeability μ. In a non-magnetic medium (a good assumption for the media of interest to us) the propagation constant is defined by[12]

$$\gamma^2 = j\omega\mu_o\sigma - \omega^2\mu_o\varepsilon \tag{5.14}$$

in which μ_o is the permeability of free space, which is a fundamental constant of nature. The permittivity of the medium can be written as

$$\varepsilon = \varepsilon_o\varepsilon_r$$

where ε_o is the permittivity of free space, again a fundamental constant, and ε_r is the dielectric constant or relative permittivity. It may be of interest to note in passing that $\mu_o = 400\pi$ nHm^{-1} and $\varepsilon_o = 8.85$pFm^{-1} so that $c = 1/\sqrt{\varepsilon_o\mu_o} = 300$ Mms^{-1}.

We saw in Fig. 5.3 that the dielectric constant of a medium can be complex. Its imaginary part accounts for losses in the medium, most often as a result of its moisture content. The losses come from energy dissipation associated with internal ionic and molecular processes in the water molecules themselves. If we write the complex dielectric constant in its standard form

$$\varepsilon_r = \varepsilon_r' - j\varepsilon_r''$$

then (5.14) becomes

$$\gamma^2 = j\omega\mu_o(\sigma + \omega\varepsilon_o\varepsilon_r'') - \omega^2\mu_o\varepsilon_o\varepsilon_r' \tag{5.15}$$

Conductivity accounts for energy absorption in the medium because of any conducting media present. Often we assume the conductivity is zero so that (5.15) can be written

$$\gamma^2 = -\omega^2\mu_o\varepsilon_o(\varepsilon_r' - j\varepsilon_r'') \tag{5.16}$$

which also follows directly from (5.14). Since this is a complex number the propagation constant γ is also complex; it can be written in the form

[12] See J.A. Richards, *Radio Wave Propagation: An Introduction for the Non-Specialist*, Springer, Berlin, 2008.

$$\gamma = \alpha + j\beta$$

which, when substituted into (5.13), shows that the field travels in the medium below the surface according to

$$E(r) = E_o \exp(-\alpha R - j\beta R)$$

It is the constant α that leads to a drop in field strength during propagation. It is referred to as the *attenuation constant* and its value is found from evaluating (5.16).

We can express the complex dielectric constant in polar, or phasor, form, so that (5.16) becomes

$$\gamma^2 = -\omega^2 \mu_o \varepsilon_o \sqrt{\varepsilon_r'^2 + \varepsilon_r''^2} \angle \tan^{-1} \frac{-\varepsilon_r''}{\varepsilon_r'}$$

Note from Fig. 5.3 that the imaginary part of the dielectric constant is considerably smaller than its real part; that is generally the case for the materials we encounter in remote sensing. Therefore the last expression can be simplified to

$$\gamma^2 = -\omega^2 \mu_o \varepsilon_o \varepsilon_r' \angle \frac{-\varepsilon_r''}{\varepsilon_r'}$$

because the tangent of a small angle is approximately the value of the angle itself in radians. Accounting for the leading negative sign by adding π to the exponent, and taking the square root, gives

$$\gamma = \omega \sqrt{\mu_o \varepsilon_o \varepsilon_r'} \angle \frac{\pi}{2} - \frac{\varepsilon_r''}{2\varepsilon_r'}$$

Converting this last polar expression back to Cartesian form gives

$$\gamma = \omega \sqrt{\mu_o \varepsilon_o \varepsilon_r'} \left[\cos\left(\frac{\pi}{2} - \frac{\varepsilon_r''}{2\varepsilon_r'} \right) + j \sin\left(\frac{\pi}{2} - \frac{\varepsilon_r''}{2\varepsilon_r'} \right) \right] = \omega \sqrt{\mu_o \varepsilon_o \varepsilon_r'} \left[\sin\left(\frac{\varepsilon_r''}{2\varepsilon_r'} \right) + j \cos\left(\frac{\varepsilon_r''}{2\varepsilon_r'} \right) \right]$$

which shows that the attenuation constant is given by

$$\alpha = \omega \sqrt{\mu_o \varepsilon_o \varepsilon_r'} \sin\left(\frac{\varepsilon_r''}{2\varepsilon_r'} \right) \approx \omega \sqrt{\mu_o \varepsilon_o \varepsilon_r'} \frac{\varepsilon_r''}{2\varepsilon_r'} = \frac{\omega}{2c} \frac{\varepsilon_r''}{\sqrt{\varepsilon_r'}} = \frac{\pi}{\lambda} \frac{\varepsilon_r''}{\sqrt{\varepsilon_r'}} \tag{5.17}$$

since $c = 1/\sqrt{\mu_o \varepsilon_o}$. Equation (5.17) describes how the electric field drops (per metre) with travel in a medium. In remote sensing we are more interested in the loss of power density. Since, from (2.7), power density is proportional to the square of the electric field, the loss of power density with transmission is described by an *absorption coefficient* κ_a which is twice the value of the attenuation constant, viz:

$$\kappa_a = \frac{2\pi}{\lambda} \frac{\varepsilon_r''}{\sqrt{\varepsilon_r'}} \tag{5.18}$$

The units for this expression are Nepers per metre. To convert them to dB per metre, which is the more normal unit in engineering, we multiply by 8.686.

Using (5.18) the loss of power density can be written

$$p(R) = p_o e^{-\kappa_a R}$$

where p_o is the power density just below the surface. We now define the *depth of penetration* δ as that value of R for which the power density has dropped to 1/e of its immediate sub-surface value. Thus

$$\delta = \frac{\lambda}{2\pi}\frac{\sqrt{\varepsilon_r'}}{\varepsilon_r''} \text{ m} \tag{5.19}$$

From (5.19) we see that the penetration depth improves with wavelength (i.e. is better at lower radar frequencies) and with reduction in the imaginary part of the dielectric constant. We note from Fig. 5.3, and similar graphs for other soil types, that the imaginary part of the dielectric constant reduces with reducing moisture content. Therefore, significant penetration of radar energy requires longer wavelengths and dry soils or sands. Employing the source data that was used to construct Fig. 5.3[13], Fig. 5.13 shows the penetration depth at L band (23.5cm) for sand as a function of volumetric moisture content[14].

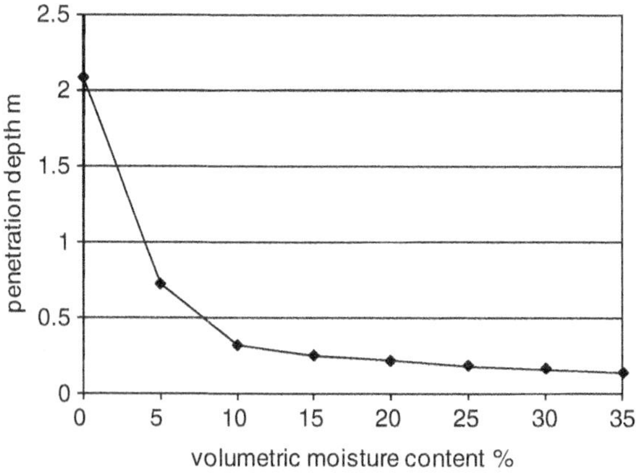

Fig. 5.13. Penetration depth for sand as a function of moisture content at L band, with a wavelength of 23.5cm

[13] J.R. Wang, The dielectric constant of soil-water mixtures at microwave frequencies, *Radio Science*, vol. 15, no. 5, 1980, pp. 997-985.

[14] See also M. Nolan and D.R. Fatland, Penetration depth as a DinSAR observable and proxy for soil moisture, *IEEE Transactions on Geoscience and Remote Sensing*, vol. 41, no. 3, March 2003, pp. 532-537, Fig. 3, for an interesting simulation of how penetration depth at X, C and L bands varies with time following a rain event.

As observed there is not much penetration at all unless the sand is very dry; even then only a few metres penetration appears possible. To give this greater perspective it could be noted that travel by one penetration depth leads to a signal reduction of 8.7 dB! Penetration to 4m leads to a loss of just over 17dB (a 50 times reduction in power density) and that is achievable only if the sand is totally dry, which happens in hyper-arid regions of the world, such as the Sahara Desert. To image below the surface two way attenuation needs to be considered. The loss therefore for imaging at 4m depth will be about 35dB! There is also loss of signal resulting from the transmission coefficient at the interface – although for a very dry medium that will be minimal (perhaps 10%). While sub-surface penetration looks almost impossible, there have been some celebrated cases of successful sub-surface imaging, including the 1981 SIR-A image recorded over the Sahara Desert in Sudan, where penetration was estimated at about 5m making sub-surface relic drainage channels and related features evident[15]. Another striking radar image of the Sahara is shown in Fig. 5.14, recorded with the multi-band, multi-polarisation SIR-C mission. It shows a hidden paleo channel of the Nile River.

In addition to loss by absorption described by (5.18), the forward travelling and backscattered energy will also be diminished if there is any appreciable scattering from inhomogeneities in the path, such as embedded gravels. We should therefore define an overall *extinction coefficient* κ_e that is the sum of the absorption coefficient and a scattering loss coefficient κ_s:

$$\kappa_e = \kappa_a + \kappa_s \tag{5.20}$$

Unless we are certain that there is significant sub-surface scattering we would normally assume that scattering loss is not as significant as absorption in sub-surface imaging.

What about penetration into water itself – such as lakes and the ocean? Water is a good conductor at microwave frequencies. The analysis is therefore different from that above in which we computed the penetration depth for materials in which the imaginary part of the dielectric constant is small. A calculation for conducting media leads us to see that the depth of penetration in sea water at L band (23.5cm) is 7mm!

5.4 Volume Scattering

5.4.1 Modelling Volume Scattering

Media such as tree canopies and sea ice contain many individual scattering sites that collectively contribute backscattered energy. Discontinuities in dielectric constant give rise to the scattering but there are so many and they are so difficult to identify and describe, that understanding how they contribute individually to backscatter is not straightforward. In the case of canopies it is the interfaces between leaves and air and twigs and air, for example, that are involved, whereas for sea ice it is air and brine inclusions in the mass of ice itself. With ice there will also be surface scattering.

Suppose we represent a scattering volume by the random set of individual scatterers illustrated in Fig. 5.15. If the density of scatterers is uniform it is evident that the volume would look much the same when viewed from any angle, in which case we could conclude that the amount of backscatter will be almost independent of, or only weakly

[15] J.F. McCauley, G.G. Schaber, C.S. Breed, M.J. Grolier, C.V. Haynes, B.Issawi, C. Elachi, and R. Blom, Subsurface valleys and geoarchaeology of the Eastern Sahara revealed by Shuttle Radar. *Science*, vol. 218, no. 4516, 1982, pp. 1004-1020.

dependent on, incidence angle. There will be no specular component as with surfaces (unless a definite surface is present as well) since the volume will look the same from above as it would at an angle.

As incident energy travels into the volume it will encounter loss in the forward direction as a result of scattering from whatever dielectric inhomogeneities are present. The scatterers themselves may also absorb some of the radiation. The scattering behaviour is what gives rise to the signal back at the radar from which we infer properties of the volume medium. It is useful to consider the scattering sites to be small compared with the wavelength of the radar signal so that they can be assumed to scatter almost isotropically (in all directions); this is another reason why the backscatter from a volume medium is almost independent of incidence angle.

Unless the volume is very lossy, in which case all forward travelling energy ultimately diminishes to zero, we need to take into account its vertical dimension. In other words, we need to recognise when analysing volume scattering behaviour that sea ice, for example, has an upper and lower boundary, just as a forest canopy has an upper and lower margin.

Fig. 5.14. Colour infrared photograph (top) and SIR-C radar image (bottom) recorded in 1995 over the Sahara Desert in Sudan. In the top right hand quadrant of the radar image a previous, ancient channel of the Nile is evident, now buried under sand; the colour composite radar image was created by displaying the C band VH cross-polar channel as red, the L band VH cross-polar channel as green and the L band co-polar HH channel as blue; since the paleo channel appears white there is good penetration at each of those wavelength/polarisation combinations (image courtesy of NASA JPL)

A simple but very helpful model of the volume scattering behaviour of a vegetation layer such as a tree canopy was devised in 1978[16]. It is based on the assumption that the dielectric property that dominates scattering is the moisture content of the vegetative matter. By assuming that the volume can be regarded as a suspension of water droplets, a credible description of volume scattering can be found in the following manner. It is based on Fig. 5.16 which shows radar energy incident on an individual resolution cell at the top "surface" of a volume of scatterers.

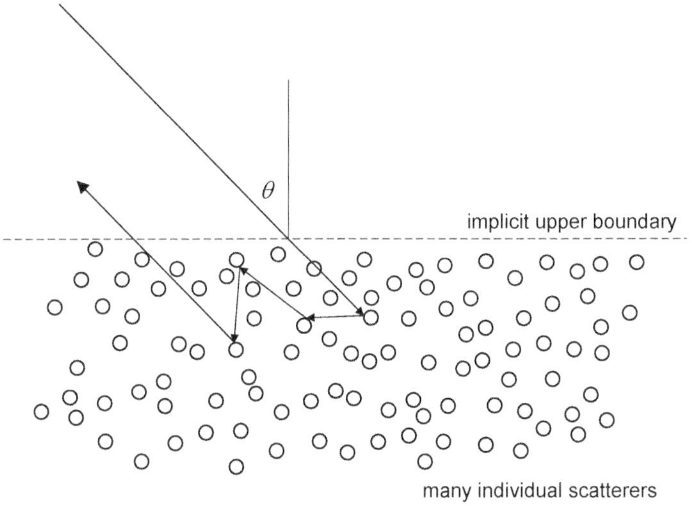

Fig. 5.15. Scattering from a "volume" of many, hard to define scatterers

Although the individual scatterers are not now delineated on the diagram assume they are identical and each has a radar cross section σ_b m^2. Further, assume that the energy an individual scatterer takes out of the forward propagating wavefront can be attributed to an extinction cross section Q_e m^2. This is an effective cross-sectional area presented to the incoming wave. The energy loss resulting from the wave encountering the single scatterer is given by the incident power density multiplied by this cross section.

Suppose there are N scatterers per unit volume in the medium; we can define

$$\sigma_v = N\sigma_b \quad \text{m}^2\text{m}^{-3} \tag{5.21a}$$

as a "volume" backscattering coefficient (i.e. radar cross section per unit volume), and

$$\kappa_e = NQ_e \quad \text{m}^{-1} \tag{5.21b}$$

as the extinction coefficient of the volumetric medium per unit of path length.

[16] E.P.W. Attema and F.T. Ulaby, Vegetation modelled as a water cloud. *Radio Science*, vol. 13, 1978, pp. 357-364.

The effective radar cross section of the infinitesimal volumetric slice *dr* shown in Fig. 5.16 is given by the product of the volume scattering coefficient of (5.21a) and the volume of the slice:

$$\sigma_v A\cos\theta dr$$

If the incoming power density is *p*, as indicated, then from the definition of radar cross section in Sect. 3.11, and ignoring for the moment any loss of power density before the slice is reached, the isotropically backscattered power from the incremental volume is

$$p\sigma_v A\cos\theta dr$$

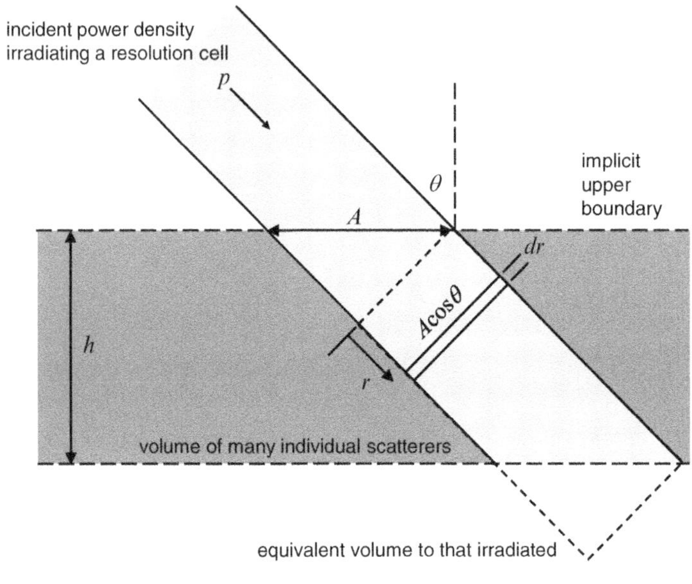

Fig. 5.16. Developing the water cloud model for a vegetation canopy

To simplify the next steps note in Fig. 5.16 that the volume and properties of the actual trapezoidal path through the medium is equivalent to that of the dotted rectangular prism, so that the geometry of the latter can be used. We now account for the loss of power density by absorption in the medium before the energy reaches the incremental slice at a distance *r* in from the implicit surface of the medium. Similarly we have to account for the comparable loss in the backscattered power as it travels back up through the medium. Thus the backscattered level of power at the surface, available for measurement by a remote sensing platform, is

$$\exp(-2\kappa_e r)p\sigma_v A\cos\theta dr$$

Integrating this last expression over the full depth of the volume gives the power backscattered from the radar resolution cell as

$$P_b = \int_0^{h\sec\theta} \exp(-2\kappa_e r) p\sigma_v A\cos\theta dr = p\sigma_v A\cos\theta \int_0^{h\sec\theta} \exp(-2\kappa_e r)dr$$

i.e.
$$P_b = \frac{p\sigma_v A\cos\theta}{2\kappa_e}[1-\exp(-2\kappa_e h\sec\theta)] \tag{5.22}$$

Equation (5.22) is the backscattered power level at the surface. We now need to turn that into the scattering coefficient for the resolution cell. From the derivation of radar cross section in Sect. 3.11 we can see that the power density received back at the radar p_r as a function of radar cross section and the power density incident on the surface p is

$$p_r = \frac{p\sigma}{4\pi R^2} = \frac{P_b}{4\pi R^2}$$

where R is the distance from the radar to the surface and P_b is the power backscattered from the resolution cell. Substituting from (5.22) in these last expressions gives the radar cross section σ of the resolution cell which, when divided by the area of the cell A, gives the backscattering coefficient[17]:

$$\sigma^o = \frac{\sigma_v\cos\theta}{2\kappa_e}[1-\exp(-2\kappa_e h\sec\theta)] \tag{5.23}$$

This assumes that any portion of the forward travelling wave that emerges from the bottom of the canopy is not subsequently reflected from some other material (such as a soil surface). While it is straightforward to consider such a composite situation it is not necessary here in our examination of the properties of volume scattering.

Fig. 5.17 shows the backscattering coefficient computed from the water cloud model for volume scattering compared with the Lambertian model for surface scattering. The Lambertian model was chosen in this comparison since it applies for the case of extreme roughness and shows the weakest dependence on incidence angle of all surface models. Even so, the volume dependence is weaker still, which is characteristic of a volume/surface behaviour comparison. Note that there is no specular component for small incidence angles in either of the curves. One would expect that for practical surfaces there would be a specular surface component as discussed in Sect. 5.3 so that it would turn upwards for smaller angles, but not so with the volume scattering curve.

Note that there has been no sense of polarisation dependence in deriving the water cloud model. That is because the scatterers have been assumed to be small and isotropic in their behaviour. In practical situations one would also expect that the *HH* and *VV* responses for true, random volumes would be comparable. If however the volumes contained scatterers that were not spatially symmetric (such as twigs, needles and branches) there will be a polarisation dependence as discussed in the next section.

[17] For a volume assumed to be composed of a very large number of identical, very small scatterers relating radar cross section and backscattering coefficient via the area of the resolution cell is acceptable. However, if the number of scatterers per resolution cell is not large, and varies from cell to cell, or if some cells contain dominant scatterers (such as hard targets), then such a relationship cannot be assumed.

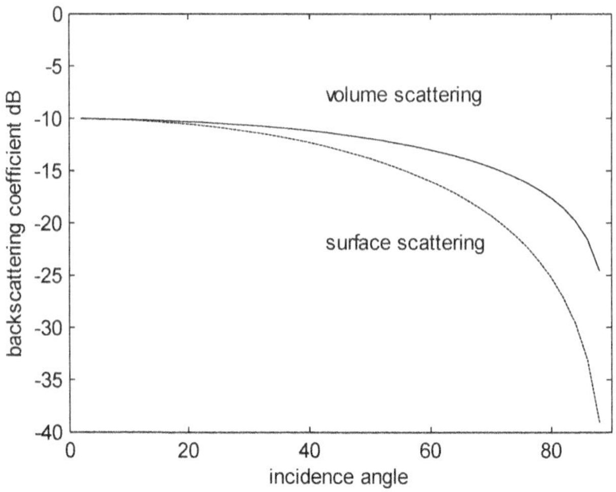

Fig. 5.17. Comparison of volume and surface backscattering showing the weaker dependence of volume behaviour on incidence angle: the surface curve here was based on the (extreme roughness) Lambertian model while the volume curve was computed with the water cloud model

5.4.2 Depolarisation in Volume Scattering

A significant feature of volume scattering is that it can lead to appreciable levels of cross polarisation, referred to as depolarisation. Depolarisation happens because scattering events lead to some degree of rotation of the polarisation vector of the incoming radiation. As a simple illustration of this consider a wave incident onto a conducting cylinder as shown in Fig. 5.18. Imagine for the moment that the cylinder is very thin even though, for purposes of illustrating its internal currents, we have shown it as slightly thick. If it were at right angles to the incoming electric field vector as seen in Fig. 5.18a then it will have no effect. In principle it looks as though it were not there. If it were perfectly aligned with the field vector as shown in Fig. 5.18b then it will scatter the incoming wave, including in the backscatter direction, with polarisation the same as that which is incident. The cylinder has maximum influence on the wave with this alignment. In essence the electric field vector induces current along the cylinder which re-radiates in the manner depicted, acting as an antenna. It is this re-radiation that we generically call scattering. The two extremes of alignment of the field and cylinder axis show, in effect, that it is the component of the electric field vector aligned with the cylinder axis that induces the current and leads to re-radiation.

Now consider the situation in Fig. 5.18c. Here the polarisation of the incoming field is at an angle to the cylinder axis. We can resolve it into components along and across the axis; only the former generates currents and thus leads to re-radiated (scattered) energy. Again, the re-radiated field is parallel to the cylinder axis and is thus "rotated" when compared with the incident field. If the incident field were a vertically polarised wave then the backscattered field will now have both vertical and horizontal components. It is not unreasonable to assume that if we were able to measure the amplitudes and relative phases of the backscattered fields then we might be able to infer something about the nature of the scatterer.

If the cylinder were not thin it will exhibit backscatter even when the incident wave is polarised orthogonal to its axis. It will also radiate, albeit with differing strengths, with components aligned with and orthogonal to its axes, but the general principle of depolarisation still applies, as it will also if the cylinder is dielectric instead of conducting.

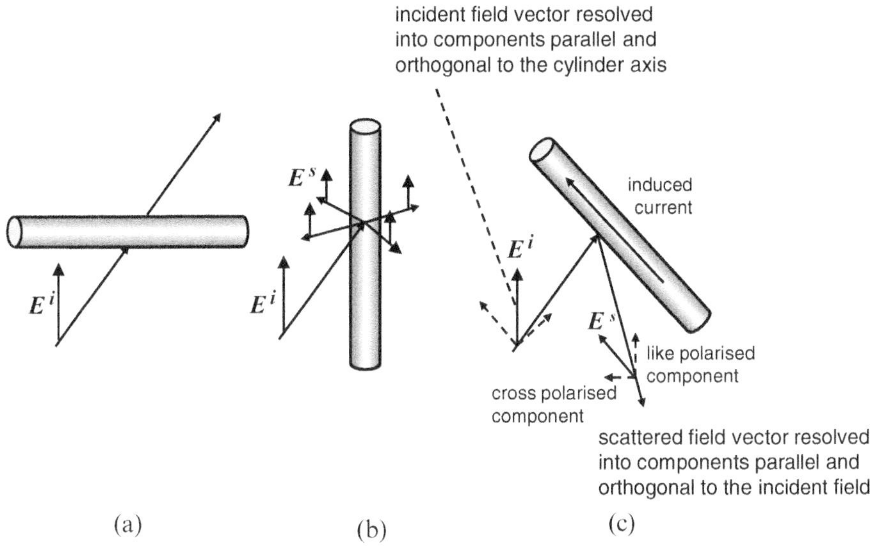

Fig. 5.18. Scattering of an incident electric field from a thin cylinder, illustrating that if the field vector is at an angle between 0° and 90° to the cylinder axis there will be a cross polarised component of the scattered field

If a volume is composed of a large collection of thin, cylinder-like elements it will exhibit strong cross-polarised behaviour, a situation that is indicative of branches and twigs in a canopy, and pine needles at shorter radar wavelengths.

It is characteristic of volume scattering that cross polarised returns are generally present and comparable in strength to co-polarised scattering, unless the wavelengths are so long that the scattering geometries have little influence. Again, except at very large angles, there is only a weak dependence of scattering coefficient on angle of incidence. Fig. 5.19 shows typical like and cross polarised returns for a forest canopy demonstrating these properties.

5.4.3 Extinction in Volume Scattering

As would be expected, and consistent with the derivation of the water cloud volume scattering model in Sect. 5.4.1, when a wave travels forward in a volumetric medium and is scattered each time it encounters a dielectric discontinuity, energy is lost from the forward travelling wavefront. The same occurs for backscattered radiation working its way back up through the medium to the sensor. Determining the extent of energy loss by scattering away from the principal directions is not straightforward, nor is the energy absorbed by the media that constitute the dielectric discontinuities. Nevertheless they can be modelled and have been incorporated into simulations of forest stands.

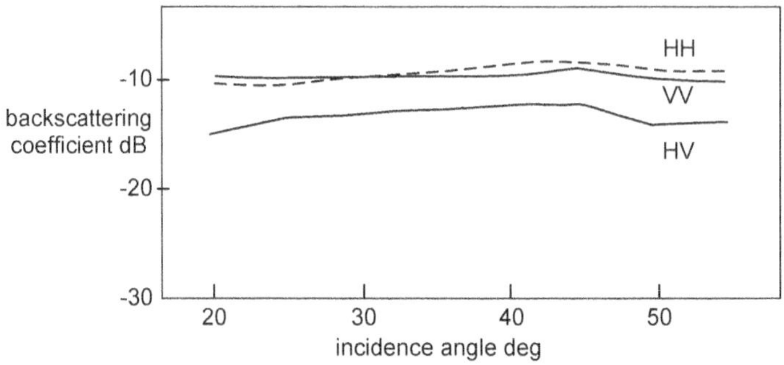

Fig. 5.19. Simulated like and cross polarised responses as a function of angle of incidence for a white spruce forest canopy at L band; based on Fig. 1 of Y. Wang, J.L. Day, F.W. Davis and J.M. Melak, Modeling L-band radar backscatter of Alaskan boreal forest, *IEEE Transactions on Geoscience and Remote Sensing*, vol. 31, no. 6, November 1993, pp. 1146-1154, ©1993 IEEE

Fig. 5.20 shows the simulated dependence of the volume attenuation (extinction) coefficient as a function of frequency for a soybean canopy in which it is assumed that the scattering elements are small compared with wavelength[18]. The most noticeable feature is the reduction in attenuation coefficient with increasing wavelength. That is the result of the reduced scattering that takes place as the size of the scattering elements reduces in comparison with wavelength. Shorter wavelengths will scatter more and thus suffer greater loss than longer wavelengths. The same effect is easily noticeable at optical frequencies. The blue sky is the result of significant scattering of the shorter optical wavelengths meaning we see energy at those wavelengths wherever we look in the sky, notwithstanding that it originates from the sun. In contrast, the longer red wavelengths don't scatter much at all. We tend only to see a reddish sky in the direction of the sun near sunset, and then only because of the longer atmospheric column than at midday. Although the dependence on wavelength is monotonic in Fig. 5.20, other canopy geometric configurations may behave differently. If structural elements such as stalks are present their scattering behaviours modify the attenuation wavelength relationship[19].

5.5 Scattering from Hard Targets

Although our interest in remote sensing tends to be largely in cover types that are distributed, such as crops, forests and soils, we frequently encounter with imaging radar individual, point-like scatterers than give exceptionally strong radar returns. They are important to understand because they can be components of composite scattering

[18] The leaves have average radii of 43mm and thickness 0.24mm, and have a gravimetric moisture content of 60%. See D.M. Le Vine and M.A. Karam, Dependence of attenuation in a vegetation canopy on frequency and plant water content, *IEEE Transactions on Geoscience and Remote Sensing*, vol. 34, no. 5, September 1996, pp. 1090-1096.
[19] *ibid.*

situations, they play a central role in scattering from urban features and, in high resolution radar, individual hard targets such as a tree can dominate a resolution cell.

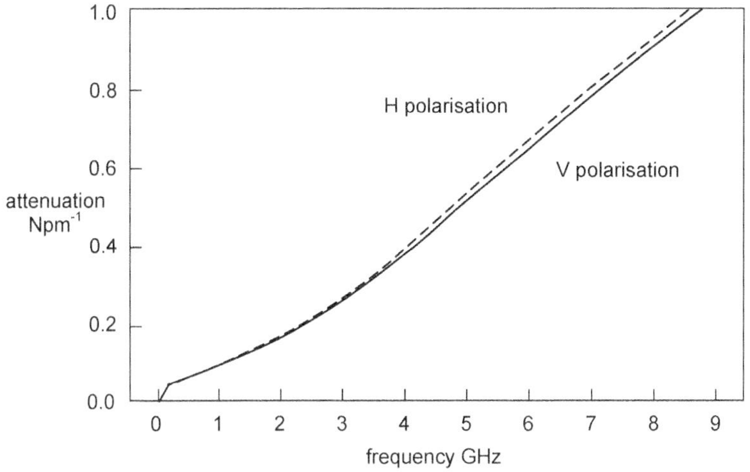

Fig. 5.20. Simulated canopy attenuation coefficient for soybeans as a function of frequency; taken from D.M. Le Vine and M.A. Karam, Dependence of attenuation in a vegetation canopy on frequency and plant water content, *IEEE Transactions on Geoscience and Remote Sensing*, vol. 34, no. 5, September 1996, pp. 1090-1096, ©1996 IEEE

Since they are discrete and not distributed we describe hard scatterers in terms of radar cross section rather than scattering coefficient. If they are the dominant scattering element in a pixel then the "scattering coefficient" of the pixel is given by dividing the radar cross section of the discrete scatterer by the size of the pixel.

One reason we encounter more hard target scattering with radar than with optical imaging is that at radar wavelengths many more surfaces appear to be smooth and are thus good reflectors. For optical imagery most surfaces are diffuse so that strong reflecting behaviour is usually not observed, except in the case of sun glint from water bodies and the occasional retro-reflector placed in a scene.

5.5.1 Facet Scattering

Occasionally, we encounter flat reflectors oriented towards the incoming radar beam, such as the house roof depicted in Fig. 5.1. If that scatterer were a rectangular conducting plate of dimensions axb m, much larger than a wavelength, then its bi-static radar cross section at an angle of incidence θ is given by

$$\sigma = \frac{4\pi}{\lambda^2}(ab)^2 \cos^2\theta \quad \text{m}^2 \tag{5.24}$$

At normal incidence, which is the situation encountered with monostatic radar, this reduces to 4π times the square of the area of the plate with dimensions expressed as a fraction of a wavelength.

5.5.2 Dihedral Corner Reflector Behaviour

Remarkably, dihedral corner reflectors, shown structurally in Fig. 4.10, occur naturally quite often and whenever there is a vertical surface adjacent to a horizontal plane. The most obvious example is the side of a building as shown in Fig. 5.1. If the building were oriented such that the corner directly faces the radar then the response will be very strong. If it is angled away then there will be no response. That is the basis of the cardinal effect treated in Sect. 5.5.5 following. The maximum radar cross section of a dihedral corner reflector is given in Table 4.1. When it is used as a model for double bounce situations, such as with the side of a building, we need to know its radar cross section at other angles. Provided the dimensions of a reflector are large compared with a wavelength then the cross section of a corner reflector not too far from bore sight is approximately

$$\sigma \approx \frac{4\pi A_e^2}{\lambda^2} \tag{5.25}$$

in which A_e is the effective area of the structure presented to the incoming beam. At 15° off bore sight it is about 3dB in error; thus over an incidence angle range of about 30-60° (5.25) can be regarded to be within 3dB of the actual value. As a function of incidence angle the radar cross section of the dihedral corner reflector from Table 4.1 can thus be shown to be

$$\sigma \approx \frac{8a^2b^2\sin^2(\theta + \pi/4)}{\lambda^2} \tag{5.26}$$

This assumes that both plates that make up the corner reflector have the same dimensions, as shown in Fig. 5.21a. The double bounce mechanism encountered in practice is more likely to be as shown in Fig. 5.21b, in which the bottom plate is a reflection of the vertical surface. It's "length" is a function of the of the angle of incidence. Using (5.25) the radar cross section of such an arrangement is given by

$$\sigma \approx \frac{16\pi a^2b^2\sin^2\theta}{\lambda^2} \tag{5.27}$$

Note the similarity of this last expression and (5.26); the latter has symmetry about an incidence angle of 45° as expected from the defined geometry of Fig. 5.21a, whereas (5.27) recognises explicitly that the cross section gets larger with angle because the horizontal projection of the vertical face monotonically increases with angle. If the model of Fig. 5.21b represented a building under a forest canopy the attenuation of the radiation as it passes through the canopy will increase with angle because of the longer path lengths; the radar cross section of (5.27) will therefore fall at larger angles.

Besides buildings, other common features that exhibit corner reflector like responses are structures over water, such as ships at sea and even oil rigs. Despite the fact that their vertical surfaces are not planar, they still behave as strong reflecting elements in the nature of corner reflectors.

A vertically standing, or near-vertically standing tree trunk also behaves like a dihedral corner reflector. It can be analysed by studying the scattering behaviour of a dielectric cylinder standing on a horizontal plane as shown in Fig. 5.22.

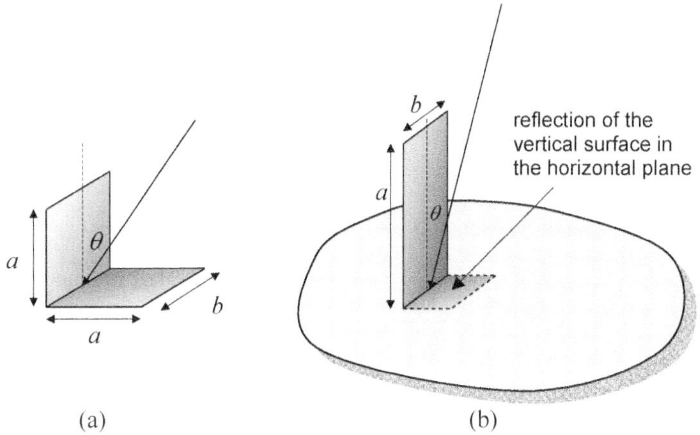

Fig. 5.21. (a) Standard dihedral corner reflector (b) projection of a vertical surface on to the horizontal plane to give a dihedral double bounce structure

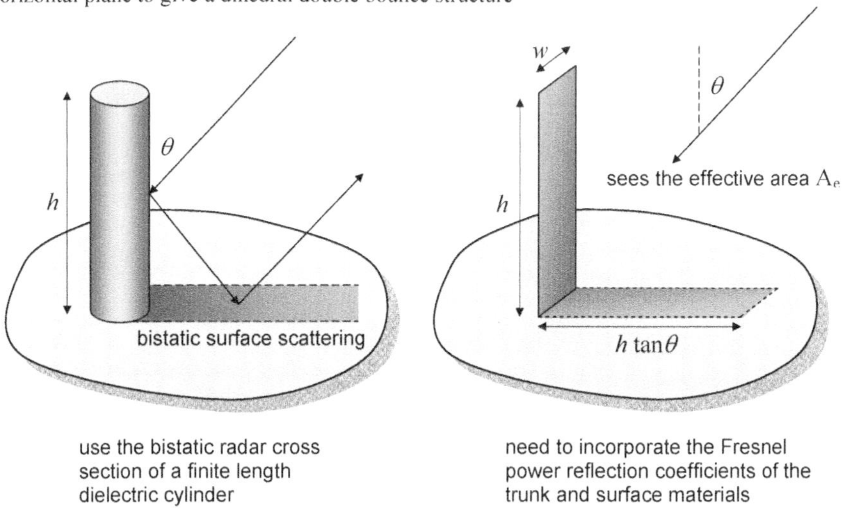

Fig. 5.22. Modelling the double bounce behaviour of a trunk standing on a horizontal surface by an equivalent dihedral corner reflector

The bistatic radar cross section of a dielectric cylinder is well known[20] and can be used to simulate a tree trunk standing on a dielectric surface. If the trunk radius y is large compared with a wavelength, its width can be approximated by a flat sheet of width[21]

[20] See G.T. Ruck, D.E. Barrick, W.D. Stuart and C.K. Krichbaum, *Radar Cross Section Handbook*, Plenum, N.Y., 1970.

$$w = \sqrt{y\lambda/2} \tag{5.28}$$

That allows us to approximate the radar cross section of a single tree trunk of height t by the expression

$$\sigma = 8\pi t^2 \rho_t^2 \rho_g^2 \sin^2\theta \frac{y}{\lambda} \tag{5.29}$$

in which ρ_t and ρ_g are the Fresnel reflection coefficients of the trunk and ground respectively, calculated from (5.3). Their squares are the Fresnel power reflection coefficients of (5.1). Equation (5.29) has been shown to underestimate the radar cross section by about 6dB or so at longer radar remote sensing wavelengths[22], but that is not important if we are interested in the general behaviour of double bounce scattering.

Since trunks rarely exist in isolation from a foliage canopy it is appropriate to add a canopy attenuation term to (5.29) to give a tree radar cross section that emulates what is observed in practice. Borrowing from the material of Sect. 5.4.1 we add an exponential decay to give as the approximate expression for the RCS of a single tree trunk

$$\sigma = 8\pi t^2 \rho_t^2 \rho_g^2 \sin^2\theta \frac{c}{\lambda} \exp(-2\kappa_e h \sec\theta) \tag{5.30}$$

where h is the depth of the canopy. Fig. 5.23 shows a plot of this expression versus incidence angle for *HH* polarisation using the parameters:

λ=0.06m (C band)	trunk dielectric constant=4
c=0.4m	ground dielectric constant = 7
t=12m	canopy depth h=5m

The extinction coefficient is varied from 0.05Npm^{-1} (low canopy absorption) to 4.5Npm^{-1} (high canopy absorption), expressed in dB in the figure. It is clear that the canopy extinction coefficient has a significant influence on the trunk response at higher incidence angles. If there were no canopy the response would not fall away at the larger angles. It is characteristic of double bounce behaviour in the presence of an attenuating canopy to peak around mid range incidence angles and to fall off at both extremes.

Fig. 5.23 was computed for the case of horizontal polarisation. Should vertical polarisation be chosen similar results would be obtained, modified only by the different behaviours of the Fresnel reflection coefficients. With a large vertical dielectric cylinder as shown there will be no or little cross polarised response. That is consistent with the behaviour of a dihedral corner reflector as seen by its scattering matrix at the end of Sect. 3.22.

---

[21] See S.D. Robertson, Targets for microwave radar navigation, *Bell System Technical Journal*, vol. 26, 1947, pp. 852-869.

[22] J.A. Richards, G-Q Sun and D.S. Simonett, L-band backscatter modelling of forest stands, *IEEE Transactions on Geoscience and Remote Sensing*, vol. GE-25, no. 4, July 1987, pp. 487-498.

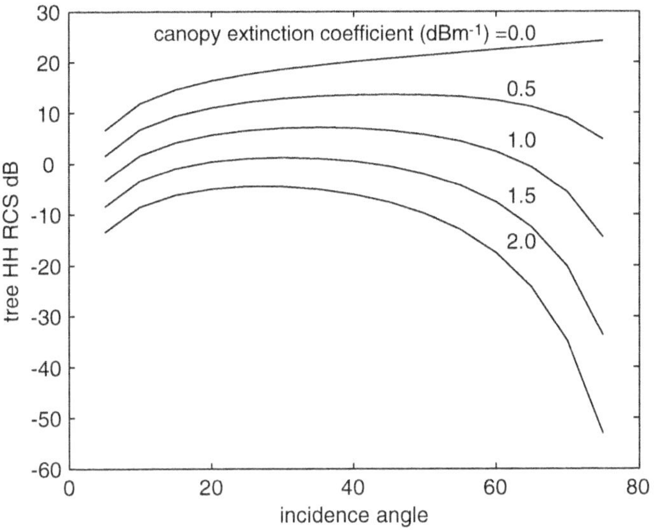

Fig. 5.23. Simulated radar cross section of a tree trunk with canopy attenuation, from (5.30)

Note the direct dependence on the ground properties in (5.30) via the reflection coefficient ρ_g. If the ground condition changes, such as a dry soil surface being replaced by water, then the radar cross section presented by the tree trunk will change accordingly. To see the magnitude of this effect assume that a dry soil surface of dielectric constant about 4 becomes flooded; water has a dielectric constant of about 81 or so. To make the calculations simple assume vertical incidence so that (5.2) can be used. With that change of dielectric constant the square of the Fresnel reflection coefficient changes from 0.11 to 0.64; that would lead to a 7.6dB increase in radar cross section. Thus a flooded forest will appear considerably brighter in radar imagery than one with a dry understory. At longer wavelengths the canopy is not very attenuating so it is even possible to observe the effect of flooding under a closed canopy[23].

An interesting composite situation that involves hard targets and dihedral reflections is radar scattering from a bridge over a river or harbour when the structure is substantially aligned with the flight path of the platform. Figure 5.24 shows an image of a region in which there are three bridges each of which appears as at least three reflections[24]. The image was recorded by a high spatial resolution X band interferometric radar at an incidence angle of 43°. Also shown is a portion of an air photo of the region for comparison.

The sketches in the figure show how the three main reflections occur for each bridge. First there is direct reflection from the bridge itself; clearly that would not be present if the side of the bridge were perfectly smooth, but generally there is enough geometric

[23] See J.A. Richards, P.W. Woodgate & A.K. Skidmore, An explanation of enhanced radar backscattering from flooded forests. *International Journal of Remote Sensing*, vol. 8, pp. 1093-1100, 1987.

[24] For a fuller description of the data set and the fusion of optical and radar data in pursuit of 3D visualisation of urbanised regions see U. Soergel, A. Thiele, H. Gross and U. Thoennessen, *Extraction of bridge features from high-resolution InSAR data and optical Images*, 2007 Urban Remote Sensing Joint Event, Paris.

detail, particularly at X band, that direct reflection will occur. We are dealing with a ground range image. Therefore the direct signal will be projected onto the ground plane displaced towards the radar as shown – a classic case of layover.

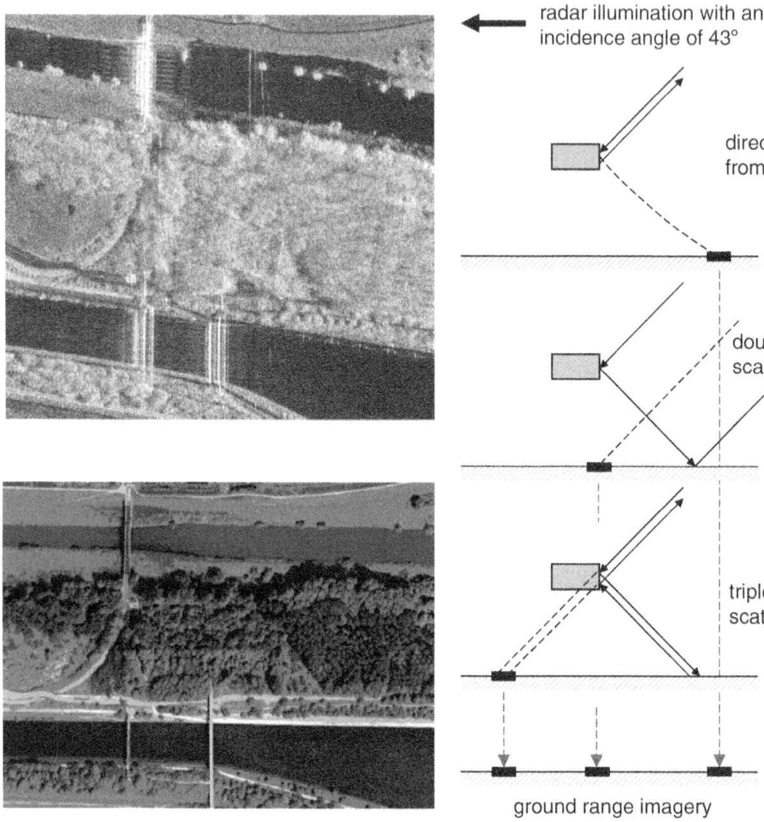

Fig. 5.24. Typical scattering from a bridge, showing how multiple reflections are formed; the imagery was taken by permission from U. Soergel, A. Thiele, H. Gross and U. Thoennessen, Extraction of bridge features from high-resolution InSAR data and optical Image", *2007 Urban Remote Sensing Joint Event*, Paris ©2007 IEEE

The second reflection is the result of double bounce dihedral behaviour involving the bridge and the water surface. Generally that would be the strongest reflection at longer wavelengths (L band) but at X band will be a little weaker owing to diffuse-like scattering from the water surface. It will not exhibit specular behaviour at those wavelengths. Where does this reflection locate on ground range imagery? In this case, with an incidence angle close to 45° the answer is straightforward. A little thought will show that the two way path travelled along the dashed line in the second sketch in Fig. 5.24, parallel to the radar rays, is approximately the same as the actual double bounce path followed, thus locating the second reflection almost directly under the bridge itself.

The third reflection is a little more complex and involves reflection from the bridge to the water, scattering back to the bridge and then reflection back to the radar. For this to have any strength the water must be a good diffuse scatterer, which we have already noted is likely to be the case at X band. Should the image have been recorded at L band this third mechanism would be weaker because of more specular behaviour at the water surface. This reflection will appear at the ground range position indicated in the third sketch which lies at the same apparent slant range position as the triple bounce reflection.

Because the spatial resolution is so high in this image (about 0.4m in range and 0.2m in azimuth) the bridge reflections show more complex detail including stanchions on the top left hand bridge and hand railings on the bottom right hand bridge. Multiple reflections of the type considered here for radar are also commonplace in optical imagery. A reflection of a bridge in calm water will often be seen when it is viewed side on.

Finally, we can derive the scattering matrix for a dihedral structure with dielectric faces, as against metallic faces described in Sect. 3.22. The metallic dihedral viewed along bore sight has a normalised scattering matrix of the form

$$\begin{bmatrix} 1 & 0 \\ 0 & -1 \end{bmatrix}$$

As seen in (5.29) and (5.30) the only factors that are polarisation dependent are the Fresnel reflection coefficients. The radar cross section for a dihedral structure therefore can be expressed in the general form

$$\sigma = A\rho_y^2\rho_x^2$$

in which A accounts for any factors that are geometrically significant, or relate to losses, and ρ_y and ρ_x are the reflection coefficients for the vertical and horizontal faces of the dihedral structure. From (3.43) we can establish $s_{PQ} \propto \rho_{yPQ}\rho_{xPQ}$ so that the normalised scattering matrix for the dielectric dihedral arrangement (such as the side of a building and the ground surface) is

$$S = \begin{bmatrix} \rho_{yHH}\rho_{xHH} & 0 \\ 0 & -\rho_{yVV}\rho_{xVV} \end{bmatrix}$$

in which the minus sign on the vertical component accounts for the change in phase between the horizontal and vertical components caused by the two reflections, as seen in Fig. 3.25 and discussed also in Sect. 8.5.1. In this expression for the scattering matrix the reflection coefficients are functions of angle so the polarimetric behaviour of the structure can be explored over a range of incidence angles. As an illustration Fig. 5.25 shows the co-polarisation plot for the side of a building with dielectric constant 4 adjacent to a soil surface with dielectric constant 5 at 30° angle of incidence[25].

5.5.3 Metallic and Resonant Elements

Metallic structures reflect radar energy and will show up in imagery if there is a component of the reflection in the backscattered direction. Although there are too many

[25] When doing these calculations it is important to recognise that the "incidence" angle for the vertical surface to use in (5.3) is 90° minus the system angle of incidence.

geometries to consider in general it is of value to examine scattering from a long, thin metallic wire, which might represent a fence line if it is in the horizontal plane.

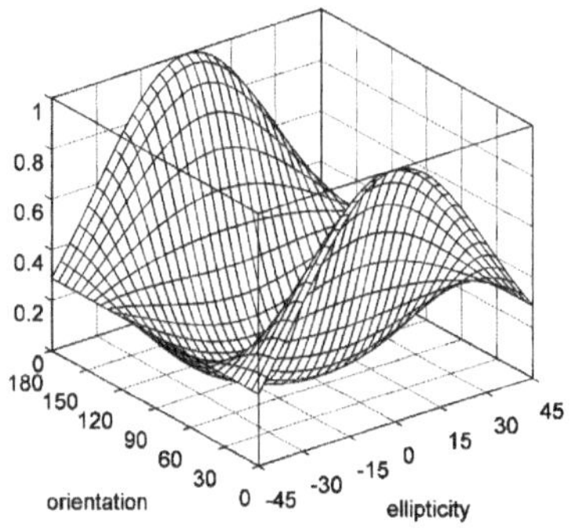

Fig. 5.25. Co-polarisation plot for a dielectric dihedral structure such as the side of a house

The backscattering radar cross section of the wire shown in Fig. 5.26 is given by[26]

$$\sigma = \frac{2\pi h^2 \cos^4\gamma}{\ln^2(0.8905\beta a\cos\Psi)+\pi^2/4)}\left[\frac{\sin(2\beta h\sin\Psi)}{2\beta h\sin\Psi}\right]$$

in which $\beta=2\pi/\lambda$. The angle γ describes the orientation of the electric field vector with respect to the cylinder axis, while the angle Ψ represents the angle with which the ray strikes the wire in the slant plane, measured against its normal, before it is scattered. While at first sight it might seem strange that there would be any backscatter except for irradiation exactly in the normal direction, a *finite* length cylinder will have backscatter in all directions in principle, albeit rapidly falling as we move away from normal irradiation. It is the square bracketed term above that determines that behaviour; note that if h goes to infinity that term will approach zero. Fig. 5.27 shows the dependence of the radar cross section of the wire on Ψ and γ. The former shows the sensitivity to alignment of the wire with the flight path of the platform while the second shows sensitivity to the polarisation of the radiation.

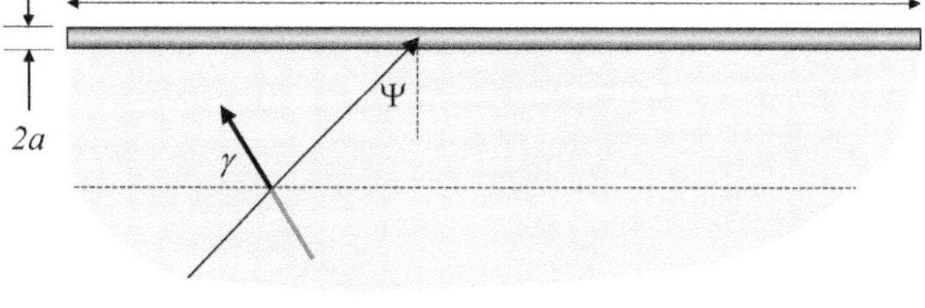

Fig. 5.26. Scattering from a long, thin horizontal wire

[26] See Ruck et al, *loc cit.*

If a metallic structure is a multiple of half wavelengths of the radiation it will have an exceptionally high radar cross section; the expression above cannot be used to show that since it relates to a wire that is long compared with a wavelength. We can however understand the effect quantitatively in the following manner. When a body such as a wire is irradiated currents are set up inside it as indicated in Fig. 5.18 ; those currents cause fields to be radiated from the object. It is those fields that represent the backscattered power density. The object is in fact behaving as though it were an antenna when it re-radiates. Antenna theory demonstrates that the most efficient radiators are those that are a half wavelength long, and then multiples of half a wavelength. Likewise passive metallic elements with those dimensions will show strong radar scattering

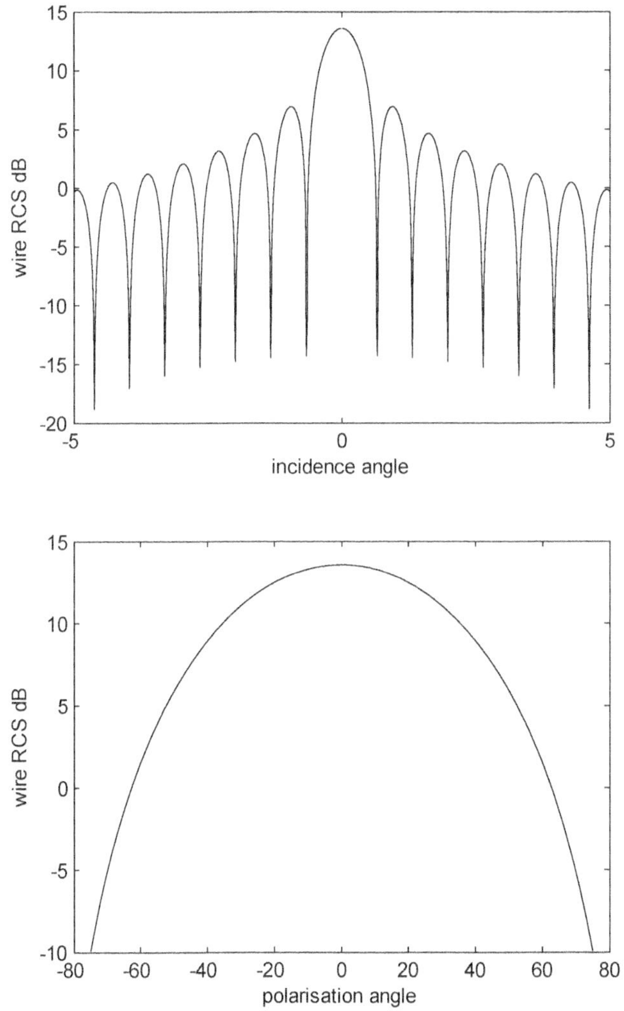

Fig. 5.27. Backscattering radar cross section of a wire 10m long and 1cm diameter at L band as a function of incidence angle (for a polarisation angle of zero) and as a function of polarisation angle (when the incidence angle is zero)

5.5.4 Bragg Scattering

The wavelengths employed in radar remote sensing are not too different from some structural periodicities often found in nature. That gives rise to a particularly interesting form of scattering that can be relevant in agriculture and underpins one of the more popular models for sea surface scattering.

Fig. 5.28 shows a sinusoidally varying surface with spatial wavelength Λ. If that structure is irradiated there will be reflections from the regularly spaced portions of the surface. Those reflections add to give the complete response from a pixel which contains the sinusoidal surface variation. If we assume the reflections are all of the same magnitude then in adding them we have to account only for the effect that some travel further than others in transmission and reception.

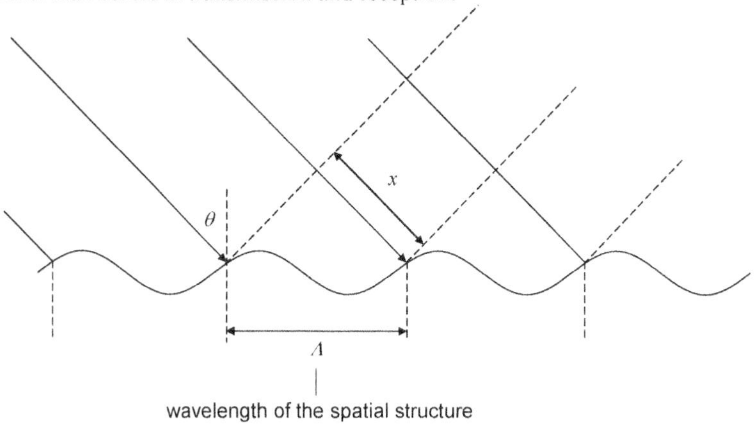

Fig. 5.28. Interaction of the radar beam with a spatially periodic structure

The additional distance x shown in Fig. 5.28 is $\Lambda\sin\theta$, giving the additional two way phase delay between the two left most rays as

$$\Delta\phi = 2\pi\frac{2x}{\lambda} = \frac{4\pi\Lambda\sin\theta}{\lambda}$$

If this additional phase is a multiple of 2π, the two waves will add in phase and re-inforce each other. If those two add in phase then so will waves reflected from other parts of the spatial periodicity. Thus, the condition for all waves to reinforce is that

$$\frac{4\pi\Lambda\sin\theta}{\lambda} = 2\pi n$$

where n is an integer, giving

$$\Lambda = \frac{n\lambda}{2\sin\theta} \tag{5.31}$$

as the condition for so-called *Bragg resonance*. While this has been developed on the basis of scattering from a sinusoidal surface any periodic repetition of scatterers aligned orthogonally to the incoming wavefront will give rise to interference among the scattered waves. A sequence of wires for example can behave this way.

Usually when we consider several scattering mechanisms within a pixel we simply add their power density contributions – effectively we add their radar cross sections or scattering coefficients. With Bragg scattering, however, the fields add. This is called coherent addition, as against non-coherent scattering when the power densities add. Coherent addition gives a much higher power density and thus scattering coefficient. To illustrate this point suppose the scattered electric field strength from each of two individual scatterers is E. If the reflections add non-coherently the total power density is proportional to $E^2+E^2=2E^2$, whereas if they add coherently the power density is proportional to $(E+E)^2=4E^2$. Fig. 5.29 shows an example of the strength of Bragg scattering believed to occur from aligned portions of circular agricultural fields in Libya.

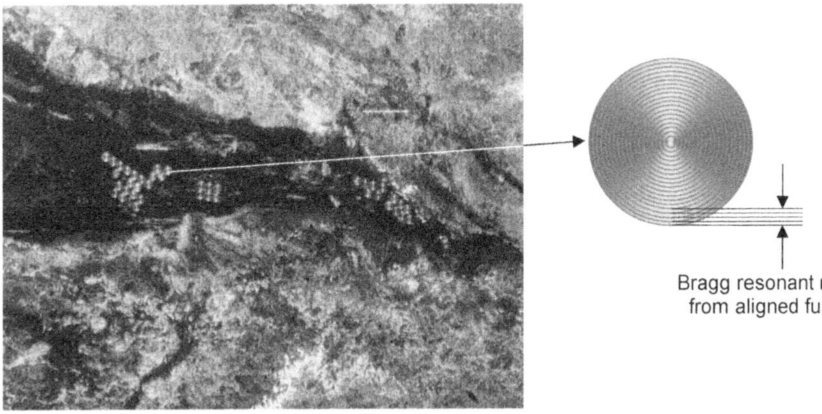

Fig. 5.29. Circular pivotal irrigated agricultural fields in Libya, demonstrating the strong returns most likely associated with Bragg Resonance; the radar illumination is from the bottom of the scene so that ploughed furrows running across the scene give the enhanced returns (from J.P. Ford, J.B. Cimino and C. Elachi, *Space Shuttle Columbia Views the World With Imaging Radar: the SIR-A Experiment*, JPL Publication 82-95, NASA, 1 January 1983)

5.5.5 The Cardinal Effect

If a region being imaged consists of a row of buildings acting like dihedral reflectors in the nature shown in Fig. 5.21 or has in it wire fence lines such as depicted in Fig. 5.26, then strong radar scattering will occur if those structural elements are aligned parallel to the platform flight line and thus orthogonally to the incoming radar beam. If they are not aligned then their radar response will be weak. Fig. 5.27 shows for example that the backscattered power from a wire will be reduced tenfold just a few degrees off broadside. It is not unusual, therefore, for urban regions of ostensibly the same housing density to show very high response when the street pattern is aligned to the flight path and low response otherwise. The same can happen if Bragg scattering occurs with, say, agricultural fields and fence lines. This is known as the *cardinal* effect because of its loose association with compass directions. Fig. 5.30 shows an image of Montreal in which the cardinal effect is evident.

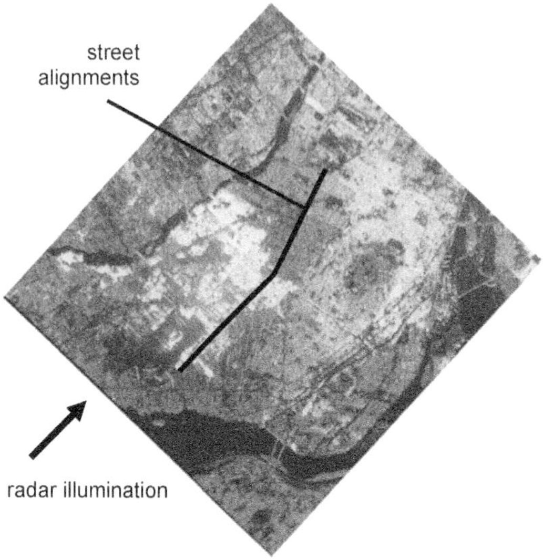

Fig. 5.30. Portion of a SIR-B image acquired over Montreal, Canada demonstrating the cardinal effect; the bright central portion of the image is where cross streets are aligned orthogonally to the incoming radar energy, whereas the portions to the north, of about the same urban density, have street patterns not orthogonal to the radar beam (from J.P. Ford, J.B. Cimino, B. Holt and M.R. Ruzek, *Shuttle Imaging Radar Views the Earth From Challenger: the SIR-B Experiment,* JPL Publication 86-10, NASA, 15 March 1986)

5.6 Composite Scatterers

In practice we often come across scattering behaviours resulting from a combination of the effects treated in the previous sections. Either there will be multiple scatterings involving the same type of element (such as from leaf to leaf in a canopy) or there will be mechanisms involving more than one scattering type. Many of these are found in the scattering behaviour of trees and forest stands, which we will use here to illustrate the effects that emerge.

When handling composite situations it is necessary to determine whether each of the scattering components that can reasonably be identified (such as the small set in Fig. 5.31) should be added coherently or non-coherently. The former can be a difficult task since it requires the scattering pathways to be described by the electric field vectors in each case, which are then combined. That might be required if there were a small number of dominant scatterers in a scene. In general, if there are many randomly dispersed scatterers within the resolution elements of a scene we can assume that the component scattering mechanisms can be combined non-coherently. That means we can add their power contributions by combining scattering coefficients or radar cross sections normalised by pixel area.

5.7 Sea Surface Scattering

Because the mechanisms for sea surface scattering are different from those generally observed with land-based features it is instructive to consider the sea as a separate scatterer type.

A perfectly flat sea will behave like a specular reflector and consequently will appear dark in monostatic radar imagery for all incidence angles except zero. Clearly, in order to receive measurable backscatter the sea surface must be made rough by some physical mechanism. The principal means for surface roughening is the formation of waves.

Fig. 5.31. Typical scattering pathways for trees and forest stands: 1 is trunk-ground corner reflector scattering; 2 is canopy-ground scattering, 3 is scattering from the ground after transmission through the canopy and 4 is canopy volume scattering

There are two broad types of wave on the surface of the ocean, both excited by the action of wind blowing across the surface, but distinguished by the mechanism that tries to restore the flat water surface against the driving effect of the wind. *Gravity waves* depend upon gravitation acting on the disturbed mass of water to counteract the effect of the wind; their wavelengths tend to be long, typically in excess of a few centimetres. On the other hand *capillary waves* have wavelengths shorter than a few centimetres and rely on surface tension to work against the disturbance caused by wind action. For both types the amplitude and wavelength is a function of wind speed, fetch (the distance over which the wind is in contact with the surface of the water) and the duration of the wind event. Capillary waves typically appear to ride on the gravity waves as depicted in Fig. 5.32.

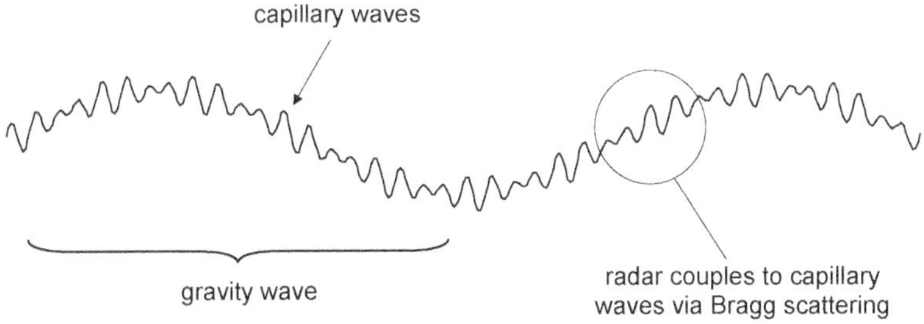

Fig. 5.32. Sea surface waveform composed of gravity and capillary waves

By their nature water waves have periodicity. They can be quite complicated in that at any time there may be a whole range of wavelengths present with more energy associated with some than others. That is summarised in the wave power spectrum of the sea state, an illustration of which is given in Fig. 5.33, which can be interpreted to mean that there are a myriad of periodicities present, some stronger than others. Importantly though there will almost certainly be available some energy at the spatial wavelength required for the

Bragg resonance mechanism discussed in Sect. 5.5.4. Bragg coupling can be used to describe the nature of sea surface imagery observed in radar remote sensing provided the gravity waves are not too large, in which case scattering from wave facets facing the radar is usually considered the most appropriate description[27]. Here, we will analyse the Bragg scattering situation by referring to the condition of (5.31) and the spectrum of Fig. 5.33.

Within the range of wavelengths relevant to capillary waves the sea surface spectrum is approximately linear on a log-log scale as represented in Fig. 5.33; the energy density increases with wind speed as indicated[28]. There is considerably more energy available at the longer sea surface wavelengths. Thus in satisfying the Bragg resonance condition of (5.31) at a given angle of incidence we would expect greater ocean returns at longer wavelengths – roughly corresponding to C band in Fig. 5.33. Extrapolating the curve to smaller wave numbers suggests there would be better backscatter still at, say, L band. However, that is not the case. The power density of capillary waves falls for wave numbers smaller than those shown so that the sea surface can appear dark at L band, particularly for incidence angles typical of space borne missions (20-40°).

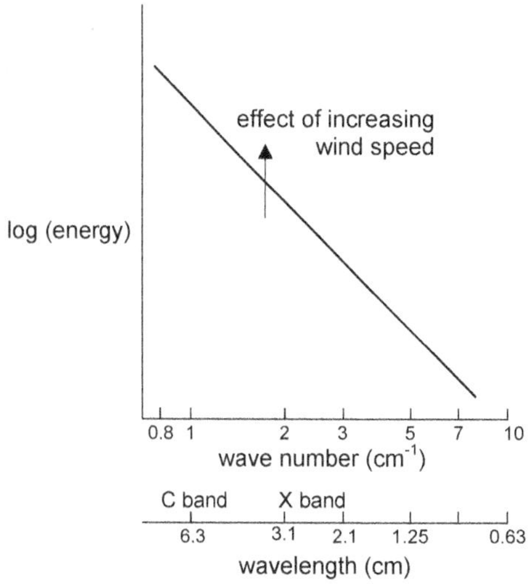

Fig. 5.33. Energy spectrum of short sea surface waves

Suppose we now decide on using C band. What incidence angles are best? Expressing the Bragg resonance condition of (5.31) in terms of sea surface wave number *k* rather than wavelength gives

$$k = \frac{4\pi \sin\theta}{n\lambda}$$

[27] See J.F. Versecky and R.H. Stewart, The observation of ocean surface phenomena using imagery from the SEASAT synthetic aperture radar: an assessment, *Journal of Geophysical Research*, vol. 87, no. C5, 3397-3430, 1982.

[28] See R.T. Lawner and R.K. Moore, Short gravity and capillary wave spectra from tower-based radar, *IEEE Transactions on Oceanic Engineering*, vol. OE-9, no. 5, 317-324, 1984, and Fig. 11.27 of F.T. Ulaby, R.K. Moore and A.K. Fung, *Microwave Remote Sensing: Active and Passive, Vol 2*, Addison-Wesley, Reading Mass., 1982

Noting from Fig. 5.33 that the greatest wave energy is at the smaller wave numbers, we see that we will get higher sea surface returns at smaller incidence angles. There will be fairly rapid fall off in radar return as the incidence angle increases.

Oceanographic mapping satellites like Seasat and ERS-1,2 employ incidence angles of around 20° to take advantage of those higher available spectral energies. Smaller angles are not used in order to avoid a specular component in the return. While angles around 20° are suitable for sea state imaging they can be problematic for land surface imaging in regions of high relief since terrain distortion is worse for smaller incidence angles, as outlined in Sect. 4.1.2.

(a)

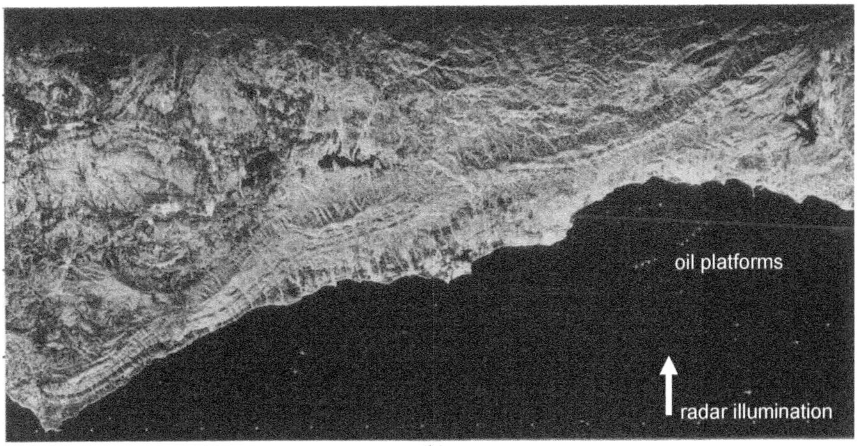

(b)

Fig. 5.34. Seasat mosaic (a) and SIR-A (b) image of the coastal region around Santa Barbara, California (from J.P. Ford, J.B. Cimino and C. Elachi, *Space Shuttle Columbia Views the World With Imaging Radar: the SIR-A Experiment*, JPL Publication 82-95, NASA, 1 January 1983)

Fig. 5.34 demonstrates these features and the importance of incidence angle. It shows a Seasat image (20°) recorded off the coast of Santa Barbara, California along with a SIR-B image (40°) of the same region. Sea surface information is only evident in the C band 20°

data, although that image also shows terrain distortion. The bright targets off the coast (which are oil drilling platforms) show well in the 40° L band imagery because of the low sea returns. They are present also in the Seasat image but masked by the strong sea surface return at the smaller incidence angle.

The dependence of radar return on incidence angle allows us to assess the modulation of backscatter that is observed across a gravity wave on which capillary waves sit, as shown in Fig. 5.32. The front slopes of the gravity waves face the radar and thus present a smaller incidence angle to the radar beam, allowing coupling to capillary waves of smaller wave numbers and thus increased energy. The back slopes show a larger incidence angle. The radar beam thus couples to larger wave number components of the sea surface spectrum; as a result they will appear considerably darker than the front slopes. Consequently, we can observe gravity waves on the sea surface as a result of Bragg resonance with the capillary waves.

Note from Fig. 5.33 that the level of backscatter increases with wind speed as is to be expected. In general, it is important to recognise that anything that affects the capillary waves, and thus their energy spectra, will lead to modulation of the radar returns. That includes rain dampening, and dampening by other mechanisms such as oil slicks. Fig. 5.35 shows a Seasat image from 1978 that includes a major oil slick. The sea surface is dark at the slick because the capillary waves have been damped by the oil.

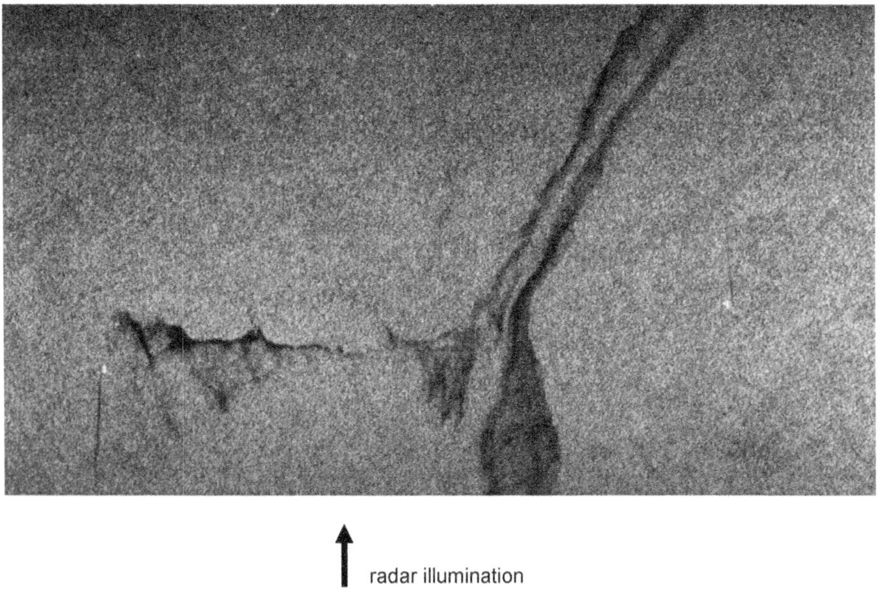

Fig. 5.35. Seasat image recorded on 3 October 1978 showing an oil slick and two ships with their (offset) wakes (from J.P. Ford et al., *Seasat Views North America, the Caribbean, and Western Europe With Imaging Radar,* JPL Publication 80-67, NASA, 1 November 1980)

Also observable in the image are two ships, both sailing in the cross track direction, but opposite to each other. Several features are noteworthy. First, the ships appear as bright spots because of the dihedral reflections caused by the sides of the ships and the ocean surface as discussed in Sect. 5.5.2. Secondly, the wakes generated by the ships are clearly visible because of their modulation of the capillary waves. Finally, the wakes are offset from their respective ship! That is because the images are formed from the Doppler history of the radar reflections as discussed in Sect. 3.6. Since the ships are moving they will have an apparent broadside position with respect to the radar platform defined by when the Doppler shift of the carrier frequency is zero. For a stationary target that happens at physical broadside. For a moving target broadside and zero Doppler are different. Because one ship in Fig. 5.35 is travelling towards the radar and the other away, their position shifts are in opposite directions.

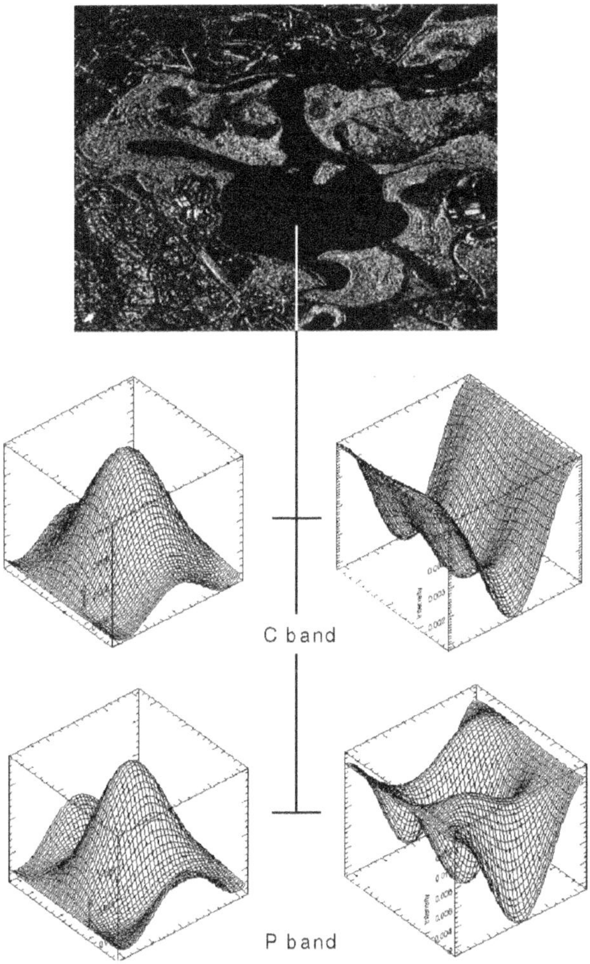

Fig. 5.36. Co and cross polarised polarisation signatures for lake water at C and P bands from an AirSAR scene of Brisbane, Australia; produced using ENVI™ (ITT Visual Information Solutions)

Finally, Fig. 5.36 shows the polarisation signatures typical of relatively calm water. They should be compared with those for a relatively smooth surface in Fig. 5.12, particularly for the case of the higher dielectric constant.

5.8 Internal (Ocean) Waves

The previous section has looked at the coupling of radar energy with waves that form on the surface of the ocean. Waves can also generate within the bulk of the ocean itself, launching on the soft boundaries that occur between water layers of differing temperatures, densities and salinities. In contrast to the much shorter surface waves that propagate on the water-air interface, so-called *internal waves* have much longer wavelengths, typically several hundreds to thousands of metres.

There is still much to be understood about internal waves. They usually exist in wave packets and appear to be generated by mechanisms that cause underwater disturbances such as river inflows, movement of water over varying bottom topography and underwater earthquakes. They express themselves in radar imagery because they modulate the capillary waves. One theory says that they have vertically circulating current patterns that sweep materials such as pollens, slicks and other debris into convergence zones that damp the capillary energy thereby causing dark bands in the imagery[29]. Fig. 5.37 shows a Seasat image of the Andaman Sea with internal waves.

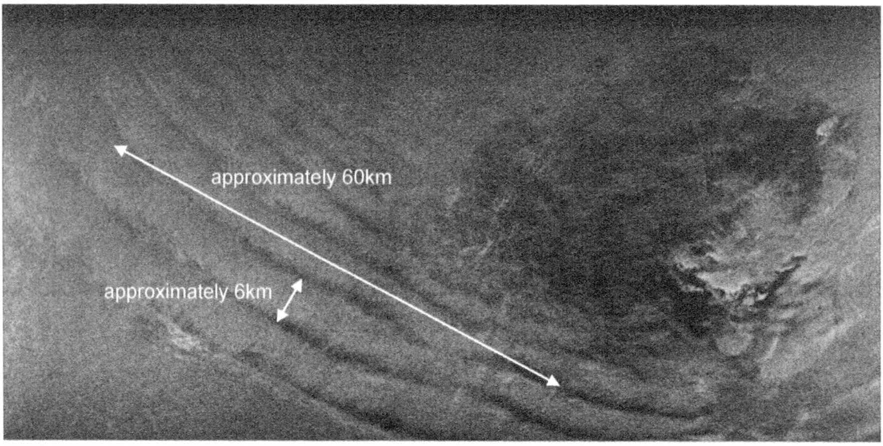

Fig. 5.37. Radar image of internal waves in the Andaman Sea; the image is about 100km across which gives an idea of the scale of the waves (from J.P. Ford, J.B. Cimino and C. Elachi, *Space Shuttle Columbia Views the World With Imaging Radar: the SIR-A Experiment*, JPL Publication 82-95, NASA, 1 January 1983)

5.9 Sea Ice Scattering

Sea ice is a particularly interesting scattering medium because its properties change with time; that leads to a change in its scattering characteristics. Newly formed ice is thin and smooth. It will appear dark in radar imagery since it behaves as a specular reflector. This will also be the case for lake ice. New sea ice can be difficult to distinguish from open water unless the water is wind roughened. If the ice is covered in a layer of moist snow[30]

[29] See W. Alpers, Theory of radar imaging of internal waves, *Nature*, vol. 314, 245-247, 1985.

[30] Dry snow – i.e. below freezing – has a very low dielectric constant and thus appears almost transparent to incident radar energy.

it will appear bright because of volume scattering from the snow and composite scattering involving the snow and the ice layer as illustrated in Fig. 5.38.

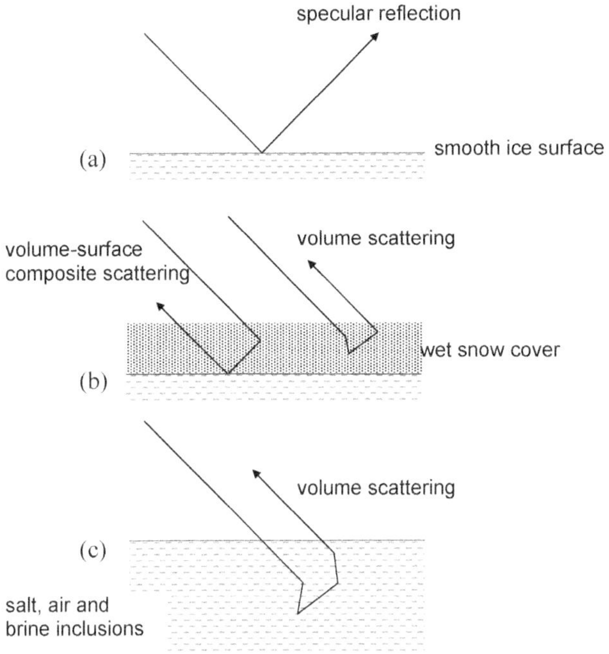

Fig. 5.38. Scattering pathways for radar backscatter from sea ice (a) specular reflection from new, smooth ice (b) volume and composite scattering involving snow cover and (c) volume scattering from within the ice itself; the surface can also be a diffuse scatterer

As sea ice ages its morphology changes. Because of temperature fluctuations and mechanical stresses caused by movements of ice floes, the surface of the ice becomes roughened with time and small pressure ridges form. As a consequence, when aged, it exhibits diffuse surface scattering behaviour, particularly at smaller incidence angles. Fig. 5.39 shows sea ice imaged at each of C, L and P bands, in which several interesting observations can be made[31]. First, C band appears to give the best range of brightness for discriminating among the ice features, especially first year (smooth) as against multi-year (surface roughened) ice. It appears that the variations in surface roughness is such that first year/multi-year differentiation is not discernable at L and P bands, along with the fact that there is likely to be penetration at those wavelengths. Pressure ridges within the multi-year ice floes however are better picked up in L band, presumably because they are rough at that wavelength compared with the smoother floe surface, whereas at C band both are rough and thus a little more difficult to differentiate; at P band both appear smooth so that the ridges are not seen.

[31] B. Scheuchl, I. Hajnsek and I. Cumming, Classification strategies for polarimetric SAR sea ice data, *Workshop on Applications of SAR Polarimetry and Polarimetric Interferometry*, Frascati, Italy, 14-16 January 2003

The dielectric constant of ice below freezing is not very high because the free water molecules that give rise to the high dielectric constant of liquid water are not present in ice[32]. Instead the water molecules are bound into the ice lattice. Typically, the dielectric constant of sea ice (3.5 - 4) is low enough that there can be transmission across its upper boundary. That component undergoes volume scattering from air bubbles, salt and brine inclusions within the bulk of the ice as shown in Fig. 5.38.

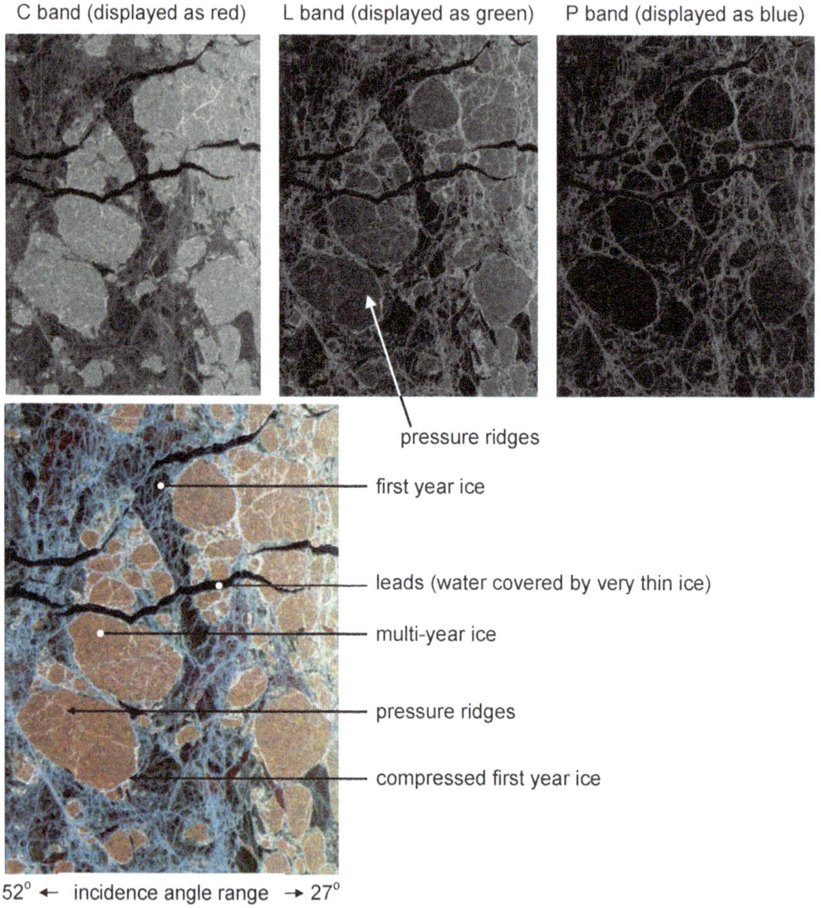

Fig. 5.39. Multi-wavelength aircraft SAR imagery of sea ice (from B. Scheuchl, I. Hajnsek and I. Cumming, Classification strategies for polarimetric SAR sea ice data, *Workshop on Applications of SAR Polarimetry and Polarimetric Interferometry*, Frascati, Italy, 14-16 January 2003, ©2003 ESA/ESRIN)

[32] See J.A. Richards, *Radio Wave Propagation: An Introduction for the Non-Specialist*, Springer, Berlin, 2008.

CHAPTER 6

INTERFEROMETRIC AND TOMOGRAPHIC SAR

6.1 Introduction

Undoubtedly, one of the more interesting applications of synthetic aperture radar imagery to emerge in the past two decades has been topographic mapping using interferometry. Because the phase angle of the backscattered signal for a given pixel is available, and phase is easily measured, it is possible to compare the phase differences of two different images of the same region and, from that comparison, find the relative locations of pixels in three dimensions: latitude, longitude and altitude, or their equivalents. In this chapter we show how that can be done, and how interferometry can also be used for change detection. The fundamental concept is extended to show how a tomographic process can be implemented, in which the vertical detail within a ground resolution cell can be resolved. The radar geometry used for interferometric applications is a special case of bistatic radar considered in Chapt. 7.

6.2 The Importance of Phase

One of the characteristics that sets radar aside from optical imaging is that we know both the amplitude and the phase of the signal backscattered from the landscape. For optical imagery we know only the intensity (radiance) which, as seen in Chapt. 2, is equivalent to amplitude squared without phase. Knowing the phases of two signals means they can be interfered as discussed in Sect. 2.17. Interference is the basis of interferometric SAR imaging.

After scattering from a particular pixel the signal received at the radar, and compressed in range and azimuth to remove the transmitted and Doppler induced chirps, can be written

$$E^r(t) = A\rho\exp(j\omega t - 2\beta R) \tag{6.1}$$

in which ω is the operating frequency of the radar – sometimes called the carrier frequency. This signal would be one of the polarisations used in a multi-polarisation radar system. It will be a function of the pixel of interest; strictly, therefore, we should write it as a function of the range and azimuth coordinates of the pixel. We will consider that detail later.

The ρ in (6.1) is the reflectivity of the pixel being imaged – i.e. its scattering characteristic that would normally be expressed as the corresponding element of the scattering matrix. As seen in Chapt. 3 it is a complex number, which indicates the effect it has on both the amplitude and phase of the incident energy when producing the backscattered signal. In general we would therefore write it as

J.A. Richards, *Remote Sensing with Imaging Radar*, Signals and Communication Technology,
DOI: 10.1007/978-3-642-02020-9_6,

$$\rho = |\rho| e^{j\phi} \equiv |\rho| \angle \phi \quad (6.2)$$

For the moment we won't have much to say about the properties of ρ, but it will be important later when we come to look at limitations in interferometry.

The factor A is a general amplitude scaling term that we will assume is the same for every pixel, and results from the inverse distance drop in signal strength during transmission along with any other factors that are not pixel specific. R is the one way distance between the radar set and the pixel, while β is the phase constant, given in Sect. 2.8, sometimes also called the wave number k. Note $\beta=2\pi/\lambda$ in which λ is the operating wavelength of the radar, given by $\lambda = c/f$ with $f = \omega/2\pi$.

We can write (6.1) as

$$E^r(t) = A\rho \exp(j\omega t - \phi_T) \quad (6.3a)$$

in which[1]

$$\phi_T = 2\beta R = \frac{4\pi R}{\lambda} \quad (6.3b)$$

is the total change in phase of the signal from when it was transmitted to when it arrived back at the radar as a result of the $2R$ path it travelled. Clearly, if two pixels or targets are at different slant ranges then they will have different total phase angles measured at the radar. The difference in their phase angles is proportional to the different distances to the targets. This is illustrated in Fig. 6.1a. It is, in principle, easy to discriminate between points a and b at the top and bottom of a topographic feature because the echoes are separated in phase. There is however an ambiguity. Because the radar works on resolving in slant range, the target at point c will appear to the radar to be at the same position as that at point a.

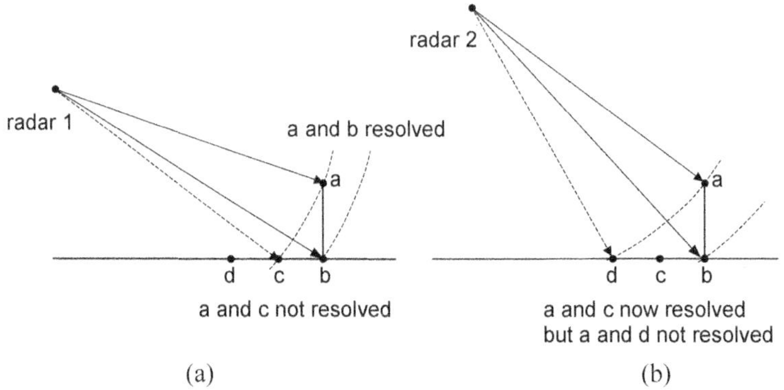

Fig. 6.1. (a) Even though the topographic variation between a and b is resolvable, there is ambiguity between a and c (b) resolving the a and c ambiguity by changing the radar position, but causing a and d ambiguity

[1] See Sect. 6.9. This expression strictly depends on the mode of operation of the interferometric radar, although that is not important at this stage

One way of resolving the a-c ambiguity is to change the viewing perspective, as in Fig. 6.1b. However, a new ambiguity has been created between points d and a. If both perspectives were used together then perhaps we might be able to resolve all such ambiguities, just like stereo vision does. That is the principle behind radar interferometry. At least two viewing perspectives are chosen. We will show now that that allows, in principle, unambiguous resolution of the landscape in three dimensions, with the exception of two further considerations: parameter uncertainties and an ambiguity in phase measurement. Phase ambiguity is a major consideration that must be resolved, as we will see shortly.

6.3 A Radar Interferometer - InSAR

Consider the geometry shown in Fig. 6.2 which is the basis of our analysis of interferometric radar, referred to commonly as *InSAR*. The two radars in this case are shown arranged horizontally at either end of a *baseline*, which approximates the situation most often encountered in practice and which we refer to generically as an *interferometer*. We treat the case of an inclined baseline in Sect. 6.8. The projection of the baseline normal to the line of sight from the radar to the target, $B_\perp$, is an important parameter; we call that the *orthogonal baseline*.

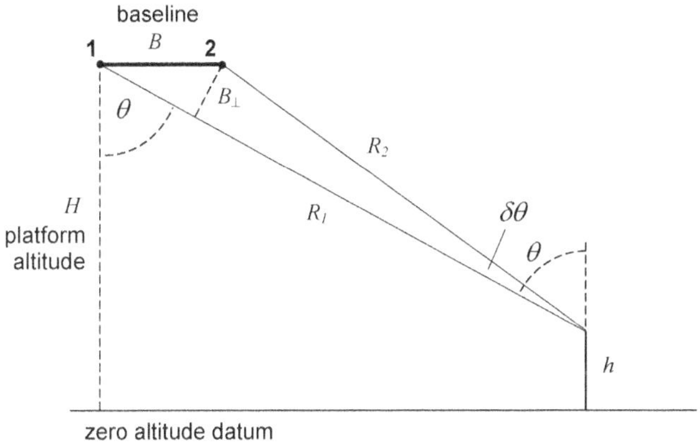

Fig. 6.2. Geometry for single baseline SAR interferometry, in which we have assumed that the look and incidence angles are the same; the platform travels out of the page

To consider the phase difference between the two radar signals we need to find the difference in the path lengths to a target, shown in the figure as sitting at a height h above the assumed zero altitude plane. It can be seen that

$$R_1 = R_2 \cos\delta\theta + B\sin\theta \tag{6.4}$$

Assuming $\delta\theta \approx 0$ this is $$R_1 = R_2 + B\sin\theta$$

so that $$\Delta R = R_1 - R_2 = B\sin\theta \tag{6.5}$$

The equivalent difference in phase angle between the two signals is, from (6.3b)

$$\Delta\phi = \frac{4\pi B \sin\theta}{\lambda} \tag{6.6}$$

We call that the *interferometric phase angle*.

The assumption of negligible difference in incidence angle $\delta\theta \approx 0$ that led to this approximation essentially means that we are considering the target to be infinitely far away from the two antennas compared with their baseline separation. While that is acceptable in the use of interferometers in radio astronomy it is a slightly poorer assumption in SAR interferometry, but is nevertheless adopted. It is sometime called the *plane wave approximation*.

Note that (6.6) is not dependent on h, which is a consequence of the plane wave assumption. However, it is a function of incidence angle which varies with target height above the datum as can be appreciated from Fig. 6.2. To find that relationship we redraw the imaging geometry simply, as shown in Fig. 6.3, in which B represents the baseline of the two-radar interferometer.

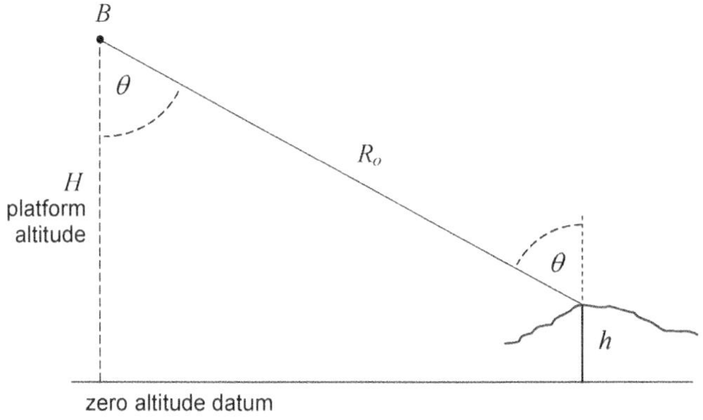

Fig. 6.3. Determining the relationship between topographic height and incidence angle; strictly the incidence and look angles would be slightly different, especially for a space borne system, but we ignore that small difference here

From Fig. 6.3 we see $$h = H - R_o \cos\theta$$

so that $$\frac{dh}{d\theta} = R_o \sin\theta \tag{6.7}$$

From (6.6) $$\frac{d(\Delta\phi)}{d\theta} = \frac{4\pi B \cos\theta}{\lambda} \tag{6.8}$$

so that, using (6.7) and (6.8),

$$\frac{d(\Delta\phi)}{dh} = \frac{d(\Delta\phi)}{d\theta}\frac{d\theta}{dh} = \frac{4\pi B \cos\theta}{\lambda R_o \sin\theta} \tag{6.9}$$

This shows how changes in terrain height result in changes to the interferometric phase angle. From Fig. 6.2 $B\cos\theta = B_\perp$ so that (6.9) is[2]

$$\frac{d(\Delta\phi)}{dh} = \frac{4\pi B_\perp}{\lambda R_o \sin\theta} = \frac{4\pi B_\perp \cos\theta}{\lambda H \sin\theta} \tag{6.10}$$

This demonstrates the dependence of the change in intereferometric phase with terrain on the three important system parameters: platform altitude H, angle of incidence (or look angle) θ and the orthogonal baseline $B_\perp$ of the interferometer. To gain some idea of the sensitivity of the system consider ERS for which H=780km, λ=0.056m and θ=23°; assume $B_\perp$ =250m, which is at the upper end of its useful range (see Sect. 6.11 for the concept of critical baseline, which limits the upper usable value of $B_\perp$). These give

$$\frac{d(\Delta\phi)}{dh} = 0.169\text{rad/m}$$

so that a full 2π cycle of phase difference corresponds to a height variation of about 37m.

If in (6.10) we call

$$\alpha_{IF} = \frac{dh}{d(\Delta\phi)}$$

the *interferometric phase factor* then the elevation of a given point at (x,y) corresponding to the phase difference at that point is given by

$$h(x,y) = \alpha_{IF}\Delta\phi(x,y) + \text{constant} \tag{6.11}$$

In principle, the constant can be found by associating the height and phase difference at one specific point, allowing the elevations at all other points then to be determined.

6.4 Creating the Interferometric Image

The difference in the phase angles of the two constituent images has to be established on a pixel by pixel basis in order to map elevation using the material in the previous section. From (6.3a) the received fields from a given pixel by each radar will be of the forms:

$$E_1(t) = \rho\exp(j\omega t - \phi_{T1}) \Rightarrow \rho\exp(-j\phi_{T1}) = e_1(x,y) \tag{6.12a}$$
$$E_2(t) = \rho\exp(j\omega t - \phi_{T2}) \Rightarrow \rho\exp(-j\phi_{T2}) = e_2(x,y) \tag{6.12b}$$

in which we have set the common amplitude factor A to unity for convenience. The signals are distinguished, as expected, by their differing phase angles. We can disregard the common time exponential terms, retaining only the phasor forms of the signals as indicated. Both are now shown as functions of x and y, signifying that they are different for pixels at different locations in the image. If we form the product

[2] We don't use tan in place of the ratio of sin and cos in (6.10) since, when the baseline is inclined, the angle of the sin term is changed – see (6.17).

$$i(x,y) = e_1(x,y)e_2^*(x,y) \tag{6.13a}$$

in which one of the images is conjugated as shown, then the result is an image with amplitude proportional to the scattering coefficient and phase being the interferometric phase difference:

$$i(x,y) = |\rho|^2 \exp[-j(\phi_{T1} - \phi_{T2})] = |\rho|^2 \exp(-j\Delta\phi) \tag{6.13b}$$

We call this the *interferogram*. Generally the pixels of the interferogram are averaged over a small neighbourhood to reduce phase noise so that the resulting elevation maps are locally smooth.

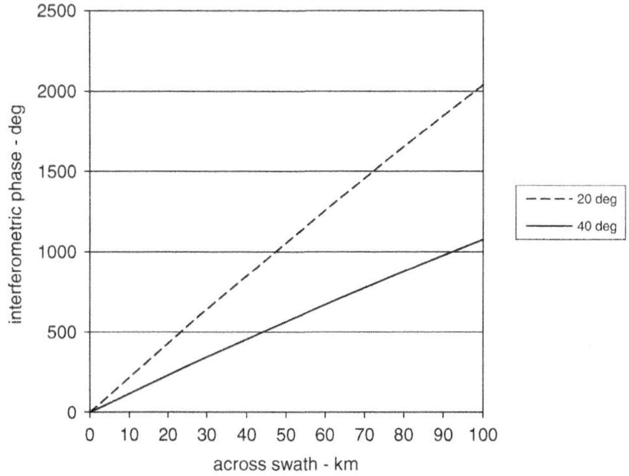

Fig. 6.4. Variation of interferometric phase in radians across a 100km swath for two different nominal incidence angles

6.5 Correcting for Flat Earth Phase Variations

In Sect. 6.3 we looked at the variation of interferometric phase difference with elevation. From (6.6) we can see that it will also vary with incidence angle across the range direction even if there were no variation in elevation. In other words, across the image swath there will be an equivalent flat earth variation in phase resulting from the corresponding change of incidence angle from near to far swath edge. That flat earth variation needs to be removed from the recorded phase difference between the two radars in the interferometer before (6.11) can be applied; otherwise the result will be biased with position across the swath.

Fig. 6.4 shows the extent of the flat earth phase variation across a 100km swath for a platform at an elevation of 800km, with a 100m baseline and for nominal incidence angles of 20° and 40°. That demonstrates the extent of correction necessary before interferometric phase can be used to deduce topography.

To illustrate this point further we use the artificial landscape[3] shown in Fig. 6.5, consisting of 500x500 pixels with a ground resolution of 20x20m, rising to a maximum elevation of 2000m. The interferometric phase variation was generated by assuming a near swath incidence angle of 20°, a baseline of 150m and a platform elevation of 800km. Fig. 6.6a shows the interferometric phase over the region. Notwithstanding the height variation in the terrain shown in Fig. 6.5 the corresponding phase change is not readily seen in the figure because it is masked by the change associated with the flat earth phase. Fig. 6.6b shows the degree of the flat earth phase variation across the swath, while Fig. 6.6c shows the effect of correcting 6.6a with 6.6b. As seen, we can now discern the variation in interferometric phase corresponding to the elevation variation, as required.

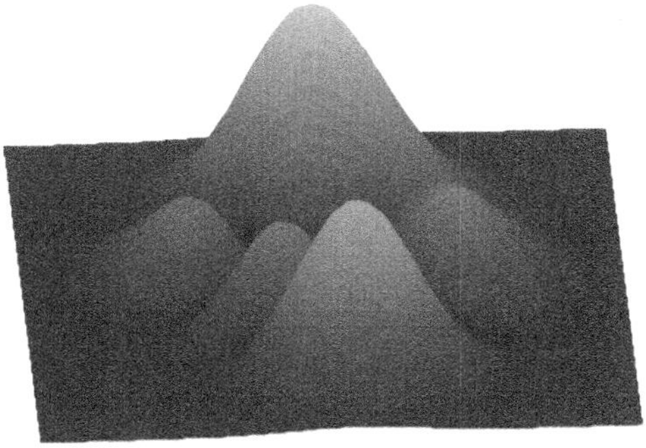

Fig. 6.5. Simulated terrain variation to assist in studying interferometric phase

6.6 The Problem with Phase Angle

Even though the physical phase angle seen in Fig. 6.6 extends over a very great range, the phase angle that results when two signals are interfered according to (6.13) is restricted to the range 0 to 2π. To see that recall that the exponential functions we use to represent sinusoids are actually just mathematical conveniences. For the interferogram of (6.13b) strictly we should write (ignoring the reflection coefficient)

$$\cos(\Delta\phi) = \mathcal{Re}\{\exp(-j\Delta\phi)\}$$

Because the cosine function is periodic, even though the interferometric phase changes substantially with terrain elevation in the manner observed in Fig. 6.6c, all we see at the radar receiver is a phase somewhere between 0 and 2π. For the example of Fig. 6.5 that means that the phase difference map between the two images of the interferometer actually produced will be as shown in Fig. 6.7. Each cycle of fringes corresponds to a 2π variation in interferometric phase, whereas in fact we want to see a smooth, non-cyclic, phase variation comparable to Fig. 6.6c.

[3] This was generated by taking the absolute value of the "peaks" function in MatLab™.

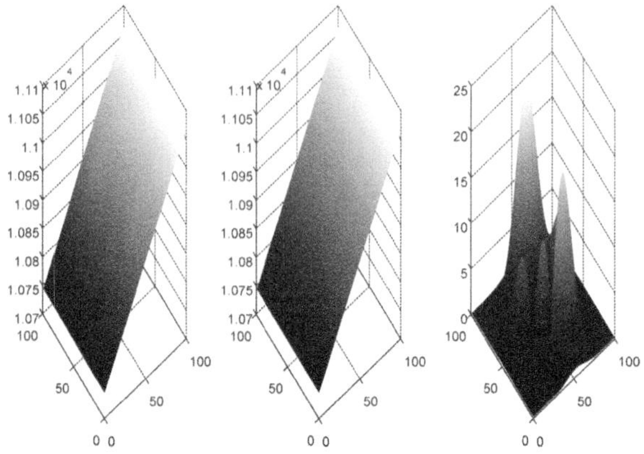

Fig. 6.6. (a) Variation of uncorrected interferometric phase, (b) flat earth phase variation, and (c) corrected interferometric phase (radians)

The question that arises therefore is how can we extract meaningful elevation information about the landscape when the phase varies cyclically in the manner of Fig. 6.7? Essentially, what we have to do is create the phase variation of Fig. 6.6c from that of Fig. 6.7. The process, known as *phase unwrapping*, can be non-trivial. It is called unwrapping because of the cyclic variation in phase with period 2π. This can be seen directly from (6.13b) by plotting $\exp(-j\Delta\phi)$ on a polar, or Argand, diagram and observing how this term changes as the phase difference increases. Fig. 6.8 shows that behaviour. As the phase difference increases we move around the polar plot cyclically, with each full cycle corresponding to the change between like shaded parts of the interference fringes seen in Fig. 6.7. In order to be able to recover the corresponding terrain height information it is necessary to roll back, or unwrap, the change in phase of Fig. 6.8.

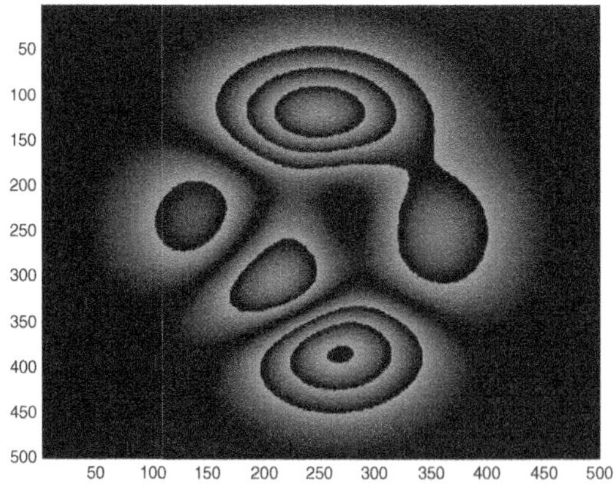

Fig. 6.7. The interferometric phase variation at the radar receiver for the landscape of Fig. 6.5

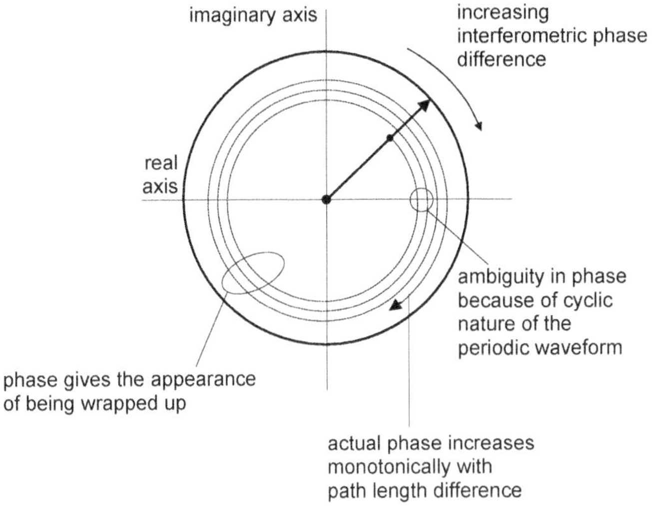

Fig. 6.8. Demonstrating how the cyclic nature of the exponential function in (6.13b) leads to ambiguity in phase

6.7 Phase Unwrapping

In principle, unwrapping the phase would appear to be straightforward. By starting where the interferometric phase difference is expected to be smallest – at the near swath edge – the phase angle difference would be tracked as we move across range. Whenever a 2π jump in phase is experienced, compensation would be made and the process continued. This is demonstrated in Fig. 6.9 using just one row of the interferogram of Fig. 6.7. The interferometric phase difference across that row is seen to have six discontinuities resulting from the phase wrapping. To unwrap the phase it is beneficial first to take its gradient across the line; as seen in Fig. 6.9 that immediately identifies the phase jumps. We then integrate along the line of interferometric phase gradient and whenever a discontinuity is encountered we add or subtract 2π, based on the sign of the discontinuity, to produce the unwrapped phase transect illustrated.

A complication arises when there is an actual jump in phase greater than 2π within the space of a single resolution cell, for example as a result of rapid changes in elevation, including layover. In such situations, which unfortunately can be common, special measures need to be taken to implement phase unwrapping. The most common are reviewed by Gens[4].

Once the phase has been unwrapped it is necessary to relate the resulting interferometric phase plot to absolute topography (referred to some datum). The simplest way to do that is via ground control points that allow at least some phase measures to be associated with elevations; the remaining phases can then be calibrated in terms of elevation.

[4] R. Gens, Two-dimensional phase unwrapping for radar interferometry: developments and new challenges, *International Journal of Remote Sensing*, vol. 24, no. 4 2003, pp. 703-710.

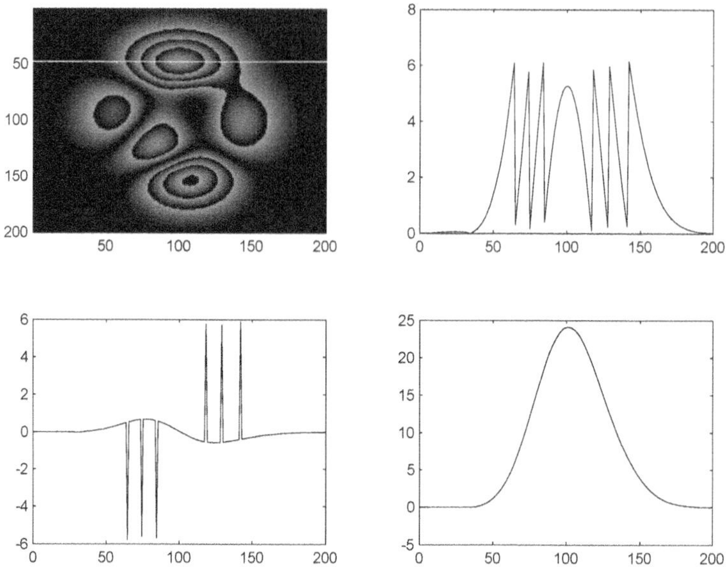

Fig. 6.9. Simple demonstration of phase unwrapping along the white transect shown on the interferogram: the top right plot shows the variation of wrapped phase along the transect, while the bottom left hand plot shows the gradient of the wrapped phase; the bottom right hand plot shows the unwrapped phase along the transect (corresponding to the topography evident in Fig. 6.5) after integrating the gradient while compensating for the phase jumps

6.8 An Inclined Baseline

We now generalise the geometry of Fig. 6.2 to the case where the baseline is inclined at an angle to the horizontal. That requires just a simple modification of the significant formulas.

Fig. 6.10 shows the general case, with important angles and distances indicated; the baseline is inclined upwards from the horizontal by the angle α. Note that the orthogonal baseline is

$$B_{\perp} = B\cos(\theta - \alpha) \tag{6.14}$$

The path length difference is

$$\Delta R = R_1 - R_2 = B\sin(\theta - \alpha) \tag{6.15}$$

so that the interferometric phase is

$$\Delta\phi = \frac{4\pi B\sin(\theta - \alpha)}{\lambda} \tag{6.16}$$

Equation (6.10) then becomes

$$\frac{d(\Delta\phi)}{dh} = \frac{4\pi B_{\perp}}{\lambda R_o \sin\theta} = \frac{4\pi B_{\perp}\cos(\theta-\alpha)}{\lambda H \sin\theta} \tag{6.17}$$

Equation (6.16) reduces to (6.6) when α=0, while (6.17) reduces to (6.10).

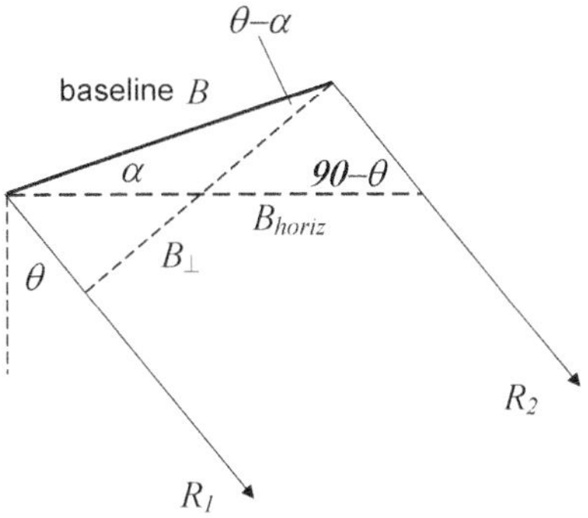

Fig. 6.10. Geometry for the case of an inclined baseline

6.9 Standard and Ping Pong Modes of Operation

In the interferometer operation outlined in Figs. 6.2 and 6.10 it is assumed implicitly that each of the two radar antennas radiate *and* receive, leading to the two way interferometric phase expression of (6.6). Some interferometers operate, however, with only one antenna transmitting and two antennas receiving; that could be the case if the antennas were both on the same platform, such as an aircraft. This configuration was adopted for the Shuttle Topography Mapping Mission (SRTM) in which the second receiving antenna was located on a 60m boom, with the primary transmitting and receiving antenna in the shuttle cargo bay.

When a single transmitting antenna is used, the time of travel of the ranging pulse from that antenna to the target is the same for both interferometer paths; it is only on the return paths that one signal travels further than the other. This is illustrated in Fig. 6.11a. In this case there is only a one way difference in phase between the paths of $\frac{2\pi}{\lambda}(R_1 - R_2)$.

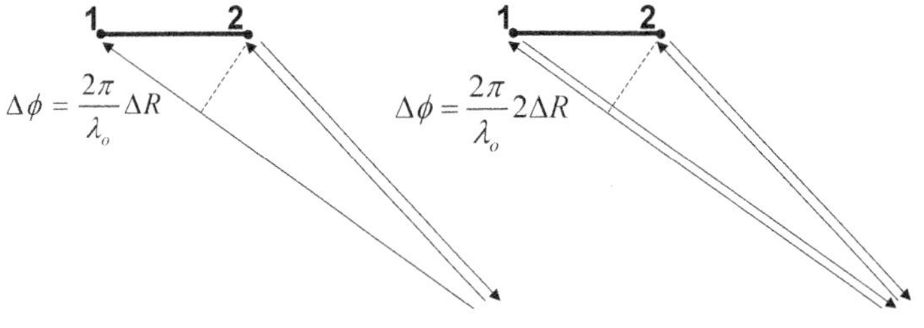

(a) (b)

Fig. 6.11. (a) Standard and (b) ping pong modes of operation, showing the differences in interferometric phase between them

Sometimes operating with a single transmitting antenna is called the *standard mode* of operation, whereas when both antennas transmit, as in Fig. 6.11b, it is called the *ping pong mode*. To account for both possibilities the interferometric phase difference of (6.6) can be written

$$\Delta\phi = \frac{2p\pi B\sin\theta}{\lambda} \tag{6.18}$$

in which p=1 for standard mode operation and p=2 for ping pong mode. Unless otherwise stated explicitly we will always assume ping pong operation in this treatment.

Fig. 6.12 shows a topographic map produced by across track interferometry on the Shuttle Radar Topography Mission (standard mode).

Fig 6.12. Digital elevation image of New Zealand produced using cross track interferometry on the Shuttle Radar Topography Mission, showing the region near Christchurch; topographic height is accentuated using colour with green at the lower elevations and white at the highest; shading is used to enhance slope information (image courtesy of NASA)

6.10 Types of SAR Interferometry

The arrangement shown in Fig. 6.2 and which has formed the basis of the development in this chapter so far has the two radar antennas arranged across the track of the platform – literally in what we have called the across track direction of the radar system in other chapters. *Cross track interferometry*, sometimes abbreviated XTI, can be achieved in two ways: either by having two antennas on the same platform, as for the SRTM mission, or by using two separate passes of a single SAR mission, such as ERS or PALSAR. Provided the landscape does not change between passes the latter arrangement constitutes a valid interferometer. XTI can therefore be subdivided into *single pass* and *repeat pass interferometry*. Clearly, repeat pass cross track interferometry always operates in the ping pong mode, whereas the single pass arrangement can operate in either ping pong or standard mode depending on the design of the system.

An interferometer can also be formed in the along track direction, parallel to the platform velocity vector. Again, *along track interferometry*, ATI, can either be single pass or repeat pass. Some aircraft systems are single pass by having antennas arranged fore and aft on the fuselage. As expected they could be either standard or ping pong as their operating mode. Repeat pass along track interferometry, in principle, requires the passes to follow the same orbital path. We will see in the following that along track interferometry is not sensitive to terrain variations, since the slant ranges are the same. It is, however, an important technology for detecting changes that occur between observations. Generally, the platforms in repeat pass ATI do not follow identical paths so that, as well as along track separation there will be some cross track separation too, leading to the detection of topographic detail as well as terrain changes. The effect of topography can be removed by using *differential InSAR*, treated in Sect. 6.13. The fundamental types of SAR interferometry are illustrated in Fig. 6.13.

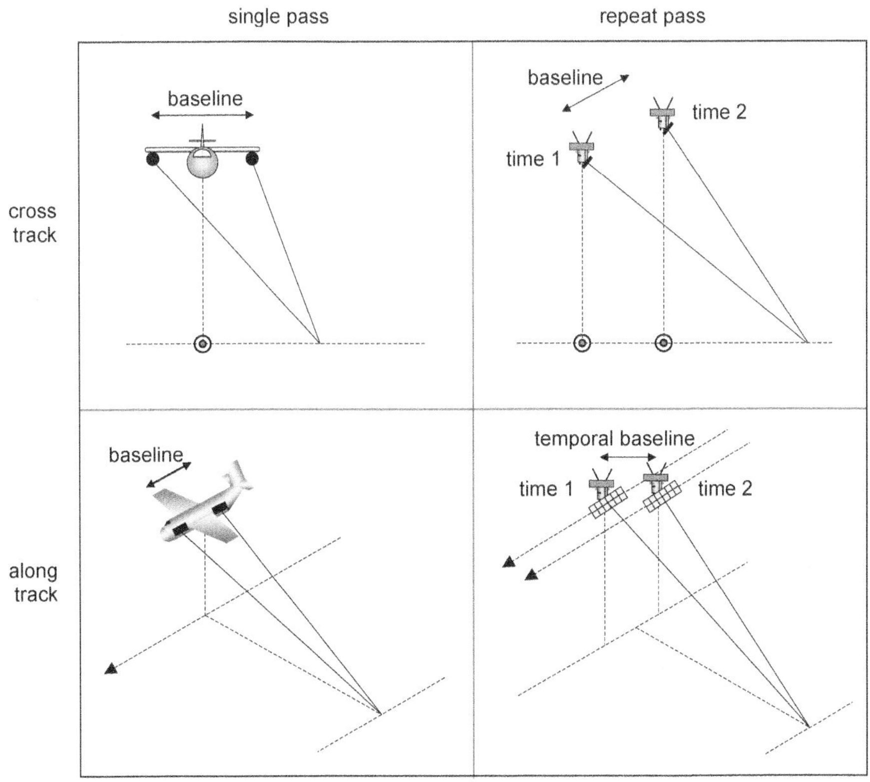

Fig 6.13. Fundamental types of SAR interferometers

6.11 The Concept of Critical Baseline

It is clear from (6.6) that the size of the baseline controls the degree of phase change with incidence angle, which in turn results from a change in topography. A larger baseline means a greater phase shift and thus potentially a more sensitive interferometer. However, there is a limit. We are interested in the change of phase with elevation from pixel to pixel. If that change exceeds 2π then we cannot readily recover the inter-pixel variation in elevation. That is demonstrated in Fig. 6.14 in which we have plotted the interferometric phase difference across a flat earth for a range of orthogonal baselines $B_\perp$ using the physical parameters of ERS[5]. The actual interferometric phase differences recorded in an ERS interferometer would be approximately the mid cell phases if the ground were perfectly homogeneous in its scattering properties.

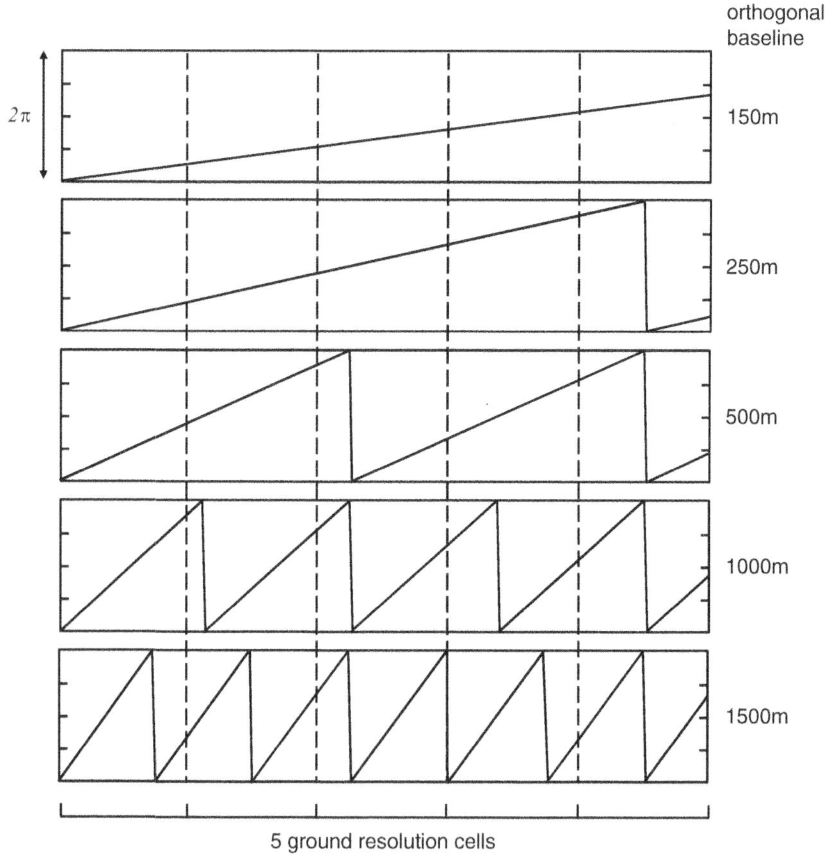

Fig. 6.14. Demonstrating how the interferometric phase difference for ERS would vary as a continuous function across the swath, over a distance equivalent to 5 ground range resolution elements

[5] Altitude 785km, near range incidence angle 23°, ground range resolution 25m, wavelength 0.056m.

As noted, for small baselines the phase varies over several resolution elements before the 2π ambiguity associated with the cyclic nature of phase becomes evident, indicating that we can readily compensate for the 2π jumps when they occur. Once the baseline exceeds 1000m however there is a phase jump *within* each ground range resolution element making it impossible understand what the flat earth variation in phase should look like. The limiting case is when there is a 2π change in phase within the distance of a single resolution cell. Even before that limit is reached it is clear that the reconstruction task is not easy.

The orthogonal baseline for which the variation in interferometric phase difference across a single ground range resolution element is 2π is called the *critical baseline*. It can be found using the plane wave approximation that led to (6.6)[6] in the following manner based on the geometry of Fig. 6.15.

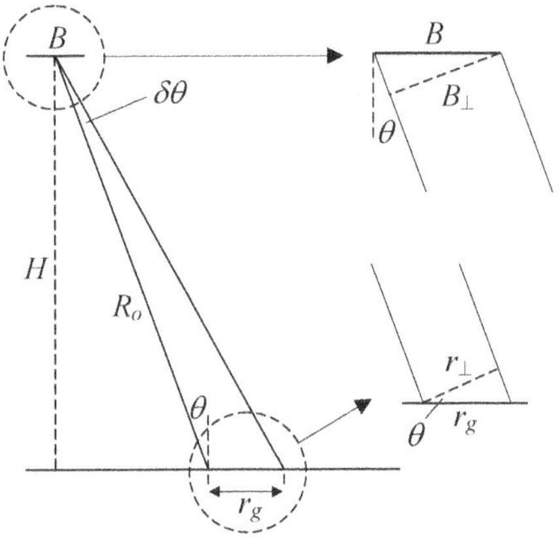

Fig. 6.15. Geometry used for calculating the critical baseline; strictly the look and incidence angles should be different but no significant error is introduced by making them the same

Using (6.6) the change in interferometric phase across the ground resolution cell is

$$\Delta\phi = \Delta\phi_1 - \Delta\phi_2 = \frac{4\pi B}{\lambda}[\sin(\theta + \delta\theta) - \sin\theta]$$
$$\approx \frac{4\pi B}{\lambda}[\sin\theta + \delta\theta\cos\theta - \sin\theta] = \frac{4\pi B}{\lambda}\delta\theta\cos\theta$$

Now
$$\delta\theta \approx \frac{r_\perp}{R_o} = \frac{r_g\cos\theta}{R_o} = \frac{r_g\cos^2\theta}{H}$$

so that
$$\Delta\phi = \frac{4\pi B r_g}{\lambda H}\cos^3\theta$$

[6] Note that (6.6) is an approximation, albeit a very good one. The results of Fig. 6.14 were generated by computing the real vectors from either side of the baseline to the earth's surface and then finding the actual interferometric phase difference.

Expressing the baseline in terms of the orthogonal baseline gives

$$\Delta\phi = \frac{4\pi B_{\perp} r_g}{\lambda H}\cos^2\theta$$

The critical baseline is given when this change in interferometric phase across the resolution cell is 2π. Thus

$$B_{\perp CRITICAL} = \frac{\lambda H}{2r_g \cos^2\theta} = \frac{\lambda R_o}{2r_g \cos\theta} \qquad (6.19)$$

It may be better to re-cast this expression in terms of the slant range resolution since the ground range resolution varies across the swath with incidence angle, while slant range resolution is a system parameter, set by the chirp bandwidth, as seen in (3.5a). Putting $r_g = r_r / \sin\theta$ in (6.19) gives

$$B_{\perp CRITICAL} = \frac{\lambda H \sin\theta}{2r_r \cos^2\theta} = \frac{\lambda R_o}{2r_r \cot\theta} \equiv \frac{\lambda H \mathcal{B}_c \sin\theta}{c\cos^2\theta} = \frac{\lambda \mathcal{B}_c R_o \tan\theta}{c} \qquad (6.20)$$

where we have used (3.5a) to express the orthogonal critical baseline in terms of the ranging chirp bandwidth $\mathcal{B}_c$.

Thus, notwithstanding the better sensitivity of phase with elevation given in (6.10) there is an upper limit on baseline set by the critical value. For ERS this is about 1030m. When compared with the plots of Fig. 6.14 we can see that that is about the point observed when there is a full 2π cycle of phase over the resolution cell. However, as the plots indicate, there may be difficulties with understanding variations in the interferometric phase with even smaller baselines. In practice, operation is not carried out above about 25% of the critical baseline, which is about the second of the plots in Fig. 6.14. We return to the concept of critical baseline later (Sect. 6.16).

6.12 Decorrelation

When the critical baseline is reached it is not possible to create a topographic map because the phase information is ambiguous, a situation referred to as *decorrelation* between the constituent images. Significant decorrelation occurs at even shorter baselines, so that operation is generally not deemed satisfactory beyond about 25% of the critical value, as noted in the previous section.

Decorrelation can also come about in other ways. Any mechanism that leads to statistical differences between the signals received by the two channels can decorrelate them. Those mechanisms include differences in the centre, or carrier, frequencies, mis-registration between the two images in range and azimuth, and noise (uncertainty) in the phase measurements on reception.

With repeat pass interferometry if regions on the ground have changed in any way between the two acquisitions that form the interferogram then the interferometric phase difference will be affected. Such changes could be fast, such as with the surface of the ocean or because of the effect of wind on vegetation canopies. Alternatively, they could be slower such as forest growth and glacial movement; generally those would not lead to decorrelation. They might also occur on a very short time scale, episodically, between acquisitions, such as ground movement or deformation resulting from earthquakes. We will be interested in detecting those types of change using along track interferometry. Any time changing phenomena can, in principle, lead to a randomising of interferometric phase for the associated pixels between passes such that, if the two images bear no correlation, interferometric information cannot be generated.

The degree of correlation, or *coherence*, between the two constituent images e_1 and e_2 of an interferometer is measured as the magnitude of the complex cross correlation between the images

$$\gamma = \frac{|< e_1 e_2^* >|}{\sqrt{< |e_1|^2 >< |e_2^*|^2 >}} \tag{6.21}$$

It will take the value 1 when the images are fully correlated over the region chosen to compute the average, and zero if there is no statistical relationship between the images, in which case they are said to be fully decorrelated.

Coherence can be expressed as the product of a number of components, each attributable to a separate decorrelating mechanism. For example we could write

$$\gamma = \gamma_{baseline}\gamma_{pixel}\gamma_{noise} \tag{6.22}$$

in which the subscripts respectively refer to the decorrelation associated with the baseline as discussed in Sect. 6.11, decorrelation caused by the pixel itself changing, or looking different from the aspects of the radars in the interferometer, and system (phase) noise. It is possible to derive explicit expressions for some of those components. Noise coherence can be expressed[7]

$$\gamma_{noise} = \frac{1}{1 + SNR^{-1}} \tag{6.23a}$$

in which SNR is the signal to noise ratio of the (two) radar receivers. For a very high receiver signal to noise ratio, which is to be expected, this term should be close to unity.

Baseline coherence is given by[8]

$$\gamma_{baseline} = 1 - \frac{2Br_g\cos^2\theta}{\lambda R_o} \tag{6.23b}$$

As expected this is a function of system parameters such as the horizontal baseline B, the ground range resolution, the slant range to the target, the operating wavelength and the incidence angle. Writing the baseline in terms of the orthogonal baseline $B = \frac{B_\perp}{\cos\theta}$ this becomes

$$\gamma_{baseline} = 1 - \frac{2B_\perp r_g\cos\theta}{\lambda R_o} \tag{6.23c}$$

[7] See H.A. Zebker and J. Villasenor, Decorrelation in interferometric radar echoes, *IEEE Transactions on Geoscience and. Remote Sensing*, vol. 30, no. 5, September 1992, pp. 950-959.
[8] *Ibid.*

While the others parameters are substantially fixed, we have control over coherence through the baseline. A large baseline leads to low coherence, while a small baseline helps keep coherence high. The orthogonal baseline at which coherence falls to zero is the critical baseline, which from (6.23b) is

$$B_{\perp crit} = \frac{\lambda R_o}{2r_g \cos\theta} \tag{6.23d}$$

This is the same as (6.19).

Decorrelation effects attributable to the pixel itself are treated in the next section in the context of detecting topographic change.

Comparison of (6.21) and (6.13a) shows that coherence is just the expected value of the magnitude of the interferogram, formed by multiplying one complex image, pixel by pixel, by the complex conjugate of the other. Likewise the interferometric phase is the argument of that product as shown in (6.13b). When we come to PolInSAR below we will generalise this concept.

6.13 Detecting Topographic Change: Along Track Interferometry

Suppose for the moment that we can set up an ideal temporal baseline; in other words the platform repeats its path exactly, with no spatial separation orthogonal to its velocity vector. That means there will be no spatial baseline of the type considered in Fig. 6.2. Topographic mapping, as treated in Sect. 6.3, is therefore not possible. However, if a feature on the landscape shifts during the two SAR acquisitions, such that there is a component of the movement in the slant range direction as illustrated in Fig. 6.16, then an interferometric phase difference will be measured for the relevant pixels, proportional to the two way change in slant range, given by

$$\Delta\phi_{change} = \frac{4\pi\Delta r_r}{\lambda} \tag{6.24}$$

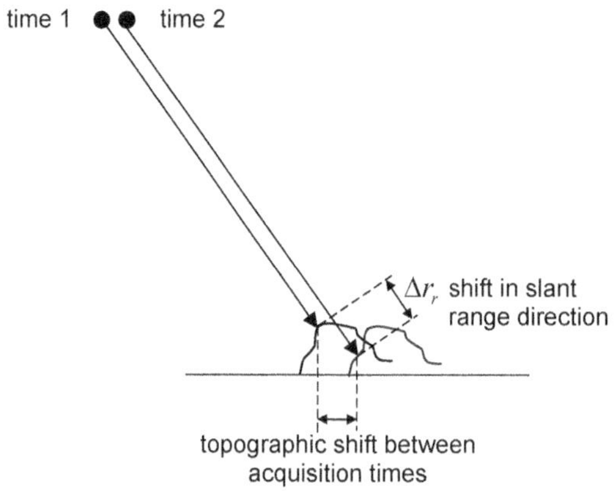

Fig. 6.16. Measuring slant range topographic variations with repeat pass along track interferometry

Note that this phase difference between the received radar signals is dependent only on the ratio of the degree of change in slant range to the operating wavelength. With ERS, for which λ=0.056m, one full cycle of phase difference is caused by a slant range shift of $\Delta r_r = \frac{\lambda}{2} = 28mm$! This should be compared with the sensitivity of 37m per cycle for topographic mapping, demonstrated in Sect. 6.3.

If there is also an across track baseline there will be a phase shift associated with topography along with the phase change related to the time variation described by (6.24). That is most often the case. We can generate an expression for the combined phase shift in the following manner.

Knowing that the total interferometric phase difference is a function of both topography and its change (usually called the *displacement phase difference*), we can write it as

$$\Delta\phi = \Delta\phi(h, \Delta r_r) \tag{6.25a}$$

To a first order this can be expanded as

$$\Delta\phi = \frac{\partial(\Delta\phi)}{\partial h}\Delta h + \frac{\partial(\Delta\phi)}{\partial(\Delta r_r)}\Delta r_r$$

We know $\frac{\partial(\Delta\phi)}{\partial h}$ from (6.10) while, from (6.24), $\frac{\partial(\Delta\phi)}{\partial(\Delta r_r)} = \frac{4\pi}{\lambda}$, so that the intereferometric phase resulting from both effects is

$$\Delta\phi = \frac{4\pi B_\perp \cos\theta}{\lambda H \sin\theta}\Delta h + \frac{4\pi}{\lambda}\Delta r_r \tag{6.25b}$$

In order to isolate the change in interferometric phase with landscape displacement between acquisitions it is necessary to remove the interferometric phase variation resulting from the underlying topography. That is done using the technique of differential interferometric SAR, or D-InSAR.

D-InSAR depends upon finding a topographic model by some other means that can be used to remove the constant topography from the ATI acquired interferogram. There are two common methods for doing that. The first entails using a pre-existing digital elevation model (DEM) to synthesise the topographic interferometric phase term in (6.25b). That can then be subtracted pixel by pixel leaving only the interferometric phase resulting from displacement between the SAR acquisitions. A second approach is to use a third SAR acquisition. Two of the SAR images are used to form an interferogram corresponding to topography alone. Its interferometric phase is then removed from the interferometric phase derived from two others of the acquisitions that maximise the effect of topographic change. Fig. 6.17 shows the mapping of topographic change by this approach.

Generally water bodies are thought not to have sufficient coherence in InSAR applications to be used as sensible targets. However, when the water forms the horizontal surface of a double bounce structure involving grasses or trees in marsh-like landscapes the associated radar cross section has a high degree of correlation between the relevant pixels of the two images in the interferometer. Through this secondary mechanism it is possible to use repeat pass interferometry to track changes in water level, as seen in Fig. 6.18.

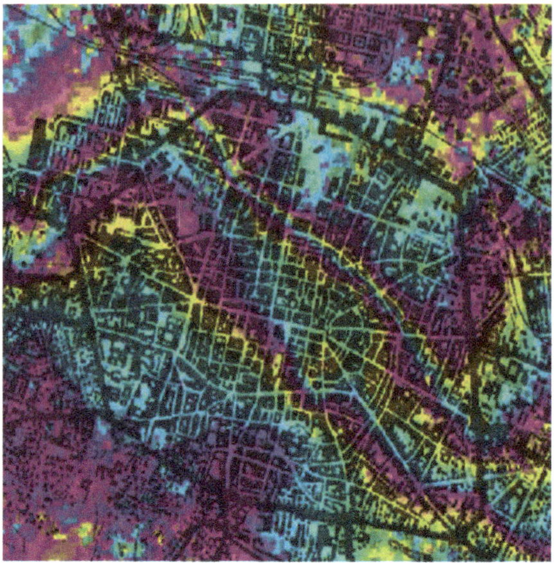

Fig. 6.17. This topographic model shows subsidence of the city of Bologna, Italy, apparently at the rate of about 1 cm per year per colour cycle shown; it was produced by D-InSAR using ERS acquisitions (image courtesy ESA/Data Processing by GAMMA)

There is an assumption implicit in the three acquisition D-InSAR approach: the two acquisitions used to synthesise the topographic interferometric phase are assumed to have no phase associated with displacement. In other words they need to have a baseline orthogonal to the platform motion and to be imaged within a time frame faster than any displacement of interest. A benefit of the DEM-based approach is that that assumption is not necessary. All that is required is the availability of a suitable DEM.

Unfortunately, interferometric phase is influenced by factors other than just topography and displacement as assumed in (6.25a). More generally we should express the phase difference in the form

$$\Delta\phi = \Delta\phi_{topo} + \Delta\phi_{disp} + \Delta\phi_{atm} + \Delta\phi_{pixel} + \Delta\phi_{noise} + \Delta\phi_{error} \tag{6.26}$$

in which $\Delta\phi_{topo}$ is the interferometric phase associated with topography, $\Delta\phi_{disp}$ is that caused by displacement and $\Delta\phi_{atm}$ is a phase difference between the acquisitions caused by variations in atmospheric dielectric constant[9]. Compensating atmospheric phase difference variations can be based on modelling[10] and the use multi-baseline interferometers[11]. $\Delta\phi_{noise}$ is a term resulting from phase noise in the radar system; that can

[9] Dielectric constant changes cause changes in the velocity of propagation of the radar energy and thus wavelength; consequently the phase delay is affected.

[10] See Z. Li., J-P Muller, P. Cross and E.J. Fielding, Interferometric synthetic aperture radar (InSAR) atmospheric correction: GPS, Moderate Resolution Imaging Spectrometer (MODIS) and InSAR integration, *J. Geophysical Research*, vol, 110, B03410, doi:10.1029/2004JB003446, 2005.

[11] See A. Ferretti, C. Prati, and F. Rocca, Multibaseline InSAR DEM reconstruction: The wavelet approach, *IEEE Transactions on Geoscience and. Remote Sensing*, vol. 37, no. 2, pt. 1, March 1999, pp. 705–715

be reduced by averaging over groups of pixels at the expense of spatial resolution. $\Delta\phi_{error}$ accounts for uncertainty in the knowledge of the platform positions and baseline. $\Delta\phi_{pixel}$ represents any change in phase between the two radar acquisitions resulting from changes in the reflectivity of the pixel being observed. Perhaps surprisingly this is an important consideration and can be the factor that limits the usefulness of repeat pass interferometry; it is therefore worth considering in a little detail.

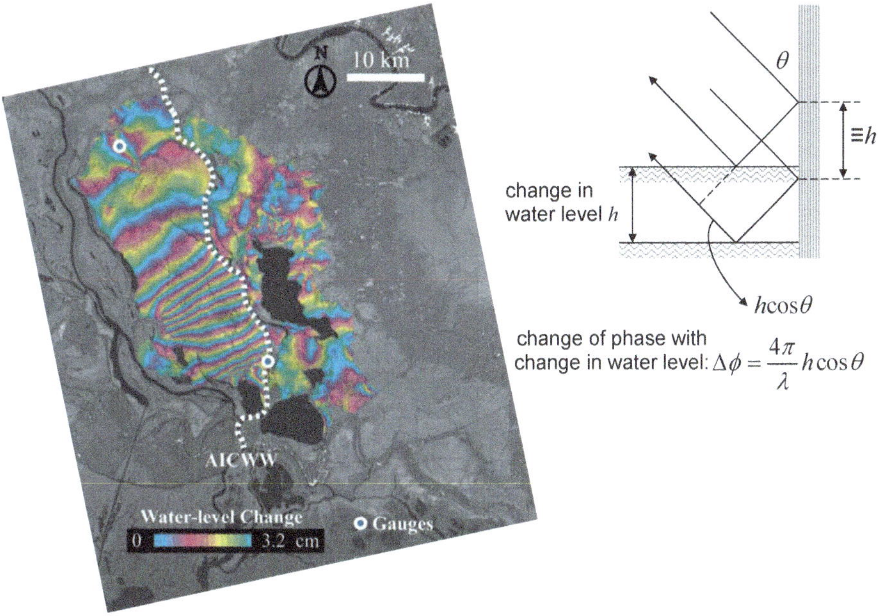

Fig. 6.18. Water level changes in swamp land mapped by repeat pass Radarsat InSAR; the inset shows how phase is affected by water level change, involving a strong reflection (from Z. Lu and O-I Kwoun, Radarsat-1 and ERS InSAR analysis over southeastern coastal Louisiana: Implications for mapping water-level changes beneath swamp forests, *IEEE Transactions on Geoscience and Remote Sensing*, vol. 46, no. 8, August 2008, pp. 2167-2184, ©2008 IEEE)

As seen in (4.4) the signal received from a given pixel, which is proportional to its reflectivity, is the sum of the fields returned from a myriad of individual scatterers within that resolution cell. It is called a coherent sum because the amplitudes and phases of those reflected fields are important is obtaining the result, as noted in (4.5). If there is a change in the amplitude of one of the component fields, because of a change in the reflectivity of the corresponding scatterer, the coherent sum will change. That could occur because of vegetation change, for instance, between acquisitions. If the angle with which a pixel is viewed changes between acquisitions because of the separation between the platform positions then the coherent sum can also change. Changes in the observed reflectivity of the pixel resulting from variations with time or viewing aspect produce errors in the interferometric phase and cause decorrelation[12], leading to the pixel coherence term γ_{pixel} in (6.22) being less than unity.

[12] A good general discussion of decorrelation will be found in H.A. Zebker and J. Villasenor, *loc cit.*

To illustrate the nature of temporal decorrelation, and that resulting from changes in viewing angle, suppose for simplicity that there are just 20 incremental scatterers in a particular resolution cell, distributed along a range line with the positions shown in Fig. 6.19a (two examples). We make the simplifying assumption that the reflectivities of those scatterers are real so we don't have to worry about the complication of changes in their phase terms with viewing angle. That will not detract from the lesson to follow. Fig. 6.19b shows the change in interferometric phase for the pixel over a very small range of incidence angles about 20°, expressed as a fraction of a wavelength of 5.6cm used in the calculations. As seen, for this toy example the change in phase with a 0.1° change in incidence angle is equivalent to 20mm, comparable to the order of displacements that (6.24) suggests are possible with along track interferometry[13]. By trebling the strength of just the first of the 20 incremental scatterers as seen in the second set in Fig. 6.19a there is an effect approximately equivalent to 2-5mm.

Pixel decorrelation can have a significant effect on the precision of any displacement measurements if not controlled. One remedy is to keep the baselines short[14] so that the chance for variations in aspect (incidence angle and any unintentional squint angle) is minimised; time variations in pixel composition are also then constrained.

Another approach to minimising pixel decorrelation is to restrict attention to parts of the scene that are assessed as having little likelihood of decorrelation. That is the basis of permanent or persistent scatterer methods[15]. Permanent scatterers are those which dominate the response of a pixel so that the pixel's properties, and especially its phase response, are moderately insensitive to angle of view and are less likely to change with time. For example, if a pixel contains an object that gives strong corner reflector behaviour, such as a building or a large tree, then its response will essentially be the radar cross section of that object; it will not be determined by the interference of many incremental scatterers. The angular dependence of its response will be that of the radar cross section of the object, which is generally weaker than that illustrated in Fig. 6.19. Because such permanent scatterers are less prone to decorrelation, accurate estimates of their elevations and rates of movement in the range direction are possible. If a large number can be identified they can be used as samples of where and how much displacement has occurred. One way to find candidate permanent scatterers is to examine coherence images. Regions associated with permanent scatterers are more likely to have high coherence because of their stability.

6.14 Polarimetric Interferometric SAR (PolInSAR)

6.14.1 Fundamental Concepts

Implicit in the expression for the interferogram in (6.13) is that the two images have the same polarisation, but that is not necessary. In principle, any polarisations could be used, provided it is possible to separate phase difference associated with the polarisations from

[13] Note that this is one look data that has not had the benefit of the multi-look averaging, that would act to smooth the variations. Note also, the results are very sensitive to the sizes and placements of the scatterers in a contrived example such as this.

[14] See P. Berardino, G. Fornano, R. Lanari and E. Sansosti, A new algorithm for surface deformation monitoring on small baseline differential SAR interferograms, *IEEE Transactions on Geoscience and Remote Sensing*, vol. 40, no. 11, November 2002, pp. 2375-2383.

[15] A. Ferretti, C. Prati and F. Rocca, Permanent Scatterers in SAR interferometry, *IEEE Transactions on Geoscience and Remote Sensing*, vol. 39, no. 1, January 2001, pp. 8-20.

the phase difference resulting from topographic effects. We could therefore generalise (6.13a) to read

$$i = e_{1,PQ}e^{*}_{2,RS} = Ie^{j\Delta\phi} \tag{6.27}$$

in which we have dropped the pixel coordinates x,y for simplicity but added subscripts implying polarisation; PQ is the polarisation state of one image and RS that of the other.

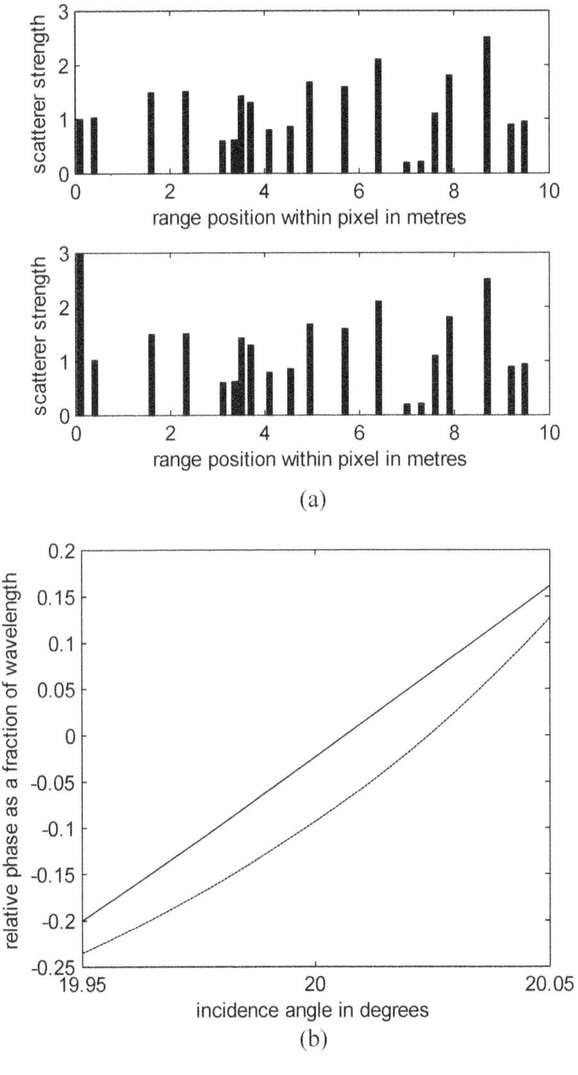

Fig 6.19. (a) Two sets of scatterers distributed across a resolution cell; the scatterer at position zero has treble the size in the second set (b) corresponding change in interferometric phase with incidence angle; the lower line corresponds to the second set in (a)

Instead of the received electric fields we could express the interferogram in terms of scattering coefficients, since they are available in the data supplied, they capture directly the polarisation states of both the incident and received fields and they are indicative of the scattering mechanisms of the pixels being imaged. Thus we could write

$$i = Ie^{j\Delta\phi} = s_{1,PQ}s^*_{2,RS}$$

to signify the complex interferogram. More generally, for fully polarised radar we could derive an interferogram-like quantity in terms of the target vectors $\mathbf{k}_1$ and $\mathbf{k}_2$ for the images that are to be interfered, since those vectors contain all the information needed for creating interference images from *any* polarisation combination. If, in addition, we average over a number of samples, or looks[16], to reduce random variations or noise then we would write a generalised interferogram in the form

$$i = \mathcal{E}(\mathbf{k}_1\mathbf{k}_2^{*T}) = \langle \mathbf{k}_1\mathbf{k}_2^{*T} \rangle \tag{6.28}$$

which from (3.50) or (3.55) will be recognised as a two image version of the covariance or coherency matrix, depending on the basis chosen for the target vectors. If we normalise this expression by the magnitudes of the single image coherences then we have a unit magnitude complex number we call the *complex polarimetric interferometric coherency* analogous to that in (6.21) for the single polarisation case; viz.

$$\gamma = |\gamma| e^{j\Delta\phi} = \frac{\langle \mathbf{k}_1\mathbf{k}_2^{*T} \rangle}{\sqrt{\langle \mathbf{k}_1\mathbf{k}_1^{*T} \rangle \langle \mathbf{k}_2\mathbf{k}_2^{*T} \rangle}} \tag{6.29}$$

The interferogram defined in (6.28) incorporates all possible polarisation combinations of the two constituent images. In practice, we would choose a particular polarisation configuration for each of the two images (often the same) and then develop the scalar interferogram. Since (6.28) is expressed in terms of the target vectors composed of all polarisations it would be interesting to know how to extract the polarisation options we are interested in from those vectors. We can do that by applying a unitary[17] filter vector **w** to the target vector to produce a modified form[18]

$$\kappa = \mathbf{w}^{*T}\mathbf{k} = w_1^* k_1 + w_2^* k_2 + w_3^* k_3 \tag{6.30}$$

If the target vectors are in the Pauli basis form of (3.49) – assuming reciprocity – then the following filter vectors generate the individual polarisation states as demonstrated:

[16] These are not necessarily the looks used for speckle reduction but can be a set of similar pixels in a neighbourhood that are assumed tacitly to come from the same cover (or scatterer) type.

[17] That is a vector whose magnitude (determined by the square root of the sum of the squares of its elements) is unity.

[18] See S.R. Cloude and K.P. Papathanassiou, Polarimetric SAR interferometry, *IEEE Transactions on Geoscience and Remote Sensing*, vol. 36, no. 5, September 1998, pp. 1551-1565.

$$\mathbf{w}_a = \frac{1}{\sqrt{2}}\begin{bmatrix}1\\1\\0\end{bmatrix},\ \mathbf{w}_b = \frac{1}{\sqrt{2}}\begin{bmatrix}1\\-1\\0\end{bmatrix},\ \mathbf{w}_c = \frac{1}{\sqrt{2}}\begin{bmatrix}0\\0\\1\end{bmatrix} \tag{6.31}$$

Thus

$$\mathbf{w}_a^{*T}\mathbf{k}_p = \frac{1}{\sqrt{2}}\left[\frac{1}{\sqrt{2}}(S_{HH}+S_{VV})\right] + \frac{1}{\sqrt{2}}\left[\frac{1}{\sqrt{2}}(S_{HH}-S_{VV})\right] + 0\text{x}2S_{HV} = S_{HH}$$

$$\mathbf{w}_b^{*T}\mathbf{k}_p = S_{VV}$$

$$\mathbf{w}_C^{*T}\mathbf{k}_p = S_{HV}$$

Note that the elements in the vectors of (6.31) are all real so that the conjugation operation in (6.30) is of no significance for this example. It will be important in cases when **w** has complex elements. Sometimes the **w** vectors are said to describe scattering mechanisms; that is because they highlight certain polarisation combinations from the target vectors. They are also referred to as polarisation (filter) vectors, which may be a better term. As this particular example shows, the result of the operation in (6.31) is to produce scattering coefficients. Thus, sometimes the κ created in (6.30) are referred to as generalised scattering coefficients.

We can choose the filter vector to be different for the two images so that the scattering coefficients are

$$\kappa_1 = \mathbf{w}_1^{*T}\mathbf{k}_1 \tag{6.32a}$$

$$\kappa_2 = \mathbf{w}_2^{*T}\mathbf{k}_2 \tag{6.32b}$$

with which we can develop a new version of the complex polarimetric interferometric coherency measure

$$\gamma = |\gamma|e^{j\Delta\phi} = \frac{\langle\kappa_1\kappa_2^*\rangle}{\sqrt{\langle\kappa_1\kappa_1^*\rangle\langle\kappa_2\kappa_2^*\rangle}} \tag{6.33}$$

We can also form an interferogram from the filtered target vectors, similar to (6.28):

$$i = \langle\kappa_1\kappa_2^*\rangle \tag{6.34}$$

As in (6.22) the complex coherence is composed of a number of components, each of which can reduce the overall coherence. For our purposes here we will decompose it into

$$\gamma = \gamma_{baseline}\gamma_{polarisation}\gamma_{pixel}\gamma_{other} \tag{6.35}$$

These include, as shown, coherence resulting from the interferometric baseline used, coherence determined by the correlation between the polarisation options chosen, coherence associated with changes in the specific region of the image of interest (sometimes called temporal coherence) and coherence associated with other factors such as noise identified earlier for single polarisation interferometry. If all the others can be maximised, then the coherence associated with the different polarisations chosen for the two images can be used as a diagnostic feature, as we will see in Chapt. 8. If the polarisations are the same for the two radars in the interferometer then the polarisation component of the coherency will be unity; that is the same as choosing $\mathbf{w}_1$=$\mathbf{w}_2$ in (6.32).

The argument, or angle, of the complex coherence, which is a polarimetric interferometric phase difference between the image pair, will be composed of the arguments of the constituent contributions:

$$\Delta\phi = \Delta\phi_{baseline} + \Delta\phi_{polarisation} + \Delta\phi_{pixel} + \Delta\phi_{other} \tag{6.36}$$

Note from (6.33) that we can write $\langle \kappa_1 \kappa_2^* \rangle = \langle \mathbf{w}_1^{*T}\mathbf{k}_1(\mathbf{w}_2^{*T}\mathbf{k}_2)^{*T} \rangle = \mathbf{w}_1^{*T}\langle \mathbf{k}_1\mathbf{k}_2^{*T} \rangle \mathbf{w}_2$, and likewise for the terms in the denominator. Since the weight vectors are constants they have been taken outside the averaging (expectation) operators. Thus the complex coherency can be written

$$\gamma = \frac{\mathbf{w}_1^{*T}\langle \mathbf{k}_1\mathbf{k}_2^{*T} \rangle \mathbf{w}_2}{\sqrt{\mathbf{w}_1^{*T}\langle \mathbf{k}_1\mathbf{k}_1^{*T} \rangle \mathbf{w}_1}\sqrt{\mathbf{w}_2^{*T}\langle \mathbf{k}_2\mathbf{k}_2^{*T} \rangle \mathbf{w}_2}}$$

or

$$\gamma = \frac{\mathbf{w}_1^{*T}\mathbf{\Omega}_{12}\mathbf{w}_2}{\sqrt{\mathbf{w}_1^{*T}\mathbf{T}_{11}\mathbf{w}_1}\sqrt{\mathbf{w}_2^{*T}\mathbf{T}_{22}\mathbf{w}_2}} \tag{6.37}$$

in which $\mathbf{T}_{11}$ and $\mathbf{T}_{22}$ are the coherency matrices[19] of each of the individual images in the interferometer and

$$\mathbf{\Omega}_{12} = \mathcal{E}(\mathbf{k}_1\mathbf{k}_2^{*T}) \equiv \langle \mathbf{k}_1\mathbf{k}_2^{*T} \rangle \tag{6.38}$$

is a new *joint image complex coherency matrix* which contains both polarimetric and interferometric information. It also implicitly contains information on the scattering properties of the pixel viewed from the perspectives of each of the radars in the interferometer – i.e. from each end of the baseline.

The coherency of (6.37) is a complex number, the phase of which contains detail on topographic effects and phase variations resulting from the polarisation differences. Its amplitude is a measure of the correlation between the two acquisitions as was the case in (6.21). As noted earlier the amplitude has an upper value of unity, so it is convenient to plot (6.37) on a complex plane that summarises the coherency for a given situation. We will develop that concept further below.

6.14.2 The $\mathbf{T}_6$ Coherency Matrix

If we write the two target vectors of (6.28) in column form $\begin{bmatrix}\mathbf{k}_1\\ \mathbf{k}_2\end{bmatrix}$, the expected value of the outer product of this vector with itself generates what has become known as the $\mathbf{T}_6$ matrix:

$$\mathbf{T}_6 = \left\langle \begin{bmatrix}\mathbf{k}_1\\ \mathbf{k}_2\end{bmatrix}\begin{bmatrix}\mathbf{k}_1^{*T} & \mathbf{k}_2^{*T}\end{bmatrix} \right\rangle = \begin{bmatrix}\langle \mathbf{k}_1\mathbf{k}_1^{*T}\rangle & \langle \mathbf{k}_1\mathbf{k}_2^{*T}\rangle\\ \langle \mathbf{k}_2\mathbf{k}_1^{*T}\rangle & \langle \mathbf{k}_2\mathbf{k}_2^{*T}\rangle\end{bmatrix} = \begin{bmatrix}\mathbf{T}_{11} & \mathbf{\Omega}_{12}\\ \mathbf{\Omega}_{21} & \mathbf{T}_{22}\end{bmatrix} \tag{6.39}$$

[19] See (3.55).

in which the component matrices on the diagonal will be recognised as the coherency matrices of the individual images in the interferometric pair. The upper right hand entry is the joint coherency matrix of (6.38). The bottom left hand entry is its conjugate transpose – $\mathbf{\Omega}_{21} = \mathbf{\Omega}_{12}^{*T}$. Since each of $\mathbf{T}_{11}$, $\mathbf{T}_{22}$ and $\mathbf{\Omega}_{12}$ are of dimension 3x3 the $\mathbf{T}_6$ matrix is of size 6x6 – hence its subscript. If the full four element version of the Pauli target vector of (3.48) were used in constructing the coherency matrix similar to (6.39) then the result would be 8x8 and the matrix referred to as the $\mathbf{T}_8$ coherency matrix.

Interestingly, if there were N radars in a multi-baseline interferometer (or a multi-static radar in the sense discussed in Chapt. 7) then (6.39) can be generalised to

$$\mathbf{T}_{3N} = \begin{bmatrix} \mathbf{k}_1 \\ \mathbf{k}_2 \\ \vdots \\ \mathbf{k}_N \end{bmatrix} \begin{bmatrix} \mathbf{k}_1^{*T} & \mathbf{k}_2^{*T} & .. & \mathbf{k}_N^{*T} \end{bmatrix} = \begin{bmatrix} \mathbf{T}_{11} & \mathbf{\Omega}_{12} & .. & \mathbf{\Omega}_{1N} \\ \mathbf{\Omega}_{21} & \mathbf{T}_{22} & & \\ \vdots & & \cdot & \\ \mathbf{\Omega}_{N1} & & & \mathbf{T}_{NN} \end{bmatrix} \quad (6.40)$$

6.14.3 Maximising Coherence

When coherence is small it is difficult to make use of the interferogram, either for interferometry as such, or as a means for understanding the landscape. As a result it is of interest to understand under what conditions complex coherence can be maximised. Clearly, all of the terms in (6.35) need to be kept high in order to achieve the best coherence possible. Good system design will optimise the coherence contributions from noise and related system properties, and a small baseline will help control the associated coherence term. However, what about coherence in general? Can it be optimised and, if so, how? To answer that question we need to know what we can change in search of maximising it.

In (6.37) the matrices $\mathbf{T}_{11}$, $\mathbf{T}_{22}$ and $\mathbf{\Omega}_{12}$ are fixed by the properties of the region being imaged; however the filter vectors $\mathbf{w}_1$ and $\mathbf{w}_2$ can be chosen in pursuit of our desired outcome. In particular, we can look to maximise the coherence of (6.37) by a careful choice of those two vectors. The optimum values of $\mathbf{w}_1$ and $\mathbf{w}_2$ come from solutions to the eigenvalue problems[20] (which share common eigenvalues ν):

$$\mathbf{B}_1\mathbf{w}_1 = \nu\mathbf{w}_1 \quad (6.41a)$$
$$\mathbf{B}_2\mathbf{w}_2 = \nu\mathbf{w}_2 \quad (6.41b)$$

in which
$$\mathbf{B}_1 = \mathbf{T}_{11}^{-1}\mathbf{\Omega}_{12}\mathbf{T}_{22}^{-1}\mathbf{\Omega}_{12}^{*T} \quad (6.42a)$$
and
$$\mathbf{B}_2 = \mathbf{T}_{22}^{-1}\mathbf{\Omega}_{12}^{*T}\mathbf{T}_{11}^{-1}\mathbf{\Omega}_{12} \quad (6.42b)$$

The maximum coherence corresponds to the square root of the dominant eigenvalue, once found. The corresponding eigenvectors $\mathbf{w}_{1opt}$ and $\mathbf{w}_{2opt}$ are the filter vectors that lead to the optimised coherence so that from (6.34) the interferogram that has maximum coherence is

$$i = \langle \kappa_{1opt}\kappa_{2opt}^* \rangle \equiv \langle (\mathbf{w}_{1opt}^{*T}\mathbf{k}_1)(\mathbf{w}_{2opt}^{*T}\mathbf{k}_2)^{*T} \rangle$$

[20] S.R. Cloude and K.P. Papathanassiou, *loc cit.*

$$
\begin{aligned}
&= \mathbf{w}_{1opt}^{*T} \langle \mathbf{k}_1 \mathbf{k}_2^{*T} \rangle \mathbf{w}_{2opt} \\
&= \mathbf{w}_{1opt}^{*T} \mathbf{\Omega}_{12} \mathbf{w}_{2opt}
\end{aligned} \tag{6.43}
$$

6.14.4 The Plot of Complex Coherence

As with any complex number the coherence of (6.29) or (6.37) can be plotted on an Argand diagram such as that shown in Fig. A1. Since the magnitude of the complex coherence has a maximum of unity, the diagram is restricted to a circle of unit radius as illustrated in Fig. 6.20. The important region of the diagram is towards its circumference since there the magnitude of the complex coherence is largest. Towards the origin signifies regions of low coherence, a situation generally not suited to interferometric applications.

For a given region of pixels in an interferometric pair of images, it is to be expected that the corresponding coherences will cluster in a particular region of the complex plane. If different groups of pixels cluster separately then the complex coherence can be used to help segment an image. We will have more to say about that in Chapt. 8.

Sometimes a particular image segment will be composed of two types of scatterer, such as a forest canopy over a diffuse soil surface. Simple modelling suggests that the resulting complex coherence of the mix of the two types lies on a straight line[21] as illustrated in Fig 6.20. More generally, complex coherence will be bounded within regions, often of elliptical shape[22] when the polarisations of the two radars are the same.

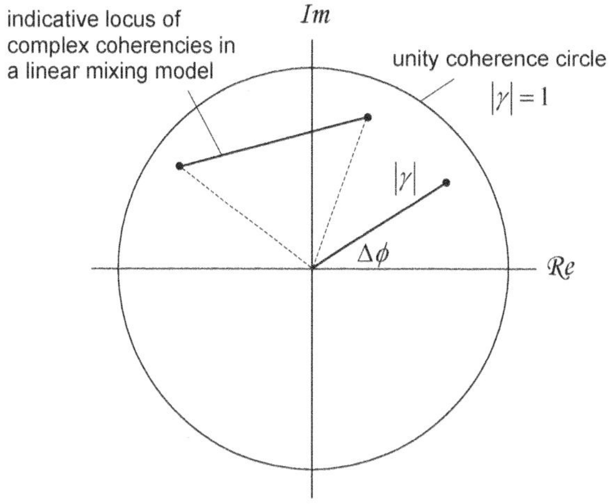

Fig. 6.20. Plot of the complex coherence and an illustration of how the coherence of a two component scatterer is likely to migrate with a change in composition

[21] K.P. Papathanassiou and S.R. Cloude, Single-baseline polarimetric SAR interferometry, *IEEE Transactions on Geoscience and Remote Sensing*, vol. 39, no. 11, November 2001, pp. 2352-2363.

[22] L. Ferro-Famil, E. Pottier and J.S. Lee, Classification and interpretation of polarimetric interferometric SAR data, *Proceeding of the International Geoscience and Remote Sensing Symposium, 2002 (IGARSS02)*, 24-28 June 2002, pp.635-637 and T. Flynn, M. Tabb and R. Carande, Coherence region shape extraction for vegetation parameter estimation in polarimetric SAR interferometry, *Proceeding of the International Geoscience and Remote Sensing Symposium, 2002 (IGARSS02)*, 24-28 June 2002, pp.2596-2598.

6.15 Tomographic SAR

Standard synthetic aperture radar generates images of the landscape in the two horizontal spatial dimensions, with detail in elevation mapped onto that two dimensional projection. Interferometric SAR using two radars deployed across track goes one step further, enabling the mapping of topography through sensitivity to elevation, but it does not permit discrimination of detail vertically, such as the internal structure of a forest. By appropriately utilising several radars with vertical separation (as passes of the same radar platform on different orbits) it is possible to identify vertical structure with the technique known as SAR tomography.

Tomography resolves vertical detail by employing a synthesised *vertical* aperture much as azimuthal detail is resolved using aperture synthesis in normal SAR. However, whereas the synthetic aperture technique used azimuthally depends on the Doppler chirp induced by platform motion, aperture synthesis vertically depends on antenna array theory as we will now demonstrate.

6.15.1 The Aperture Synthesis Approach

Consider the arrangement shown in Fig. 6.21. Several flight lines of the same platform (or even different platforms if their imaging characteristics are compatible) are used to image the landscape from different altitudes. After the images have been formed (i.e. after range and azimuth compression) the set of measurements for each pixel is used to resolve vertical detail in the manner developed below. Some pixels will not necessarily exhibit vertical structure, such as simple surfaces, but if the region is a volumetric or composite scatterer there will be vertical detail that may be of interest. In understanding how that can be revealed we concentrate on a single pixel and imagine it is being irradiated simultaneously by the set of radars shown in the figure. In particular, we concentrate on a position within the pixel volume at a height g above the datum. There is an assumption here that the incident radiation can penetrate any intervening volume to allow the structure of interest to be seen, albeit partially.

In the figure we have made the unrealistic assumption that the flight lines are so arranged that when they are projected on to a line orthogonal to the line of sight to the target they are uniformly spaced. Such an assumption simplifies our analysis and generates results that are generally applicable. After we have looked at the fundamental properties of tomography we will examine what happens if the flight lines are not uniformly spaced – which is of course what happens in practice. Because we are looking at the sloped multi-baseline assumption we also project the vertical detail of the pixel onto a parallel sloped line as shown. The point g within the pixel volume we also measure along that projection rather than along the vertical. To clarify the variables involved the vertical plane containing the set of radar beams – which we might call the *orthogonal* (to the) *slant plane* – is redrawn as shown as shown in the figure.

What we will be looking for in the first instance is the condition under which we can bring all the radar beams into focus vertically on the spot g. When we understand how to do that we will know how much vertical resolution is available to us.

In Fig. 6.21 we have shown N radars (with N odd for convenience) over a total separation L_T orthogonal to the line of sight. As in interferometry, that is referred to as the orthogonal baseline. Here we will call it the *tomographic aperture*. Consider first the radar which is located at the general position z within the discrete array of radars, for which

$$z = \frac{nL_T}{N-1} \text{ with } -(N-1)/2 \le n \le (N-1)/2 \tag{6.44}$$

The distance between that radar and the point g in the target pixel – the slant range – is

$$R = \sqrt{R_o^2 + (z-g)^2}$$

in which R_o is the slant range from the centre of the array to the base of the pixel. Since z and g will be very small by comparison to R_o we can approximate the last expression as

$$R = R_o + \frac{(z-g)^2}{2R_o}$$

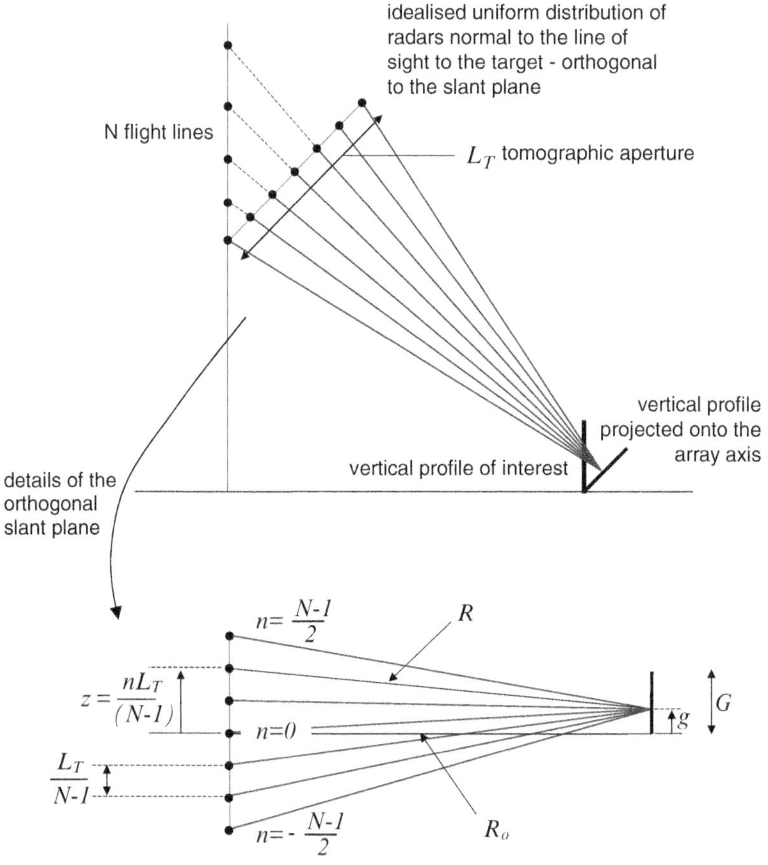

Fig. 6.21. Idealised SAR tomographic arrangement in which a vertical array of radars is used to synthesise high resolution vertically

The equivalent two way phase delay associated with that distance is

$$\begin{aligned}\phi(z) &= \frac{4\pi}{\lambda}\left[R_o + \frac{(z-g)^2}{2R_o}\right]\\ &= k\left[2R_o + \frac{(z-g)^2}{R_o}\right]\end{aligned} \tag{6.45}$$

in which λ is the radar operating wavelength and k is the corresponding wave number (sometimes written as the phase constant β). As with SAR interferometry we are not interested in absolute phases but in the phase difference between the return signal of a given radar and a reference beam. If we choose the radar at z=0 (n=0) as the reference, then from (6.45) the difference in phase of the radar at location z is

$$\begin{aligned}\Delta\phi(z) &= k\left[2R_o + \frac{(z-g)^2}{R_o}\right] - k\left[2R_o + \frac{(g)^2}{R_o}\right]\\ &= \frac{k}{R_o}\left[z^2 - 2zg\right]\end{aligned} \tag{6.46}$$

Subtracting the phase corresponding to the z=0 radar is called *de-ramping*, and is one of the steps involved in tomographic processing[23]. Using (6.44) we can re-write (6.46) as

$$\Delta\phi(n) = \frac{k}{R_o}\left[d^2n^2 - 2dng\right]$$

in which $d = L_T/(N-1)$ is the spacing between the flight lines.

The combined signal received by the set of radars from the element of the pixel at elevation g can be expressed

$$s(t) = \sum_{n=-(N-1)/2}^{(N-1)/2} s_n e^{j\omega_o t} e^{-j\Delta\phi(n)}$$

in which ω is the radar operating frequency and the amplitudes s_n account for the reflectivity of the pixel at the elevation seen by the nth radar along with any factors during transmission that change the signal levels. It is reasonable to assume that the s_n are all the same – i.e. independent of n – so that the amplitudes can be ignored, as can the exponential function of the carrier frequency, to leave the received signal as function of g

$$s(g) = \sum_{n=-(N-1)/2}^{(N-1)/2} e^{-j\Delta\phi(n)} = \sum_{n=-(N-1)/2}^{(N-1)/2} \exp\left[-j\frac{k}{R_o}(d^2n^2 - 2dgn)\right] \tag{6.47}$$

The first term in the exponent of (6.47) is not a function of g and is known explicitly for each radar in the array. When processing the images acquired by the platforms or by

[23] See A. Reigber and A. Moreira, First demonstration of SAR tomography using multibaseline L band data, *IEEE Transactions on Geoscience and Remote Sensing*, vol. 38, no. 5, pt. 1, September 2000, pp. 2142-2152.

separate passes of the same platform it is possible to remove it in each case before the sum in (6.47) is taken, leaving as the composite signal

$$s(g)=\sum_{n=-(N-1)/2}^{(N-1)/2}\exp\left[j\frac{2kdgn}{R_o}\right] \tag{6.48}$$

If we put $\psi=\frac{2kdg}{R_o}=\frac{2kL_Tg}{R_o(N-1)}$ then (6.48) can be written in expanded form as

$$s(g)=e^{-j\frac{(N-1)}{2}\psi}\ldots e^{-j2\psi}+e^{-j\psi}+1+e^{j\psi}+e^{j2\psi}\ldots e^{j\frac{(N-1)}{2}\psi} \tag{6.49}$$

This is a geometric progression with first term $e^{-j\frac{(N-1)}{2}\psi}$ and common ratio $e^{j\psi}$. The sum over N terms is

$$s(g)=e^{-j\frac{N-1}{2}\psi}\frac{(1-e^{jN\psi})}{(1-e^{j\psi})}$$

$$=e^{-j\frac{N-1}{2}\psi}\frac{e^{j\frac{N}{2}\psi}(e^{-j\frac{N}{2}\psi}-e^{j\frac{N}{2}\psi})}{e^{j\frac{\psi}{2}}(e^{-j\frac{\psi}{2}}-e^{j\frac{\psi}{2}})}$$

$$=\frac{\sin N\psi/2}{\sin\psi/2}$$

Substituting for ψ this is

$$s(g)=\frac{\sin\frac{NkL_Tg}{(N-1)R_o}}{\sin\frac{kL_Tg}{(N-1)R_o}} \tag{6.50}$$

The magnitude of this expression is plotted as a function of g in Fig.6.22 for several different values of N, but keeping L_T constant at 300m. The slant range was chosen as 5km and the wavelength as 0.235 (L band), values used by Reigber et al[24].

Several observations can be made from this figure:

1. For each value of N the maximum signal occurs for g=0. In other words the radar array is focussed at the base of the pixel. From (6.49) the maximum is N, but we have normalised the graphs by dividing by N to make the comparisons more meaningful. By "focussing" we mean that most of the backscattered signal reaching the radar comes from that point. The side lobes evident in the diagram will also contribute small amounts of energy from other heights too, but essentially we regard the principal contribution to the return for this example as coming from the pixel's properties at zero elevation.
2. Other maxima also occur – in other words the radar array will focus at other elevations as well. From (6.49) it is easily seen that the maxima are given by $\psi=2m\pi$ where m is an integer. Substituting for k and ψ gives a condition on g for a maximum:

[24] *ibid.*

$$g_{\max} = m\lambda R_o(N-1)/2L_T = m\lambda R_o/2d \tag{6.51}$$

in which $d = L_T/(N-1)$ is the spacing (sample interval) between the flight lines, as seen in Fig. 6.21. With m=1 this shows that the radar will also focus at g=3.92m elevation for N=3, g= 7.83m for N=5 and g=19.6m for N=11. The first two are seen in Fig. 6.23 in which a greater range of elevations is shown and the results are only to one side of the principal maximum. Taking the case of N=11, the radar will also receive signal from any elements of the pixel in the vicinity of 20m elevation. If the detail reaches that elevation then an ambiguous signal is received. In order to avoid elevation ambiguities (6.51) can be re-cast (with m=1) to give a criterion on the selection of the flight line spacing for a specified maximum elevation G. Elevation ambiguity will be avoided if

$$d \le \lambda R_o/2G \tag{6.52}$$

which is sometimes referred to as the *elevation ambiguity criterion*. Thus for a maximum pixel height of 20m a sample (flight line) spacing no greater than 28.5m is needed, given the other parameters chosen for this example.

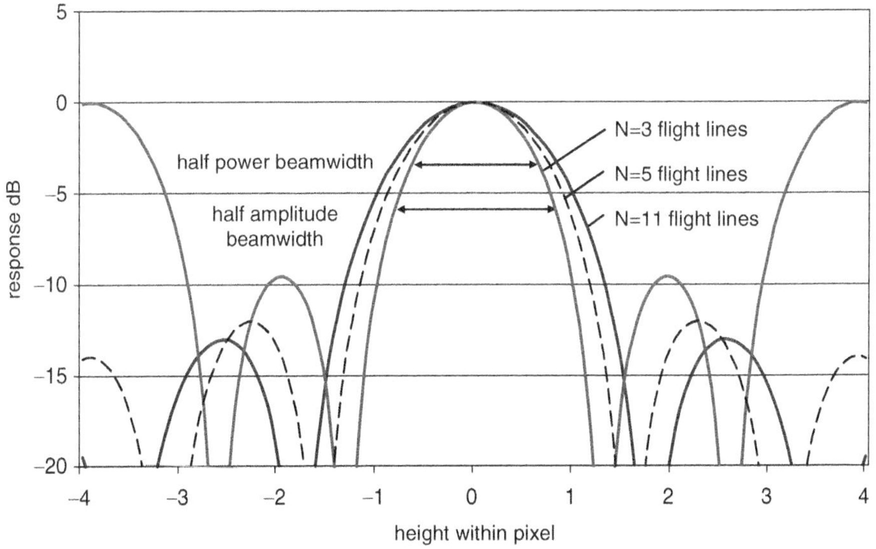

Fig. 6.22. Received radar signal with 3, 5 and 11 vertical flight lines.

3. The main lobe, focussed on a specific elevation (so far in this case at g=0), has a half power width of $\Delta g = 2g_1$ where g_1 is the height for which the lobe has dropped to $1/\sqrt{2}$ of its peak value; it is a solution to

$$\frac{\sin N\psi_1/2}{\sin\psi_1/2} = \frac{N}{\sqrt{2}}$$

with $\psi_1 = \dfrac{2kL_T g_1}{R_o(N-1)}$. The denominator changes more slowly than the numerator and for the range of g of interest its argument is small, permitting the approximation

$$\sin N\psi_1/2 = \frac{1}{\sqrt{2}} N\psi_1/2$$

This transcendental equation has the solution $N\psi_1/2 \approx 1.39$[25] which gives

$$\Delta g = \frac{1.39}{\pi}\frac{N-1}{N}\frac{\lambda R_o}{L_T} \approx \frac{0.35\lambda R_o}{L_T} \tag{6.53}$$

This is the effective vertical resolution of the array. The half width of the received *signal* as against the half *power* width is the solution to the transcendental equation

$$\sin N\psi_1/2 = \frac{1}{2} N\psi_1/2$$

which requires $N\psi/2 \approx 1.90$. This gives[26]

$$\Delta g = \frac{1.9}{\pi}\frac{N-1}{N}\frac{\lambda R_o}{L_T} \approx \frac{N-1}{N}\frac{\lambda R_o}{2L_T} \approx \frac{\lambda R_o}{2L_T} \tag{6.54}$$

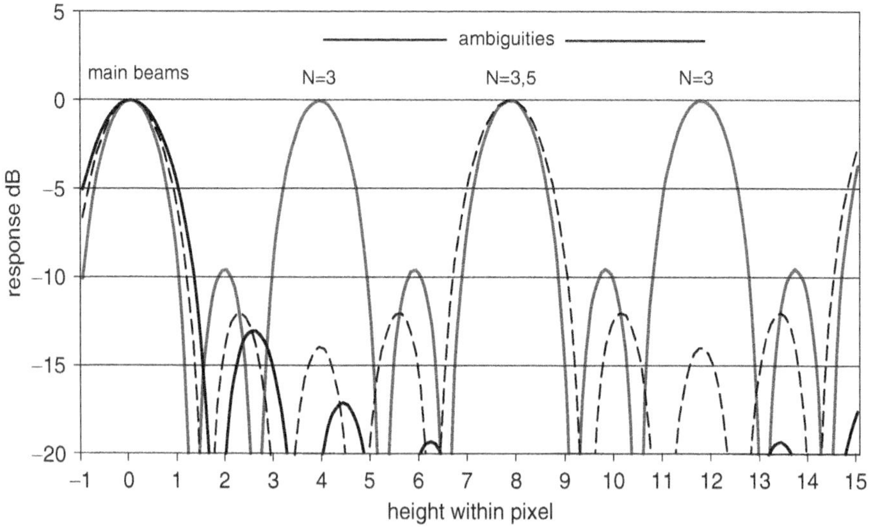

Fig. 6.23. Expanded vertical coverage showing the height ambiguities that result from too few flight lines

[25] See R.W.P. King, *The Theory of Linear Antennas*, Harvard UP., Cambridge, Mass., 1956.
[26] See also Reigber and Moreira, *loc cit.* for this same result, derived slightly differently.

We now need to consider how to focus at other elevations within the pixel volume. Again, we need to find conditions such that the sum in (6.49) has its maximum value of N, but with $g \neq 0$. To do that we add an incremental phase angle ϕ to ψ in (6.49) such that the received signal is

$$s(g) = \sum_{n=-(N-1)/2}^{(N-1)/2} e^{jn(\psi+\phi)} = \frac{\sin N(\psi+\phi)/2}{\sin(\psi+\phi)/2}$$

By appropriately choosing ϕ we can focus the array at a desired value of g. The condition for maximum signal is that the exponent is zero for each term, which requires

$$\phi = -\psi = -\frac{2kL_T g}{R_o(N-1)} \quad (6.55)$$

Effectively, what we are doing here is zeroing out the phase angle associated with the non-zero value of g by adding that value of ϕ. Thus by stepping through g from 0 to G in increments of Δg, we can ascertain the appropriate incremental phases to add to focus the radar successively up through the volume of the pixel. While that is technically acceptable there is a more elegant approach based on a Fourier transform understanding of the vertical focussing process, treated in the next section.

6.15.2 The Fourier Transformation Approach to Vertical Resolution

In Fig. 6.21 we analysed the situation where all the radars in the vertical array were illuminating a single vertical position within the pixel volume and receiving the corresponding echoes. We now consider a different approach. We look at just one of the radars irradiating and receiving echoes from a discretised model of the pixel volume as shown in Fig. 6.24 – we envisage the vertical detail of the volume being broken into N samples $\rho(l)$, l=0…N-1 each of which corresponds to one half width of the focussed array defined by (6.54).

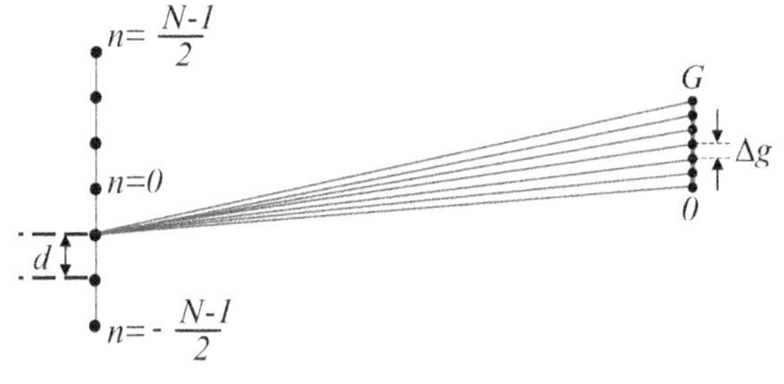

Fig. 6.24. Basis of the Fourier transform approach to tomographic focussing

We can use the development that led (6.48) to help find the signal received by the single radar[27]; however now it is important to recognise that the reflectivity of the pixel will vary with height – indeed that is what we are interested in finding – so that the signal received by the nth radar in the array is given by

$$s(n)=\sum_{l=0}^{N}\rho(l)\exp\left[j\frac{k}{R_o}2ndl\Delta g\right] \qquad -\frac{N-1}{2}\leq n\leq\frac{N-1}{2}$$

Putting $d=L_T/(N-1)$ and substituting from (6.54) for Δg, this last expression becomes

$$s(n)=\sum_{l=0}^{N}\rho(l)\exp\left[j\frac{4\pi}{\lambda R_o}n\frac{L_T}{N-1}l\frac{N-1}{N}\frac{\lambda R_o}{2L_T}\right]$$

i.e.
$$s(n)=\sum_{l=0}^{N}\rho(l)\exp\left[j\frac{2\pi}{N}nl\right] \qquad -\frac{N-1}{2}\leq n\leq\frac{N-1}{2} \tag{6.56}$$

Equation (6.56) is the expression for the discrete inverse Fourier transform[28] of the pixel reflectivity as a function of elevation $\rho(g)$ when represented by the set of vertical samples $\rho(l)$ [29]. The *set* of received signals $s(n)$ for all n, in (6.56) are the complete set of samples of the discrete Fourier transform of the pixel reflectivity with height. To recover the vertical detail of the pixel all that needs to be done is to perform an Fourier transform on the set of signals $s(n)$ received for that pixel; the transform is

$$\rho(l)=\sum_{n=-(N-1)/2}^{(N-1)/2}s(n)\exp\left[-j\frac{2\pi}{N}nl\right] \tag{6.57}$$

In practice the radar signal received on each flight line would be range and azimuth compressed to form the set of images, which would then be registered to each other. The set of measurements available for each pixel (one from each flight line) then form the Fourier spectrum of the pixel vertical profile, which is recovered by applying the Fourier transform (based on the Fast Fourier Transform algorithm).

6.15.3 Unevenly Spaced Flight Lines

Clearly the situation depicted in Fig. 6.21 will not be achieved in practice because the flight lines are time sequenced passes of the same platform or possibly other compatible platforms. Instead, the flight lines are likely to be unevenly spaced so that the set of

[27] This is after de-ramping and assuming that the quadratic factor in (6.47) has been compensated for. Incidentally the de-ramping in this case is based on a reference point in pixel elevation, rather than within the radar array. Nevertheless (6.47) is still the end result.

[28] See E.O. Brigham, *The Fast Fourier Transform and its Applications*, 2nd ed., Prentice Hall, Englewood Cliffs, N.J., 1988 or J.A. Richards and X. Jia, *Remote Sensing Digital Image Analysis*, 4th ed., Springer Verlag, Berlin, 2006.

[29] Because of the two sided nature of n the exponent in (6.56) can be positive or negative without affecting the result. Some authors call (6.56) the discrete Fourier transform and (6.57) the discrete inverse Fourier transform.

samples incorporated into the Fourier transform of (6.57) will not be uniformly spaced, thus affecting the integrity of the operation. The simplest means for obtaining a uniform spacing over the (orthogonal) tomographic aperture of Fig. 6.21, is to use the available irregularly spaced set of samples (flight lines) to estimate a set of samples on a uniform spacing by using an appropriate resampling technique[30]. If some gaps are especially large then a form of infilling will be required. One such method is based on the assumption that there will be a dominant scattering centre somewhere in the vertical profile from which synthetic flight paths can be established[31].

Another consideration that can arise when using unevenly spaced flight lines is which one to choose as the reference when de-ramping using reference phase subtraction. A simple solution is to use the average slant range to the array from the pixel position[32].

6.15.4 Polarisation in Tomography.

In the development of SAR tomography above there has been no explicit mention of polarisation since, in principle, any polarisation is suitable. It is possible to build tomographic pixel elevation profiles for a range of polarisation configurations – the benefit of doing so is that the scattering properties with elevation may be polarisation sensitive. That would certainly be the case for a forested pixel. To demonstrate the combination of tomography and polarisation Fig. 6.25 shows the analysis of vertical structure with several polarisations along an azimuth line in an L band airborne radar image with 13 flight lines. It is also shown in colour composite form using the Pauli display basis[33]. The vertical structure is readily evident and comparable in scale with the features on the ground. The association of scattering properties with polarisation is as expected and is particularly evident in the elevated HV scattering from the forest foliage (green in the colour image).

There is one point concerning polarisation over which care needs to be taken; that relates to the removal of the quadratic phase term between (6.47) and (6.48). It is important to keep track of that phase because relative phase is significant among polarisations.

6.15.5 Polarisation Coherence Tomography

When the term tomography is applied to SAR interferometry it implies a procedure for understanding the vertical structure within a pixel. The technique just considered does so through using a multiple baseline configuration to create vertical spatial discrimination.

Another approach is to postulate a model of the vertical structure and see whether the parameters of the model can be found from radar measurements. That is the approach adopted with polarisation coherence tomography (PCT)[34]. It uses interferometric polarimetric data to understand simple vertical structures via the properties of the complex coherence. This only requires a two radar (single baseline) arrangement such as

[30] See G. Fornano, F. Serafino and F. Soldovieri, Three-dimensional focussing with multipass SAR data, *IEEE Transactions on Geoscience and Remote Sensing*, vol. 41, 2003, pp. 507-517.
[31] See Reigber and Moreira, *loc cit.*
[32] *ibid.*
[33] The Pauli basis displays S_{HH}-S_{VV} as red, S_{HV} as green and S_{HH}+S_{VV} as blue.
[34] See S.R. Cloude, Polarisation Coherence Tomography, *Radio Science*, vol. 41, RS4017, 2006

that shown in Fig. 6.2 but the radars operate with different polarisations as we will ultimately see.

As with Fig. 6.21 we have a problem in that the vertical detail is resolved along a sloped line orthogonal to the line of sight. We need to correct that before we can apply the PCT approach. That is done through the artifice of *range spectral filtering*, which we consider first.

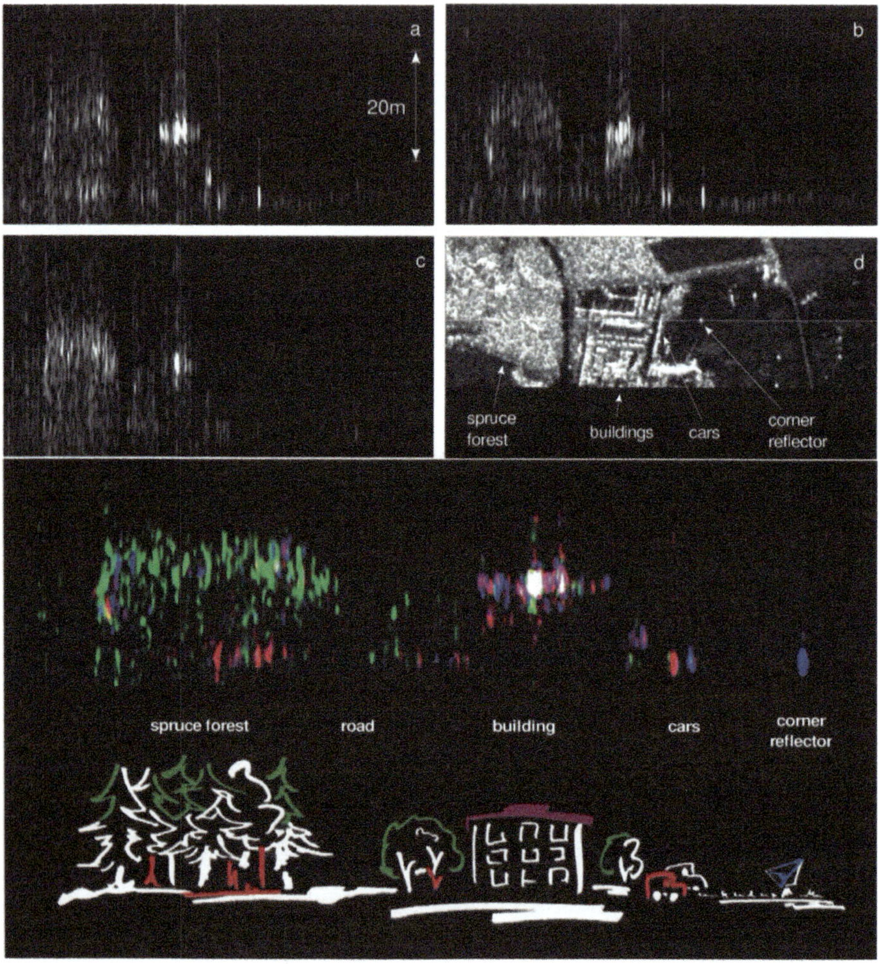

Fig. 6.25. Vertical detail versus azimuthal distance for a test site in Oberpfaffenhofen, Germany, demonstrating the efficacy of SAR tomography and the additional information available from adding a polarimetric dimension; a, b and c in the top image correspond to HH, VV and HV polarisation respectively; the transect shown in d has been used to form the colour composite in the lower image, which represents elevation versus azimuth position but in the Pauli display basis (from A. Reigber and A. Moreira, First demonstration of airborne SAR tomography using multibaseline L-Band data, *IEEE Transactions on Geoscience and Remote Sensing*, vol. 38, no 5, pt.1, September 2000, pp. 2142- 2152, ©2000 IEEE)

Fig. 6.26 allows us to compute interferometric phase as a function of position and height within a pixel. We concentrate on the point at position (r,h) and see how the interferometric phase varies with incidence angle. If one end of the interferometric baseline subtends an incidence angle of θ_2 and other an incidence angle of θ_1 the two way differences in phase (not yet the interferometric phase) between the point of interest and the origin shown are

$$\Delta\phi_2 = \frac{4\pi}{\lambda}(h\cos\theta_2 - r\sin\theta_2) = \frac{4\pi f}{c}(h\cos\theta_2 - r\sin\theta_2) \qquad (6.58a)$$

$$\Delta\phi_1 = \frac{4\pi}{\lambda}(h\cos\theta_1 - r\sin\theta_1) = \frac{4\pi f}{c}(h\cos\theta_1 - r\sin\theta_1) \qquad (6.58b)$$

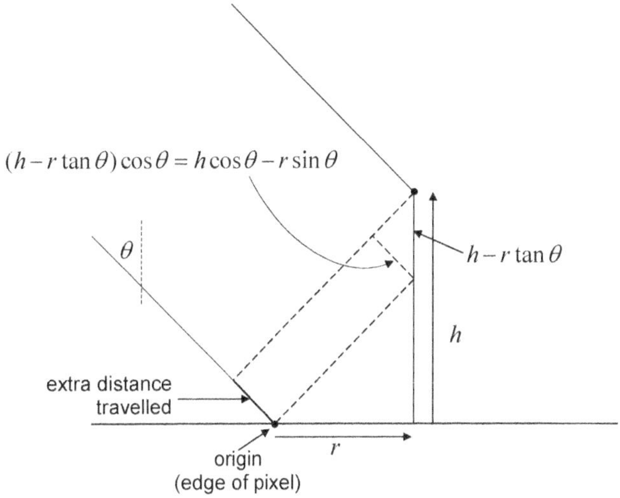

Fig. 6.26. Geometry for calculating the effect on phase of a vertical offset within a pixel

The resulting difference in phase, from pixel edge to elevated detail, between the two radars – i.e. the interferometric phase – is

$$\Delta\phi = \Delta\phi_2 - \Delta\phi_1 = \frac{4\pi}{\lambda}[h(\cos\theta_2 - \cos\theta_1) - r(\sin\theta_2 - \sin\theta_1)]$$
$$= \frac{4\pi}{\lambda}[-2h\sin\frac{\theta_2+\theta_1}{2}\sin\frac{\theta_2-\theta_1}{2} - 2r\cos\frac{\theta_2+\theta_1}{2}\sin\frac{\theta_2-\theta_1}{2}]$$

Putting $$\theta = \frac{\theta_2+\theta_1}{2} \text{ and } \frac{\Delta\theta}{2} = \frac{\theta_2-\theta_1}{2} \qquad (6.59)$$

we have $$\Delta\phi = \frac{4\pi\Delta\theta}{\lambda}(-h\sin\theta - r\cos\theta)$$

We can change the sign of this expression since we do not know which of θ_1 and θ_2 is the larger, leaving the dependence of interferometric phase on incidence angle as

$$\Delta\phi = \frac{4\pi\Delta\theta}{\lambda}(h\sin\theta + r\cos\theta) \tag{6.60}$$

Thus the interferometric phase not surprisingly varies (i) with height – which is what we are interested in – and (ii) range position in the pixel – which is a nuisance because we are interested in the vertical reflectivity profile of the pixel. We need to ask ourselves now if there is anyway we can compensate for the range variation. Although not immediately obvious at this stage we can achieve this goal by shifting the carrier (centre) frequency in (3.3) of the second radar by the small amount[35]

$$\Delta f = f\frac{\Delta\theta}{\tan\theta} \tag{6.61}$$

in which θ is the average incidence angle over the two radars in the interferometer and $\Delta\theta$ is the difference in their incidence angles at the point of interest, as shown in (6.59). Making this change in (6.58a) gives

$$\begin{aligned}\Delta\phi_2 &= \frac{4\pi(f+\Delta f)}{c}(h\cos\theta_2 - r\sin\theta_2)\\ &= \frac{4\pi}{\lambda}(h\cos\theta_2 - r\sin\theta_2) + \frac{4\pi}{\lambda}\frac{\Delta\theta}{\tan\theta}(h\cos\theta_2 - r\sin\theta_2)\end{aligned}$$

Subtracting (6.58b) to give the interferometric phase, and using (6.60) we have

$$\Delta\phi = \frac{4\pi\Delta\theta}{\lambda}(h\sin\theta + r\cos\theta) + \frac{4\pi}{\lambda}\frac{\Delta\theta}{\tan\theta}(h\cos\theta - r\sin\theta)$$

in which we have used the approximation $\theta_2 \approx \theta$. Taking the tangent inside the brackets at the right hand side gives

$$\Delta\phi = \frac{4\pi\Delta\theta}{\lambda}(h\sin\theta + r\cos\theta) + \frac{4\pi\Delta\theta}{\lambda}(h\frac{\cos^2\theta}{\sin\theta} - r\cos\theta)$$

i.e.

$$\Delta\phi = \frac{4\pi\Delta\theta}{\lambda\sin\theta}h \tag{6.62}$$

which is effectively the same result as (6.10). Here we have the interferometric phase insensitive to position in the pixel and dependent only on elevation, as required. Having made that correction we can now proceed to consider polarisation coherence tomography.

The first step in PCT is to assume that the backscattered power can be represented by a vertical profile function *f*(*h*) the shape of which accounts for the vertical distribution of scattering material and the loss of energy by absorption and scattering in that medium (similar to the assumptions for the water cloud model of Sect. 5.4.1). Because we are dependent on phase in interferometry it is important to account for the phase associated

---

[35] See Sect. 16.6.

with each incremental scatterer as the energy penetrates and scatters from inside the column of material. We do that by attaching an exponential (interferometric) phase term to the vertical scattering profile so that we could write the average interferogram from the two radars s_1 and s_2 as

$$\left\langle s_1 s_2^* \right\rangle = e^{j\Delta\phi_{topo}} \int_0^{h_v} f(h) e^{jk_h h} dh \tag{6.63}$$

in which it is assumed that the scattering medium extends from 0 to h_v in elevation. It could be a vegetation canopy, for example. The exponential term outside the integral accounts for interferometric phase associated with the surface in the absence of any pixel vertical detail and can be obtained from (6.10).

The spatial phase constant (wave number) is given from (6.62) as

$$k_h = \frac{d(\Delta\phi)}{dh} = \frac{4\pi\Delta\theta}{\lambda \sin\theta} = \frac{4\pi B_\perp}{\lambda R_o \sin\theta} \tag{6.64}$$

where $B_\perp$ is the orthogonal baseline of the interferometer.

If the two radar signals are identical there will be no surface interferometric phase and no baseline so that (6.63) becomes

$$\left\langle s_1 s_1^* \right\rangle = \int_0^{h_v} f(h) dh$$

As a result, the interferometric complex coherence for a single channel at each radar can be written, similar to (6.29), as

$$\gamma = |\gamma| e^{j\Delta\phi} = \frac{\left\langle s_1 s_2^* \right\rangle}{\sqrt{\left\langle s_1 s_1^* \right\rangle \left\langle s_2 s_2^* \right\rangle}} = \frac{e^{j\Delta\phi_{topo}} \int_0^{h_v} f(h) e^{jk_h h} dh}{\int_0^{h_v} f(h) dh} \tag{6.65}$$

The next step in PCT is to assume a profile $f(h)$ that matches what is expected of the medium of interest. The simplest is a constant between the lower and upper limits as might be expected for a uniform density, lossless forest canopy. Thus if $f(h)=A$ for $0 \le h \le h_v$ and zero otherwise then (6.65) becomes

$$\begin{aligned} \gamma &= e^{j\Delta\phi_{topo}} \frac{\int_0^{h_v} e^{jk_h h} dh}{\int_0^{h_v} dh} \\ &= e^{j\Delta\phi_{topo}} e^{jk_h h_v/2} \frac{\sin(k_h h_v/2)}{k_h h_v/2} \\ &= e^{j\Delta\phi_{topo}} e^{jk_h h_v/2} \operatorname{sinc}(k_h h_v/2) \end{aligned} \tag{6.66}$$

Therefore the amplitude of the complex coherence for a uniform, lossless canopy, irrespective of the polarisations chosen, is a sinc function of half the product of the canopy depth and the vertical wave number; that function is shown in Fig. 6.27. It is taken out to arguments beyond which coherence falls to zero simply to illustrate its sinc-like behaviour. In practice it is unlikely we would be interested in coherences less than about 0.5, so the range of the function would be much less than shown here.

For a given vertical wave number, which is set by system parameters as shown in (6.64), the magnitude of (6.66) allows an estimation of the canopy depth h_v. There is also an interesting lesson in (6.66) for simple topographic mapping with interferometric radar, through an inspection of the phase angle terms. Suppose the region in which we were interested for mapping topography is overlain by a vegetation layer that extends from the surface to a height h_v . The second phase term shows that any interferometric phase expected to be associated with the surface will be affected by the canopy, an effect known as *vegetation bias* in SAR interferometric mapping. From (6.64) we can see that to keep it small the baseline should be made as small as possible but, from (6.10), we see that the sensitivity of the interferometer is then reduced.

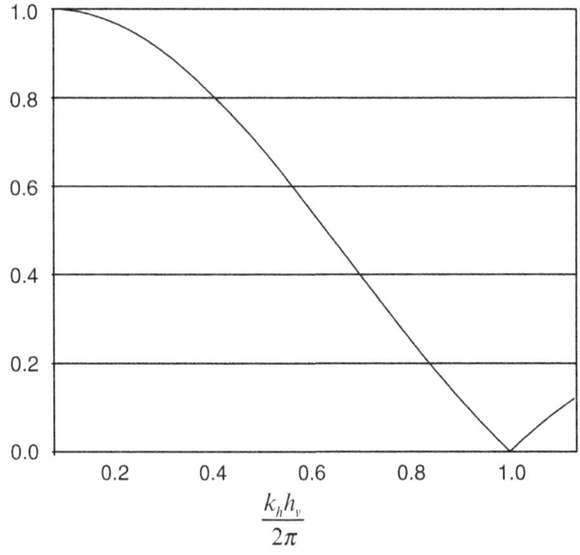

Fig. 6.27. Magnitude of the complex coherence for a uniform, lossless canopy of height h_v

Note that we have not had to use different polarisations so far; the height can be inverted, in principle, from the complex coherence with the simple slab model of the vegetation canopy. In other words a single measurement of the magnitude of the complex coherence is enough to allow a value of canopy height to be estimated.

Now consider another simple profile, but one that can be used to describe a lossy canopy. An exponentially decreasing vertical profile function with elevation downwards, as depicted in Fig. 6.28, signifies that more backscattering occurs from the top layers of the canopy and less from the lower layers as a result of loss. The energy loss is the result of absorption by the material that composes the canopy and scattering of energy away from the forward and backward paths travelled by the rays from and to the radar.

Even though we used (6.61) to allow us just to consider vertical variations and not horizontal displacement when computing interferometric phase with vertically structured pixels, it is nevertheless important to recognise that the path travelled by the rays in the lossy canopy is slanted by the incidence angle of the radar system. Therefore the effective canopy extinction vertically has to account for the real, longer path travelled per unit of vertical distance. If κ_e is the actual one way power extinction coefficient of the canopy then the equivalent two way vertical extinction coefficient can be written

$$\xi = 2\kappa_e \sec\theta \tag{6.67}$$

in which θ is the local incidence angle. Writing the vertical profile function as

$$f(h) = Ae^{\xi h}$$

where the exponent is positive since canopy penetration is in the negative h direction, then the complex coherence of (6.65) becomes

$$\gamma = \frac{e^{j\Delta\phi_{topo}} \int_0^{h_v} e^{\xi h} e^{jk_h h} dh}{\int_0^{h_v} e^{\xi h} dh}$$

$$= e^{j\Delta\phi_{topo}} \frac{\xi}{\xi + jk_h} \frac{e^{(\xi + jk_h)h_v} - 1}{e^{\xi h_v} - 1} \tag{6.68}$$

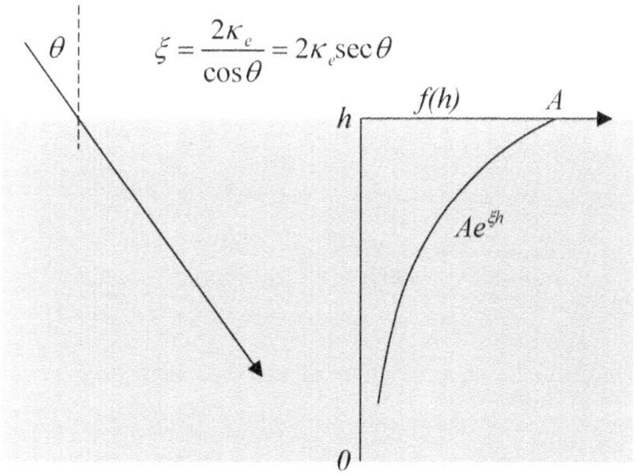

Fig. 6.28. Exponentially decreasing vertical profile function representing a lossy canopy

This expression has two unknowns – the canopy depth h_v and the power extinction coefficient κ_e (via ξ). They need to be estimated from the recorded data. Since the coherence is complex it's amplitude and phase provide the two necessary measurements for that to be done. Also, since it is complex, its phase again adds to the topographic interferometric phase to give a vegetation bias. Fig. 6.29 shows the magnitude of the complex coherence as a function of canopy depth and power extinction coefficient, using the same parameters as Cloude[36], viz k_h=0.1567, θ=45°, 0≤h≤40m, but with κ_e=0dB/m, 0.25dB/m, 0.5dB/m and 0.75dB/m.

Several interesting observations can be made from this graph. First, note that when the canopy is lossless the coherence is the sinc function of Fig. 6.27 because then the profile is constant with height. At the other extreme of very high canopy attenuation the coherence is high and independent of height except for shallow canopies. That is because most of the incident radiation is absorbed and backscattered by the uppermost parts of the canopy.

Again with such a simple vertical scattering profile extending to the earth's surface we can, in principle, determine the two parameters without resort to multi-polarisation radar. So let's now go to the next stage of a more complicated vertical structure $f(h)$. It would be possible to take an arbitrary $f(h)$ and find the appropriate complex coherence through a numerical evaluation of (6.65). However, if we want to develop an inversion algorithm to permit the vertical detail to be characterised from recorded coherence data then it is better to be able to model the general $f(h)$ with a set of (so-called) basis functions that lend themselves to an analytical evaluation of (6.65). With the right set of functions we should be able to derive inversion formulas.

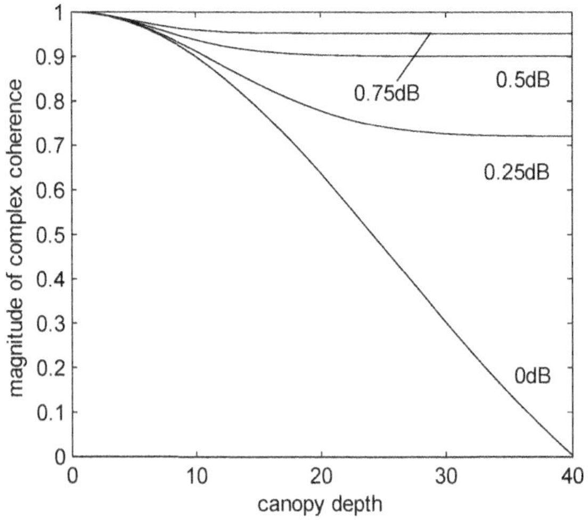

Fig. 6.29. Magnitude of the complex coherence as a function of canopy depth and power extinction coefficient

There are many sets of functions that might be used to represent $f(h)$ including simple polynomials, Chebyshev polynomials and sets of exponential functions. An appealing set of basis functions that have been shown to be of value in PCT are the Legendre

[36] S.R. Cloude, Polarisation coherence tomography (PCT): A tutorial introduction, in http://earth.esa.int/polsarpro/Manuals/3_PCT_Training_Course.pdf

polynomials[37]. The first few of these polynomials[38] in terms of an independent variable x are

$$\begin{aligned} P_0(x) &= 1 \\ P_1(x) &= x \\ P_2(x) &= \frac{1}{2}(3x^2 - 1) \\ P_3(x) &= \frac{1}{2}(5x^3 - 3x) \\ P_4(x) &= \frac{1}{8}(35x^4 - 30x^2 + 3) \end{aligned} \tag{6.69}$$

The Legendre polynomials of (6.69) are shown plotted in Fig. 6.30.

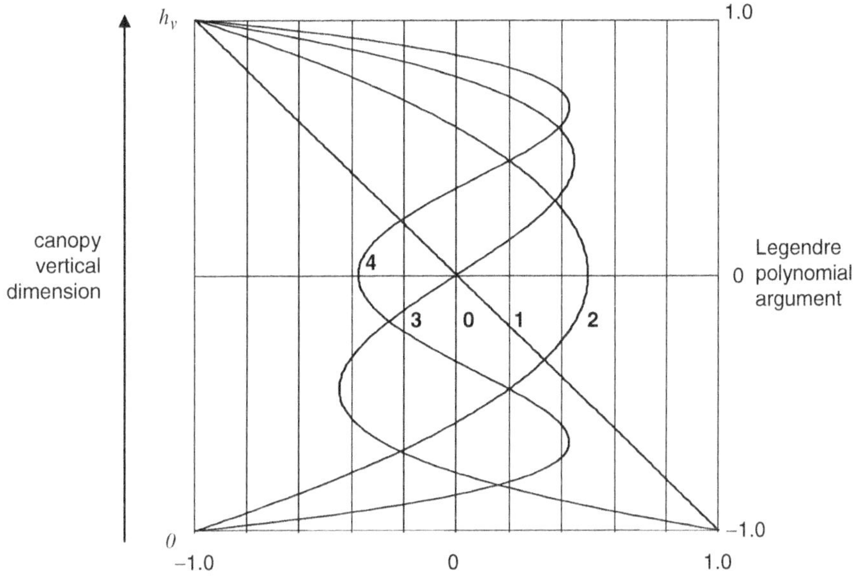

Fig. 6.30. The first five Legendre polynomials plotted vertically to illustrate that they can be used as a set of basis functions with which to model the vertical structure profiles of a pixel

An arbitrary function defined over the interval [-1,1] can be represented by a weighted set of Legendre polynomials

$$f(x) = \sum_{n=0}^{\infty} a_n P_n(x)$$

in which the expansion coefficients a_n are all real. With well behaved functions, such as might be expected for the vertical scattering properties of a pixel, it is possible to truncate that infinite series with little error, so that the function might be approximated

[37] See Cloude, *loc cit.*

[38] For a larger set of Legendre polynomials see Wikipedia or WolframMathWorld™.

$$f(x) = a_0 P_0(x) + a_1 P_1(x) + a_2 P_2(x) ... + a_n P_n(x) \tag{6.70}$$

The coefficients in the Legendre model are given by

$$a_n = \frac{2n+1}{2} \int_{-1}^{1} f(x) P_n(x) dx$$

In order to apply the Legendre model directly to the vertical profile function $f(h)$ we need first to map the independent variable h to the range [-1,1]. We do that by introducing the change of variable

$$x = 2\frac{h}{h_v} - 1 \tag{6.71a}$$

It is also helpful to define a new vertical profile function

$$f(x) = f(h) - 1 \tag{6.71b}$$

With these two substitutions the complex coherence in (6.65) becomes

$$\gamma = \frac{e^{j\Delta\phi_{topo}} \frac{h_v}{2} e^{jk_h h_v/2} \int_{-1}^{1} [1 + f(x)] e^{jk_h h_v x/2} dx}{\frac{h_v}{2} \int_{-1}^{1} [1 + f(x)] dx}$$

We now express the modified profile $f(x)$ by the truncated Legendre series to give

$$\gamma = \frac{e^{j\Delta\phi_{topo}} e^{jk_v} \int_{-1}^{1} [1 + \sum_{m=0}^{n} P_m(x)] e^{jk_v x} dx}{\int_{-1}^{1} [1 + \sum_{m=0}^{n} P_m(x)] dx}$$

in which we have put $k_v = k_h h_v / 2$. Cloude[39] shows that this can be written

$$\gamma = e^{j\Delta\phi_{topo}} e^{jk_v} (f_0 + a_{10} f_1 + ... + a_{n0} f_n) \tag{6.72}$$

with

$$a_{m0} = \frac{a_m}{1 + a_0} \tag{6.73}$$

and in which the first four constituent functions are

$$f_0 = \frac{\sin k_v}{k_v}$$

[39] Cloude, *loc cit.*

$$f_1 = j\left\{\frac{\sin k_v}{k_v^2} - \frac{\cos k_v}{k_v}\right\}$$

$$f_2 = 3\frac{\cos k_v}{k_v^2} - \left\{\frac{6-3k_v^2}{2k_v^3} + \frac{1}{2k_v}\right\}\sin k_v$$

$$f_3 = j\left\{\left\{\frac{30-5k_v^2}{2k_v^3} + \frac{3}{2k_v}\right\}\cos k_v - \left\{\frac{30-15k_v^2}{2k_v^4} + \frac{3}{2k_v^2}\right\}\sin k_v\right\} \tag{6.74}$$

If we can estimate the unknowns in (6.72) from the recorded radar data – i.e. the expansion coefficients $a_{m0,}$ the topographic phase $\Delta\phi_{topo}$ and the baseline-height product k_v, – then we can reconstruct the vertical profile function $f(h)$ for the pixel of interest. That is not a simple task in general[40]. Here we illustrate the simpler case where we assume that the vertical structure can be adequately represented by truncating (6.72) to just the first two terms

$$\gamma = e^{j\Delta\phi_{topo}} e^{jk_v}(f_0 + a_{10}f_1) \tag{6.75}$$

There are now only three unknowns to be determined – $\Delta\phi_{topo}$, h_v and a_{10}. In reality the latter is a combination of two unknowns as seen in (6.73) but that turns out not to be important. If we make the reasonable assumption that $\Delta\phi_{topo}$, h_v are not polarisation dependent then neither are the functions in (6.74). That leaves the expansion coefficient a_{10} in (6.75) as the only term that could depend on polarisation. This is where polarisation comes into PCT. Suppose we have two different polarisation configurations with which we estimate complex coherence. Denote them by superscripts p1 and p2 respectively so that the two measurements yield

$$\gamma^{p1} = e^{j\Delta\phi_{topo}} e^{jk_v}(f_0 + a_{10}^{p1}f_1) \tag{6.76a}$$
$$\gamma^{p2} = e^{j\Delta\phi_{topo}} e^{jk_v}(f_0 + a_{10}^{p2}f_1) \tag{6.76b}$$

in which we now have four unknowns $\Delta\phi_{topo}$, h_v, a_{10}^{p1} and a_{10}^{p2} but also four measurements in the amplitudes and phases (or real and imaginary parts) of the two measured coherences thereby, in principle, allowing the unknowns to be determined. We can estimate those unknowns in the following manner.

First, form the function

$$\gamma^{pd} = \gamma^{p1} - \gamma^{p2} = e^{j\Delta\phi_{topo}} e^{jk_v}(a_{10}^{p1} - a_{10}^{p2})f_1$$

Since f_1 is imaginary, as seen in (6.74), it is convenient to write this last expression as

$$\gamma^{pd} = \gamma^{p1} - \gamma^{p2} = je^{j\Delta\phi_{topo}} e^{jk_v}(a_{10}^{p1} - a_{10}^{p2})Im(f_1)$$

---

[40] See S.R. Cloude, Polarisation Coherence Tomography, *Radio Science*, vol. 41, RS4017, 2006 and S.R. Cloude, Polarisation coherence tomography (PCT): A tutorial introduction, in http://earth.esa.int/polsarpro/Manuals/3_PCT_Training_Course.pdf for available methods.

By looking at the argument (phase angle) of $-j\gamma^{pd}$ we can find the composite phase

$$\Phi = \Delta_{topo} + k_v = \arg(-j\gamma^{pd}) = \arg[e^{j\Delta\phi_{topo}} e^{jk_v}(a_{10}^{p1} - a_{10}^{p2}) Im(f_1)] \qquad (6.77)$$

because the a_{10} and $Im(f_1)$ are real numbers that don't contribute to the overall phase. Using the definition of Φ in (6.77) we can write (6.76a) as

$$\gamma^{p1\Phi} = e^{-j\Phi}\gamma^{p1} = (f_0 + a_{10}^{p1} f_1) \qquad (6.78)$$

Since f_1 is imaginary the real part of this expression is

$$\mathcal{Re}(\gamma^{p1\Phi}) = f_0 = \frac{\sin k_v}{k_v} \qquad (6.79)$$

Because we know Φ from (6.77) we can now find k_v and then we can determine $\Delta\phi_{topo}$. Thus two of the unknown parameters have now been found. From (6.78) we can evaluate a_{10}:

$$a_{10}^{p1} = \frac{Im(\gamma^{p1\Phi})}{Im(f_1)}$$

We actually don't need to go any further since we now have enough information for constructing the vertical profile function $f(h)$ for the pixel. From (6.70), truncating at the second term, we have

$$f(x) = a_0 P_0(x) + a_1 P_1(x)$$

so that from (6.71), and the definitions of the first two Legendre polynomials in (6.69), we have

$$f(h) = 1 + a_0 + a_1\left\{2\frac{h}{h_v} - 1\right\}$$

Dividing throughout by $1+a_0$ gives the final expression for the vertical profile

$$f(h) = 1 + a_{10} + \frac{2a_{10}}{h_v} h \qquad (6.80)$$

which, as expected with the level of approximation (truncation) used, is linear, the sign and slope of which depends on the coefficient a_{10}.

What has been achieved here is an identification of the vertical structure profile of the pixel along with the topographic phase and the height of the vegetation layer. To do that required two polarisations with a single interferometer. Although this has generated a very simple linear approximation to whatever the actual profile might be, it is clear from the increasingly complex shapes of the Legendre polynomials in Fig. 6.30, that incorporating more terms in (6.75) will allow more complex vertical profiles to be identified, at least in principle. As noted above that is not as easy as the analysis just outlined for a linear variation, but has been demonstrated for quadratic shapes[41].

6.16 Range Spectral Filtering and a Re-examination of the Critical Baseline

The frequency shift we used in (6.61) to remove the horizontal dependence of pixel properties in tomography arises from the fact that the two radars in an interferometer subtend slightly different incidence angles at the same spot on the ground; that effectively gives rise to a relative incremental wave number shift between the radars[42]. We will now describe that effect and see how it can be used to generate (6.61), leading to the procedure called range spectral filtering. We will also use it to verify the expression for critical baseline of (6.20) from a different perspective.

Consider the two radars separated by an orthogonal baseline $B_\perp$ in the interferometer of Fig. 6.31. For generality we have shown the ground to be sloped upwards away from the radar at an angle ϑ. The two-way phase angles for each of the two radars are

$$\phi_1 = \frac{4\pi}{\lambda}\frac{Y}{\sin(\theta-\vartheta)} = \frac{4\pi f}{c}\frac{Y}{\sin(\theta-\vartheta)} \tag{6.81a}$$

$$\phi_2 = \frac{4\pi}{\lambda}\frac{Y}{\sin(\theta+\delta\theta-\vartheta)} = \frac{4\pi f}{c}\frac{Y}{\sin(\theta+\delta\theta-\vartheta)} \tag{6.81b}$$

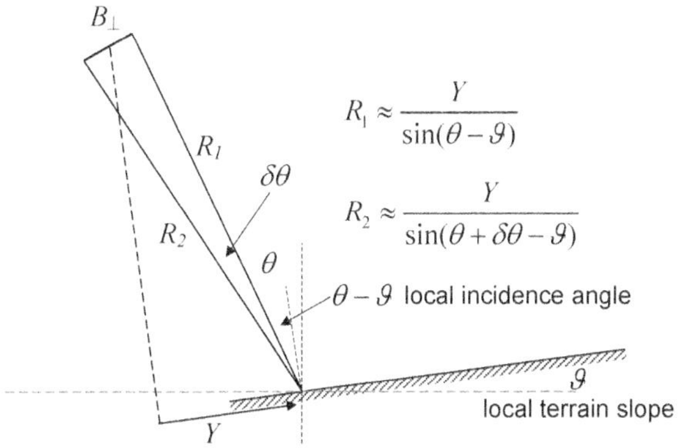

Fig. 6.31. Geometry for computing range spectral filtering; the orthogonal baseline of the interferometer is assumed negligible compared with platform altitude

[41] See Cloude, *Radio Science*, *loc cit.*

[42] See C. Prati and F. Rocca, Improving slant range resolution with multiple SAR surveys, *IEEE Transactions on Aerospace Systems*, vol. 29, 1993, pp. 135-144 and F. Gatelli, A.M. Guanieri, F. Parizzi, P. Pasquali, C. Prati and F. Rocca, The wave number shift in SAR interferometry, *IEEE Transactions on Geoscience and Remote Sensing*, vol. 32, 1994, pp. 855-865 and S.R. Cloude, Polarisation coherence tomography (PCT): A tutorial introduction, in http://earth.esa.int/polsarpro/Manuals/3_PCT_Training_Course.pdf

Under what circumstance can the phase angles of (6.81a) and (6.81b) be the same? To pre-empt the answer let's add a small frequency offset Δf to f in (6.81b) and equate the phases[43]. First note that for $\delta\theta$ small $\sin(\theta+\delta\theta-\vartheta)\approx\sin(\theta-\vartheta)+\delta\theta(\cos\theta-\vartheta)$ so that we are looking for a value of Δf that gives rise to

$$\frac{4\pi(f+\Delta f)Y}{c[\sin(\theta-\vartheta)+\delta\theta\cos(\theta-\vartheta)]}=\frac{4\pi fY}{c\sin(\theta-\vartheta)}$$

i.e.
$$(f+\Delta f)\sin(\theta-\vartheta)=f[\sin(\theta-\vartheta)+\delta\theta\cos(\theta-\vartheta)]$$

$$\Delta f\sin(\theta-\vartheta)=f\delta\theta\cos(\theta-\vartheta)$$

so that
$$\Delta f=\frac{f\delta\theta}{\tan(\theta-\vartheta)} \qquad (6.82)$$

Equation (6.82) can be interpreted in two ways. First, it indicates the frequency offset that would have to be added to the carrier (centre) frequency of one of the radars in the interferometer so that the two-way phases are the same. That is the compensation we used at (6.61) to remove the horizontal variation of phase across a pixel when interested in resolving intra-pixel vertical detail. That process is called *range spectral filtering*.

We can also interpret (6.82) as an amount by which the centre frequency of the ranging chirps, used to achieve range resolution, will be offset between the two radars on reception. As a result, the chirp spectra (see Fig. D.4) received by the two radars will overlap, as illustrated in Fig. 6.32. It is only the common region that can be used to achieve range resolution. The range resolution is thus degraded because of that smaller effective chirp bandwidth and will be given from (3.5b) by

$$r_g=\frac{c}{2(\mathcal{B}_c-\Delta f)\sin(\theta-\vartheta)}$$

in which $\mathcal{B}_c$ is the transmitted chirp bandwidth. Note that as the frequency offset approaches the bandwidth of the transmitted chirp the resolution deteriorates badly. The extreme is when $\Delta f=\mathcal{B}_c$. Under what conditions will that occur? To answer that examine their ratio. There will be range resolution while so ever

$$\frac{\Delta f}{\mathcal{B}_c}=\frac{f\delta\theta}{\mathcal{B}_c\tan(\theta-\vartheta)}\leq 1$$

Since $\delta\theta=\dfrac{B_\perp}{R_o}$ (where $R_o\approx R_1\approx R_1$) this gives

$$\frac{\Delta f}{\mathcal{B}}=\frac{fB_\perp}{R_o\mathcal{B}_c\tan(\theta-\vartheta)}\leq 1$$

Thus to avoid the complete loss of range resolution the orthogonal baseline of the interferometer must satisfy

[43] See also Sect. 4.5.1 of D. Massonnet and J-C Souyris, *Imaging With Synthetic Aperture Radar*, Taylor and Francis, Roca Baton, Florida, 2008.

$$B_{\perp} \le \frac{R_o \mathcal{B}_c \tan(\theta - \vartheta)}{f}$$

which, in the limit, is the critical baseline of (6.20):

$$B_{\perp crit} = \frac{R_o \mathcal{B}_c \tan(\theta - \vartheta)}{f} = \frac{R_o \lambda \mathcal{B}_c \tan(\theta - \vartheta)}{c}$$

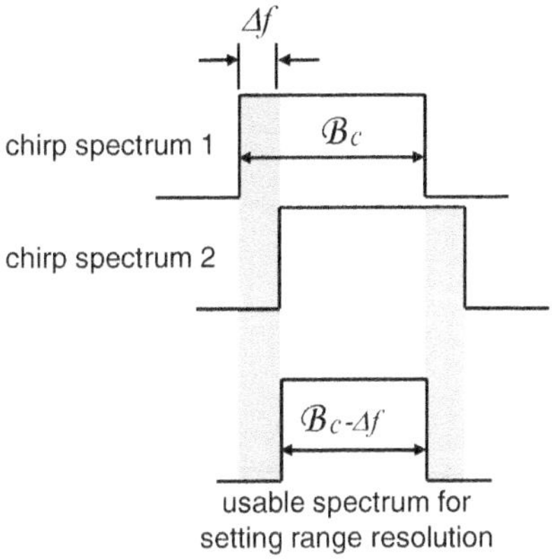

Fig. 6.32. Chirp spectra from either side of the baseline showing the overlap caused by the frequency shift associated with the different look angles, and its impact on the usable bandwidth for determining range resolution

CHAPTER 7

BISTATIC SAR

7.1 Introduction

The monostatic radar configuration, in which the transmitter and receiver are collocated, seems so logical that we are not led naturally to contemplate other arrangements. As noted in Chapt. 2 monostatic radar is tantamount to using a torch or flashlight to see objects when it is dark. In this visual example it is not necessary to collocate the source of illumination and the receiver (our eyes). The energy source can be located in a roof light or lamp. We then see objects through bistatic light scattering. The same situation occurs with standard optical remote sensing; the source of energy – generally the sun – is located spatially quite separate from the sensor.

Clearly we could do the same with radar. The source of irradiation can be located in a different position from the receiver. However, since radar uses time delay to ascertain range information there needs to be some form of communication and synchronisation between the transmitter and receiver; nevertheless, a bistatic radar configuration is certainly technologically feasible[1]. There are advantages in such an arrangement. In defence applications it is advantageous to have a separate receiver since transmissions from a radar make it liable to detection; having the receiver in a different position renders its location silent in a radio sense. To improve its security such a system could even operate with a satellite based transmitter and an aircraft based receiver. Also, just as we can gain more information about an object by viewing it from different perspectives so bistatic radar, in principle, might yield better data for discriminating targets, including those of interest in remote sensing. Interestingly, it is also possible to have a forward looking radar system in the bistatic mode, provided the transmitter is off-axis[2].

A bistatic radar can use transmitters of opportunity[3] (just as the sun is a convenient source in optical imaging). Sources of microwave energy designed for other purposes, including telecommunications, navigation and positioning (as for example GNSS[4]) can be used to irradiate a target. The scattered energy can then be detected by a radar receiver. In the case of GNSS the transmissions are time encoded (which is the very basis of GNSS) so that it is possible to synchronise scattered signals with those transmitted.

---

[1] In the radar literature a bistatic radar is sometimes called *passive* since the receiver is not accompanied by a transmitter and hence is undetectable. That should not be confused with passive imaging in microwave remote sensing (using the earth's thermal emission as a source), even though the connotations are similar.

[2] See X. Qiu, D. Hu and C. Ding, Some reflections on bistatic SAR of forward-looking configuration, *IEEE Geoscience and Remote Sensing Letters*, vol. 5, no. 4, October 2008, pp. 735-739.

[3] See H.D. Griffiths, From a different perspective: principles, practice and potential of bistatic radar, *Proc IEEE International Conference on Radar 2003*, Adelaide, 3-5 Sep 2003, pp. 1-7.

[4] GNSS (Global Navigational Satellite System) is now widely used to describe satellite based navigation and positioning systems such as the US GPS, the Russian Glonass and the forthcoming European Galileo program.

J.A. Richards, *Remote Sensing with Imaging Radar*, Signals and Communication Technology,
DOI: 10.1007/978-3-642-02020-9_7,

We can generalise further. We need not restrict our attention to a single transmitter and a single receiver. Monostatic and bistatic configurations are special cases of the multiple input, multiple output (MIMO) radar systems summarised in the next section.

As with the other radar configurations treated in this book we develop our analysis here based on rectilinear flight paths and flat earth assumptions so that the expressions derived strictly apply only to airborne systems. Nevertheless they are good approximations for spacecraft operation as well, unless the transmitter and receiver are widely spaced. Deriving expressions for range and azimuth resolutions for bistatic radar can be a little complicated, so we develop the essential concepts in this chapter by building up from simple, special cases to the most general bistatic situation later on. In doing so we will derive general formulas for resolution that can be used for any radar topology.

7.2 Generalised Radar Networks

In principle we could have as many transmitters and as many receivers as we like. The manner in which they interact defines a number of identified radar configurations. While we will focus just on bistatic radar in this chapter it is likely that some of the more general configurations will feature in future remote sensing radar sensor networks.

Although the definitions and nomenclature are still developing, Fig. 7.1 shows the set of radar network topologies currently recognised[5]. This is drawn in terms of the rays that connect the transmitters and receivers rather than in terms of physical layouts. Monostatic and bistatic radars are included to show where they sit in the hierarchy. The important differentiator is the set of pathways between the transmitters and receivers. Even though those paths contain the target of interest, it is the number of paths that intersect with the target (with which we diagnose its properties), and how their signals are processed, that define the different radar configuration types and their subsets.

The netted radars shown in Fig. 7.1c and d come in several forms. They can be used to provide different perspectives of and thus information on the target of interest. Information fusion procedures would be employed to integrate the information available from each of the radars. Alternatively, they can be used as a set of cooperating radars to provide enhanced areal coverage.

The multistatic radar of Fig. 7.1e normally consists of a single transmitter and a set of cooperating receivers although some multistatic radars use more than one transmitter. A multistatic radar can also be established using a monostatic configuration with a second receiver[6]. In principle, the interferometers of Chapt. 6 are multistatic radars. Effectively the standard mode uses one transmitter and two receivers while the ping pong mode uses two transmitters and two receivers. The tomographic SARs of Sect. 6.15 can also be regarded as multistatic or netted radars.

Multiple input, multiple output (MIMO) radars networks are a generalisation of multistatic radars. Shown in Fig. 7.1f they consist of a set of transmitters and a set of

---

[5] There is another form of radar not included here called *secondary radar*. It is widely used in air traffic control and depends upon having a cooperative target. The target (aircraft) carries a receiver and re-transmitter (together called a transponder) which detects an incoming radar pulse. It then transmits a signal back to the radar set. This has the advantage that the signal level received at the radar can be larger than that through passive scattering from a target. Correspondingly, the radar transmitter power (and antenna) can be much smaller since the level of signal received is, again, not the result of passive scattering. The return signal can also carry information about the aircraft and its position. The secondary radar principle is similar to that of active radar calibrators (ARCs) of Sect. 4.2.3.

[6] See A. Moccia, N. Chiacchio and A. Capone, Spaceborne bistatic synthetic aperture radar for remote sensing applications, *Int. Journal of Remote Sensing*, vol. 21, 2000, pp. 3395-3414.

receivers[7]. Each receiver detects scattered energy from the target originating from every transmitter as seen in the path diagram of Fig. 7.1g. The received signal is therefore quite complex. Although not shown here explicitly the paths could also include monostatic ones – in other words the receivers could be collocated with the transmitters. MIMO radars are further subdivided. If the transmitters are well separated, as are the receivers, there is no correlation between the signals and the network is referred to as a *statistical MIMO radar*. If the system is designed to have the transmitting antennas closely arranged so that they look like an array antenna, and similarly for the receivers, the network is referred to as a *coherent MIMO radar* because the signal set transmitted is designed to achieve specified performance objectives through signal processing.

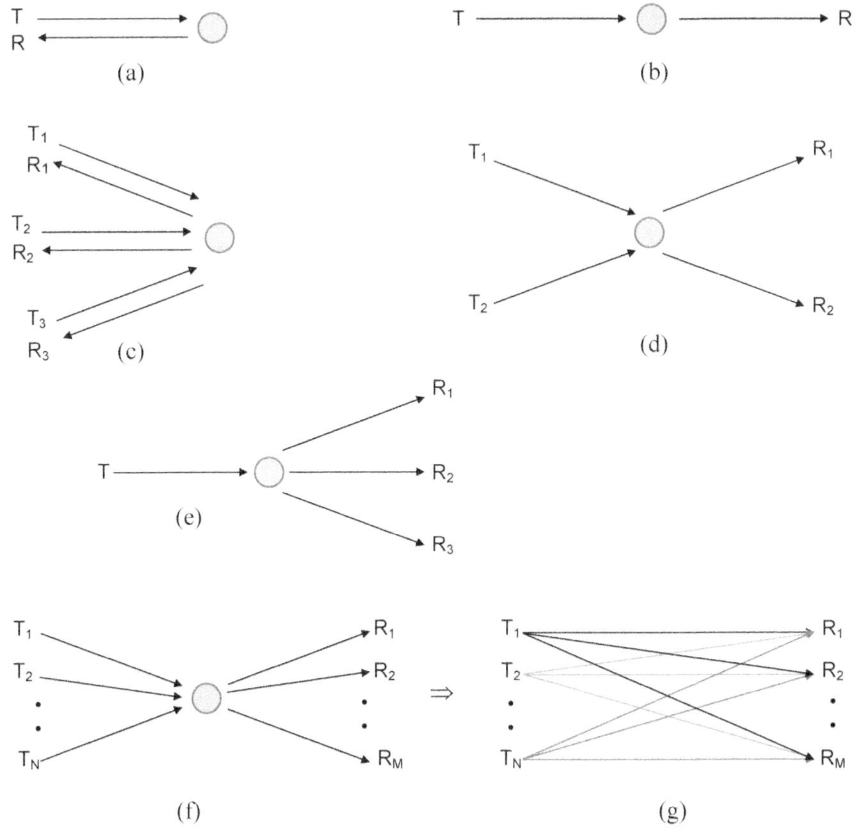

Fig. 7.1. (a) Monostatic radar (b) bistatic radar (c) monostatic netted radar (d) bistatic netted radar (e) multistatic radar (f) MIMO radar and (g) signal paths for the MIMO radar

[7] See K.W. Forsythe and D.W. Bliss, Chapter 2 MIMO radar: concepts, performance enhancements, and applications, in J. Li and P. Stoica, *MIMO Radar Signal Processing*, Wiley, Hoboken, N.J., 2009.

7.3 Analysis of Bistatic Radar

Consider the particular bistatic arrangement shown in Fig. 7.2, in which the transmitter and receiver are on platforms travelling out of the page parallel to each other. They are separated in the cross track direction by a baseline B and the rays from the transmitter to the target and the receiver to the target have an angular separation β; this is called the *bistatic angle*. There are two other significant angles; one subtended by the transmission path and the other by the scattering or reception path. We call the former the incidence angle, in keeping with the monostatic radars of the previous chapters, and the latter the observation or scattering angle. This configuration is geometrically similar to that used in cross track interferometry in Chapt. 6 but here the baseline can be large and one platform transmits and one receives, although as noted in Fig. 7.1 in some applications both platforms might transmit and receive.

We will consider the alternative case when the transmitter and receiver are separated in the along track direction after we have analysed the cross track situation. We will then consider the case of an arbitrary baseline and, finally, examine the most general configuration.

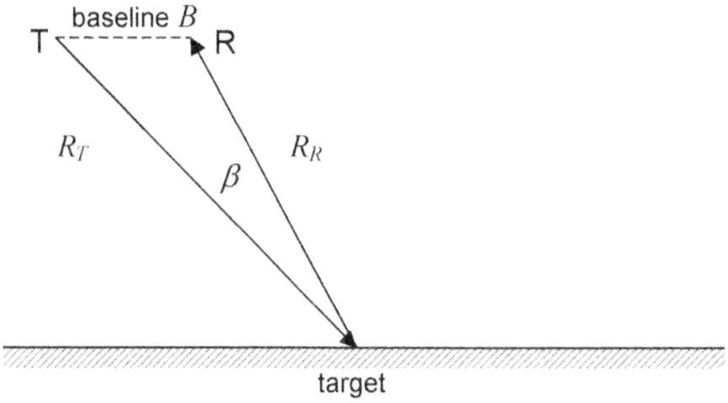

Fig. 7.2. Definition of physical parameters used in bistatic radar

7.3.1 The Bistatic Radar Range Equation and the Bistatic Radar Cross Section

Equivalently to (3.33) for monostatic radar the radar range equation for bistatic radar can be derived in the following manner. A transmitter power of P_t radiated on an antenna with gain G_t will generate a power density at the target of

$$p_i = \frac{P_t G_t}{4\pi R_T^2} \text{ Wm}^{-2}$$

This is assumed to be scattered isotropically by the target in the direction of the receiver which, for this purpose, is described by a *bistatic radar cross section* σ_B m^2. The power produced at the receiver is

$$P_r = A_r \frac{1}{4\pi R_R^2} \sigma_B \frac{P_t G_t}{4\pi R_T^2}$$

in which A_r is the aperture of the receiving antenna which, if expressed in terms of its gain, gives

$$P_r = \frac{G_r\lambda^2}{4\pi}\frac{1}{4\pi R_R^2}\sigma_B\frac{P_tG_t}{4\pi R_T^2} = \frac{P_tG_tG_r\lambda^2}{(4\pi)^3(R_TR_R)^2}\sigma_B \tag{7.1}$$

The inverse squares of the distances in the denominator are interesting to explore. For a given separation between the transmitter and the receiver, calculation will demonstrate that the worst case received power level is when the target is mid way between the two – which is exactly as it is for monostatic radar. If the target is closer to either the transmitter or receiver the received power will be higher.

The bistatic radar cross section (or equivalently the bistatic scattering coefficient for a resolution cell) will be a function of both the scattering angle and the incidence angle. It will also be a function of polarisation and can thus be expressed in the form of a matrix as in (3.36). We can also use a scattering matrix for multipolarisation bistatic radar. This is taken up in Sect. 7.10.

7.3.2 Bistatic Ground Range Resolution

Recall from Sect. 3.2 and Fig. 3.6 that the minimum resolvable detail on the ground is determined by the duration of the transmitted ranging pulse which, in a chirp-based pulse compression radar, is determined by the chirp bandwidth – see (3.5). The same consideration sets the ground range resolution for a bistatic radar. Fig. 7.3 shows transmitted rays to two spots on the ground, in adjacent ground resolution cells, and their reflections to the receiver. We have assumed that the transmitter and receiver are sufficiently far away that the rays can be drawn parallel and that they are both to the same side of the target. The receiver, alternatively, could be on the other side of the target; that situation just requires the scattering angle to be regarded as negative – in our diagram the incidence and scattering angles are defined as positive anticlockwise from the vertical.

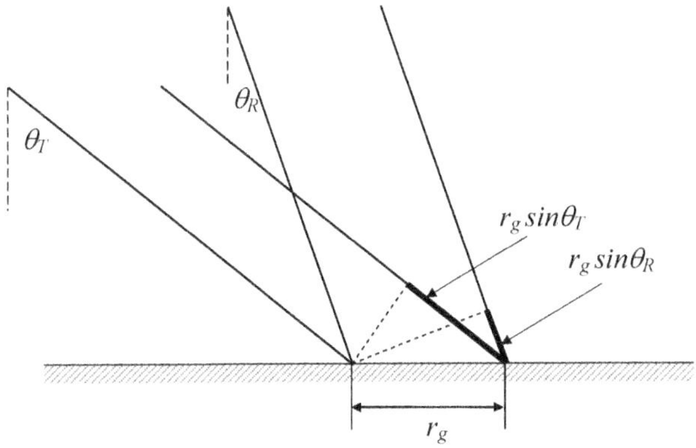

Fig. 7.3. Geometry for calculating the ground range resolution of bistatic radar

The right hand incident ray and its reflection each travel further than the left hand incident and reflected rays. The additional distances are shown in heavier lines. The reflection from the right hand resolution cell, or pixel, arrives at the receiver $r_g(\sin\theta_T + \sin\theta_R)/c$ seconds later than the reflection from the left hand pixel.

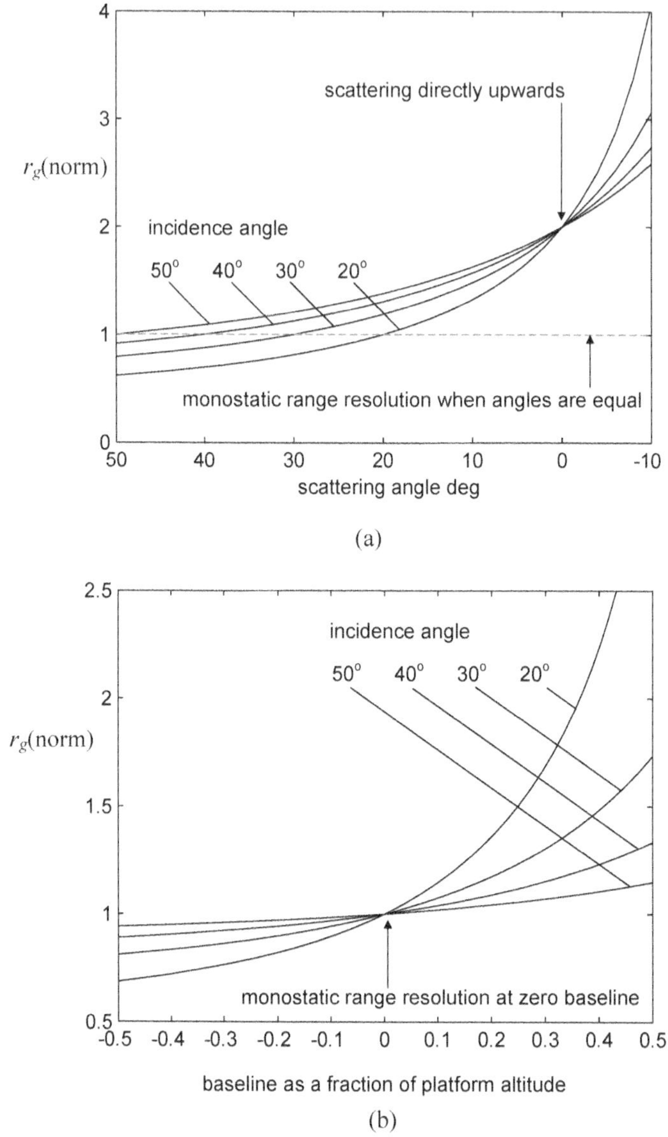

Fig. 7.4. Range resolution of cross track bistatic radar for a range of incidence angles, normalised to the monostatic case (a) as a function of scattering angle and (b) as a function of normalised baseline

In order to resolve the bistatic echoes that time difference must be no smaller than the duration of the compressed ranging chirp. The limit of resolution is when the time difference equals the compressed chirp width τ, giving the ground range resolution as

$$r_g = \frac{c\tau}{\sin\theta_T + \sin\theta_R} = \frac{c}{\mathcal{B}_c(\sin\theta_T + \sin\theta_R)} \tag{7.2}$$

where $\mathcal{B}_c$ is the chirp bandwidth. There are some interesting special cases of this expression. If $\theta_R = \theta_T$ then it reverts to (3.5b) for monostatic radar. If $\theta_R = -\theta_T$ $r_g \rightarrow \infty$ showing there is no range resolution. Both of these can be seen easily by appropriate adjustments to Fig. 7.3. If $\theta_R < \theta_T$ the ground range resolution is better than achievable with monostatic radar. To demonstrate these points further, Fig. 7.4a shows the ground range resolution for a range of incidence and scattering angles, normalised to the monostatic case; effectively this plots $2\sin\theta_T/(\sin\theta_T + \sin\theta_R)$. The dependence on relative baseline is shown in Fig. 7.4b, computed by noting from Fig. 7.2 that the scattering angle is related to platform altitude H, baseline B and incidence angle θ_T by $\theta_R = \tan^{-1}(\tan\theta_T - B/H)$.

Now consider the along track bistatic configuration of Fig. 7.5. This could be established by having the platform carrying the receiver either preceding or following the transmitter platform in orbit. Generally the transmitter radiates to the side, as in the monostatic radar technology considered in Chapt. 3, so that the receiver has to squint forward or backwards to see the scattered signal. It is possible of course for the transmitter or both to squint but we won't examine those variations here. Note that the incidence and scattering angles are now not in the same plane. This configuration goes by several names: we will call it an along track baseline here, but it is also known as a tandem configuration or sometimes a squint configuration since either or both of the transmitter and receiver have to squint to ensure simultaneous coverage of the same spot on the ground.

To find the ground range resolution for this case attention is focussed on the plane that contains the broadside distance from the target to the satellite path and the reflected ray. The expanded detail in Fig. 7.5 shows that the reflected path difference is $r_g\sin\chi$. Since from Fig. 7.3 the transmitted path difference is $r_g\sin\theta_T$ then the ground range resolution for the along track bistatic configuration is

$$r_g = \frac{c}{\mathcal{B}_c(\sin\theta_T + \sin\chi)} \tag{7.3}$$

From the geometry of the shaded triangle in Fig. 7.5 we can see that

$$\sin\chi = \tan\theta_T \cos\theta_R \tag{7.4}$$

which we can use to find ground range resolution in terms of the scattering angle, if that were of interest. In the along track case it is more appropriate simply to express resolution in terms of the baseline separation. If both platforms are at altitude H and there is a baseline B the scattering angle at transmitter broadside can be shown to be $\theta_R = \cos^{-1}[(\sec^2\theta_T + (B/H)^2)^{-1/2}]$. From (7.4) we can then get χ which we substitute into (7.3) to find how the range resolution varies with baseline; that is shown in Fig. 7.6. Note that the range resolution for this configuration is always poorer than the monostatic equivalent.

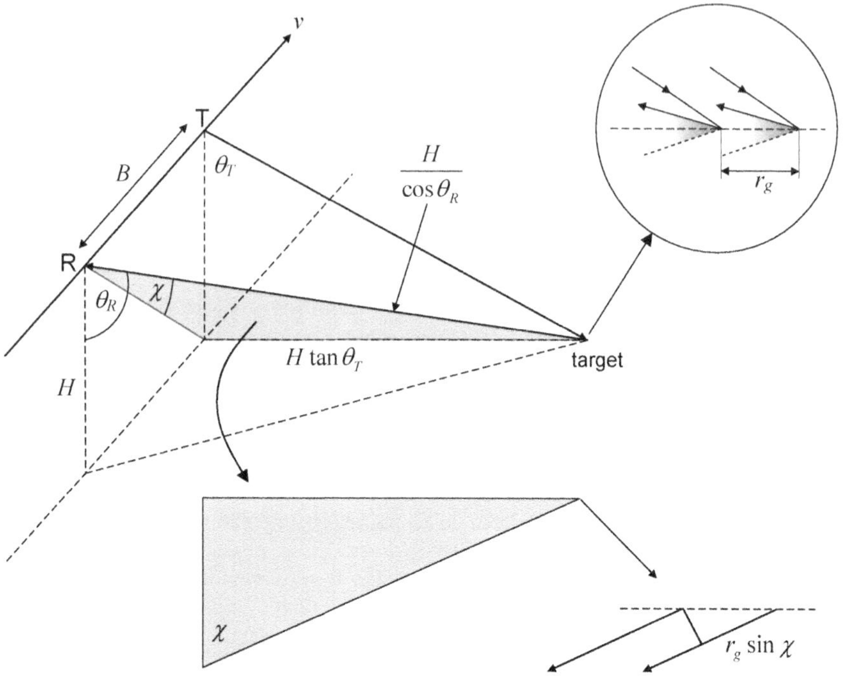

Fig. 7.5. Along track bistatic radar with a squinting receiver following the transmitter in orbit; the shaded plane containing the reflected ray and the projection of the transmitted ray on the ground is used to derive the expression for ground range resolution

Having looked at the two special cases of an across track baseline and an along track baseline it is straightforward to look at the situation in which the platforms are on parallel tracks but at different altitudes with the receiver lagging (or leading) the transmitter. This is shown in Fig. 7.7, in which the inclined baseline is shown resolved into its three Cartesian components.

The relative positions of the platforms are described by

X their along track separation
Y their cross track separation
Z their altitude separation

Other significant parameters are the altitude and incidence angle of the transmitting platform, H_T and θ_T respectively.

The plane in which scattering takes place is shaded; by reference to Fig. 7.5 it can be seen that the ground range resolution is given by (7.3), although χ is different from that case, so (7.4) doesn't apply. From the scattering plane geometry in Fig. 7.7 we see that

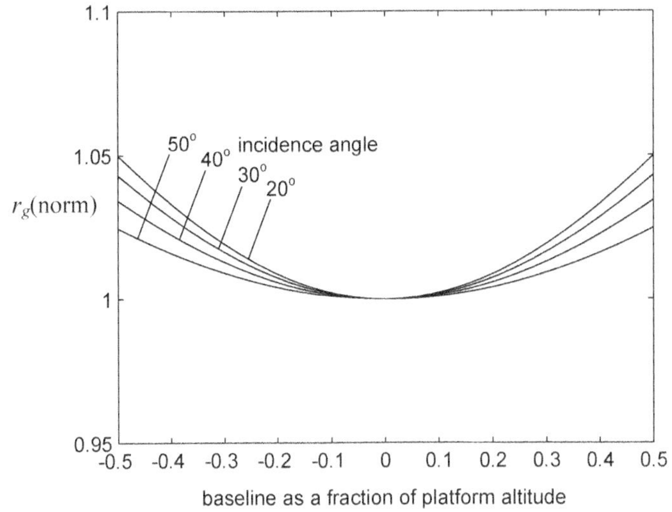

Fig. 7.6. Range resolution of along track bistatic radar as a function of normalised baseline for a range of incidence angles, normalised to the monostatic case; the negative baseline corresponds to the receiver leading the transmitter for the sketch of Fig. 7.5.

$$\sin\chi = \frac{H_T \tan\theta_T - Y}{H_R \sec\theta_R} \tag{7.5}$$

where θ_R is the scattering angle (at the receiver) for the instant when the transmitter is broadside of the target. Equation (7.5) replaces (7.4) in (7.3) for this general case.

We can now consider special cases:

1. For monostatic SAR Y=0, $\theta_R=\theta_T=\theta$ and $H_T=H_R$, so that (7.5) gives $\sin\chi=\sin\theta$; (7.3) then degenerates to (3.5b).
2. For cross track bistatic SAR X=0, Z=0, $Y=B$; also $H_T=H_R=H$. Equation (7.5) then becomes

$$\sin\chi = \frac{H\tan\theta_T - B}{H\sec\theta_R}$$

   From the view along the velocity vector shown in Fig. 7.7 and Fig. 7.12, with Z=0 and $Y=B$, we can see that the numerator in the last expression is simply $H\tan\theta_R$, so that we have $\sin\chi = \sin\theta_R$. The ground range resolution is then given by (7.2) as required.
3. For along track bistatic SAR Y=0, Z=0, $X=B$ and again $H_T=H_R$. Equation (7.5) then reduces to (7.4), as required.

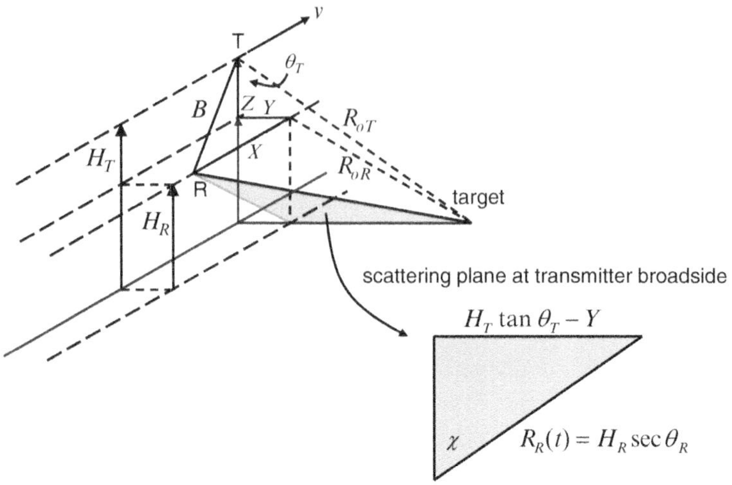

Fig. 7.7. Geometry for calculating the ground range resolution with an inclined baseline; the transmitter is at its broadside position for the calculation of range resolution

7.3.3 Bistatic Azimuth Resolution

Consider the cross track bistatic configuration of Fig. 7.2, but viewed from above as the platforms are approaching a point target, as shown in Fig. 7.8. Using the same development and approximations as for monostatic radar in Sect. 3.6, and assuming both platforms are travelling at the same velocity relative to the target, we can see that

$$R_T(t) \approx R_{oT} + \frac{(vt)^2}{2R_{oT}}$$

and

$$R_R(t) \approx R_{oR} + \frac{(vt)^2}{2R_{oR}}$$

in which R_{oT} and R_{oR} are the slant ranges at broadside for the transmitter platform and receiver platform respectively.

The phase delay associated with the total path followed by the ranging pulses (from transmitter to target to receiver) is

$$\phi(t) = \frac{2\pi}{\lambda}\left[R_{oT} + R_{oR} + \frac{(vt)^2}{2}\left(\frac{1}{R_{oT}} + \frac{1}{R_{oR}}\right)\right]$$

so that the motion induced Doppler frequency is given by

$$\omega(t) = \frac{d\phi(t)}{dt} = \frac{2\pi v^2 t}{\lambda}\left(\frac{R_{oT} + R_{oR}}{R_{oT}R_{oR}}\right)$$

The chirp rate is therefore

$$b = \frac{2\pi v^2}{\lambda}\left(\frac{R_{oT} + R_{oR}}{R_{oT}R_{oR}}\right) \text{ rad s}^{-1}\text{s}^{-1}$$

or

$$\beta = \frac{v^2}{\lambda}\left(\frac{R_{oT} + R_{oR}}{R_{oT}R_{oR}}\right) \text{ Hzs}^{-1}$$

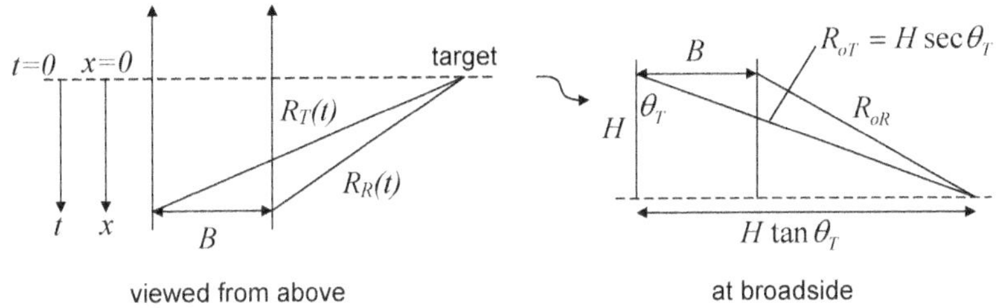

Fig. 7.8. Cross track bistatic SAR geometry: the along track dimensions are exaggerated for clarity; the beamwidth of the transmitting antenna, which has no squint, is illuminating the target

To find the azimuth resolution we need next to find the chirp bandwidth, which is the product of the chirp rate and chirp duration. As with monostatic radar the duration of the chirp is set by the time the point target is irradiated; that time is given by the azimuth beamwidth of the transmitting antenna projected onto the ground divided by the transmitter platform velocity, giving the azimuth chirp length as

$$T_a = \frac{L_a}{v} = \frac{\lambda R_{oT}}{v l_a}$$

in which l_a is the azimuth length of the transmitting antenna and L_a is the azimuth beamwidth at the ground. The chirp bandwidth is therefore

$$\mathcal{B}_c = \beta T_a = \frac{v}{l_a}\frac{R_{oT} + R_{oR}}{R_{oR}}$$

If the Doppler induced azimuth chirp is compressed in the usual way for synthetic aperture radar by correlating it against a replica of itself, then the compressed version has a duration equal to the reciprocal of the chirp bandwidth. As in Sect. 3.6, using the platform velocity, that is equivalent to a travel in azimuth over that period of:

$$r_a = \frac{1}{\mathcal{B}_c} v = l_a \frac{R_{oR}}{R_{oT} + R_{oR}} \qquad (7.6)$$

which is the azimuth resolution of the bistatic radar. Note that if $R_{oR}=R_{oT}$ (7.6) reduces, as expected, to the azimuth resolution of monostatic SAR given by (3.8).

Equation (7.6) can be re-written in terms of the platform altitude, baseline separation and incidence angle of the transmitter as

$$r_a = l_a \frac{\sqrt{1 + (\tan\theta_T - B/H)^2}}{\sec\theta_T + \sqrt{1 + (\tan\theta_T - B/H)^2}} \tag{7.7}$$

Using this expression Fig. 7.9 shows how azimuth resolution varies with baseline. Unfortunately, in the bistatic case the positions of the platforms affect the resolution, whereas for monostatic SAR the theoretical azimuth resolution is independent of platform position. Note that it is theoretically possible for the azimuth resolution to be better than the monostatic equivalent, as seen in the figure. That is when the receiver is on the side of the transmitter furthest from the transmitter. However, because of the correspondingly longer path from transmitter to target to receiver, such an arrangement results in less available power at the receiver, limiting the ability to measure smaller radar cross sections or scattering coefficients before noise becomes a problem.

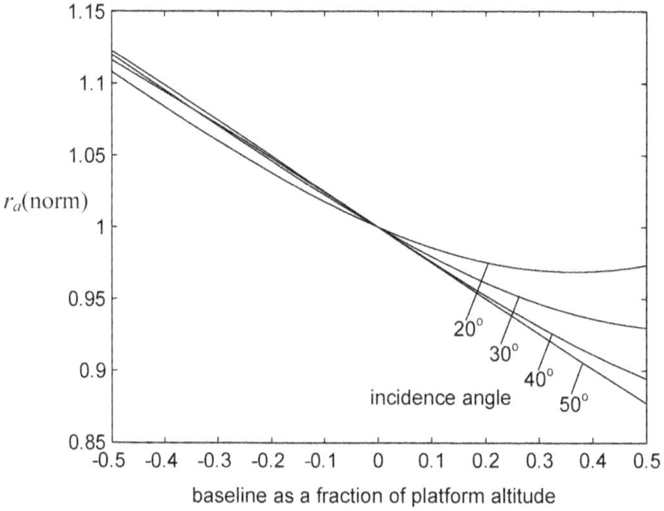

Fig. 7.9. Azimuth resolution, normalised to the monostatic case, for cross track bistatic SAR, as a function of incidence angle, and baseline to platform altitude ratio

We now consider azimuth resolution when the bistatic configuration is in the along track mode as in Fig. 7.5 in which the transmitter has no squint angle, and the receiver squints forward. Fig. 7.10 shows a slant plane view of the radar just acquiring a target. The transmitter and receiver are in the same slant plant; that makes the analysis a little easier.

We can derive an expression for azimuth resolution by following the same procedure as for monostatic radar and the across track bistatic configuration, with one exception. Because the baseline B can be quite large, as determined by the separation in orbit of the platforms, we cannot use the simple two term power series approximation to the square root in calculating the slant range from the target to the receiver $R_R(t)$. Because of that we will not approximate the transmitter-target slant range either, even though that would have been acceptable.

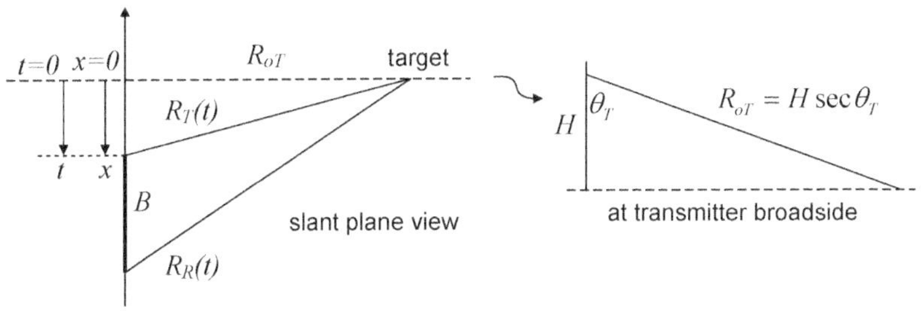

Fig. 7.10. Slant plane view of along track bistatic SAR: again the along track dimensions are exaggerated for clarity; the beamwidth of the transmitting antenna, which has no squint, is reaching the target

From Fig. 7.10 the total range is

$$R(t) = R_T(t) + R_R(t)$$

with $R_T(t) = \sqrt{R_{oT}^2 + (vt)^2}$ and $R_R(t) = \sqrt{R_{oT}^2 + (vt+B)^2}$. This gives the total phase change from transmitter to target to receiver as

$$\phi(t) = \frac{2\pi}{\lambda} R(t) = \frac{2\pi}{\lambda}\left[\sqrt{R_{oT}^2 + (vt)^2} + \sqrt{R_{oT}^2 + (vt+B)^2}\right] \tag{7.8}$$

The corresponding Doppler induced frequency shift of the radar carrier frequency is

$$f(t) = \frac{1}{2\pi}\frac{d\phi(t)}{dt} = \frac{1}{\lambda}\frac{dR(t)}{dt} \tag{7.9}$$

As the transmitter passes broadside the change in frequency will be linear (as was the case for monostatic SAR) emulating a chirp that can be compressed to provide high azimuth resolution. It is likely that the frequency variation will also have higher order dependences on time; in its most general form it could be written

$$f(t) = f_o + \beta t + \beta_2 t^2 + \beta_3 t^3 \ldots \tag{7.10}$$

in which f_o is the carrier (or centre) frequency of the radar and β is the linear chirp rate – see (3.11b) for the monostatic equivalent. To find β we evaluate $\left.\frac{df(t)}{dt}\right|_{t=o}$; thus

$$\beta = \frac{1}{\lambda}\left.\frac{d^2R(t)}{dt^2}\right|_{t=0} \tag{7.11}$$

Substituting from (7.7) gives

$$\beta = \frac{v^2}{\lambda R_{oT}} + \frac{v^2}{\lambda R_{oRt}} - \frac{v^2B^2}{\lambda R_{oRt}^3} \tag{7.12a}$$

in which
$$R_{oRt} = \sqrt{R_{oT}^2 + B^2} \tag{7.12b}$$

is the receiver slant range at t=0, i.e. at transmitter broadside. Proceeding as for the cross track case, the azimuth chirp bandwidth is

$$\mathcal{B}_c = \beta T_a = \beta \frac{\lambda R_{oT}}{v l_a}$$
$$= \frac{v R_{oT}}{l_a}\left[\frac{1}{R_{oT}} + \frac{1}{R_{oRt}} - \frac{B^2}{R_{oRt}^3}\right]$$

Finally the azimuth resolution is given by the platform velocity multiplied by the compressed chirp duration (the reciprocal of the chirp bandwidth) to give

$$r_a = l_a\left[1 + \frac{R_{oT}}{R_{oRt}} - \frac{R_{oT}B^2}{R_{oRt}^3}\right]^{-1}$$

This can be simplified:

$$r_a = l_a\left[1 + \frac{R_{oT}R_{oRt}^2 - R_{oT}B^2}{R_{oRt}^3}\right]^{-1}$$

i.e.
$$r_a = l_a\left[1 + \frac{R_{oT}^{\;3}}{R_{oRt}^3}\right]^{-1} \tag{7.13}$$

Note that if B=0, $R_{oRt} \equiv R_{oT}$ and this reduces to the monostatic formula. We re-write this in terms of the baseline to altitude ratio and the transmitter incidence angle; note in Fig. 7.10 that $R_{oT} = H \sec\theta_T$ so that $R_{oRt} = H\sqrt{\sec^2\theta_T + (B/H)^2}$ giving

$$\frac{R_{oT}}{R_{oRt}} = \frac{\sec\theta_T}{\sqrt{\sec^2\theta_T + (B/H)^2}}$$
$$= [1 + \cos^2\theta_T (B/H)^2]^{-1/2}$$

Equation (7.13) then becomes

$$r_a = l_a\left[1 + [1 + \cos^2\theta_T (B/H)^2]^{-3/2}\right]^{-1} \tag{7.14}$$

Fig. 7.11 shows how azimuth resolution varies in this tandem situation as a function of incidence angle and baseline-to-altitude ratio compared with the monostatic case. As with range resolution the azimuth resolution in the along track configuration is never better than the monostatic value.

Finally we can consider the azimuth resolution for the inclined baseline case of Fig. 7.7, but shown more generally in Fig. 7.12. The transmitter to target and target to receiver distances respectively are

$$R_T(t) = \sqrt{R_{oT}^2 + (vt)^2} \tag{7.15a}$$

$$R_R(t) = \sqrt{R_{oR}^2 + (vt + X)^2} \tag{7.15b}$$

The broadside slant range for the receiver R_{oR} is related to the broadside slant range for the transmitter R_{oT} by

$$R_{oR}^2 = R_{oT}^2 + Z^2 + Y^2 - 2R_{oT}(Y\sin\theta_T + Z\cos\theta_T) \tag{7.16}$$

That expression comes from looking along the flight direction and projecting the transmitted and received rays onto the plane orthogonal to the velocity vector at the position of the target.

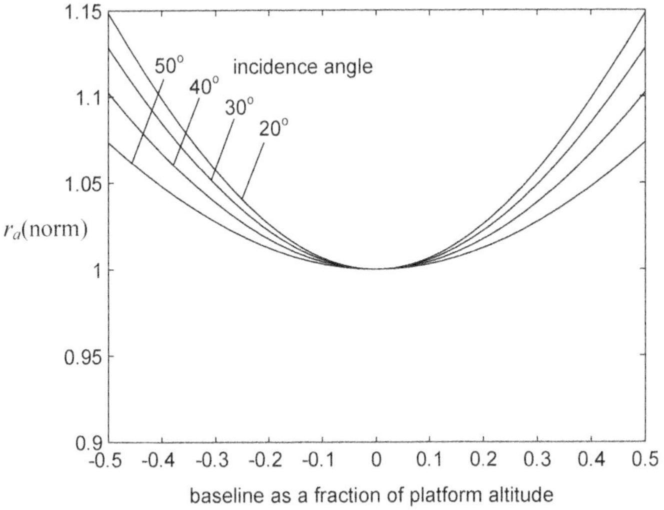

Fig. 7.11. Azimuth resolution, normalised to the monostatic case, for along track bistatic SAR, as a function of incidence angle, and baseline to platform altitude ratio; the transmitter radiates normal to its velocity vector

Since (7.15b) is of the same form as $R_R(t)$ in (7.8) we can use the results for the along track baseline special case by noting now that R_{oT} and R_{oR} are different and putting X in place of B. Using (7.11) with the appropriate substitutions the linear chirp rate of the motion induced Doppler frequency in this general case is

$$\beta = \frac{v^2}{\lambda R_{oT}} + \frac{v^2(R_{oR}^2 + X^2)^{-1/2}}{\lambda} - \frac{v^2X^2(R_{oR}^2 + X^2)^{-3/2}}{\lambda}$$

From (7.16) we note

$$R_{oR}^2 + X^2 = R_{oT}^2 + B^2 - 2R_{oT}(Y\sin\theta_T + Z\cos\theta_T) \tag{7.17}$$

As before, this is the square of the receiver to target slant range when the transmitter is at broadside; call it R_{oRt}^2. Again noting that the azimuth chirp bandwidth is

$$\mathcal{B}_c = \beta T_a = \frac{\beta \lambda R_{oT}}{v l_a}$$

then the azimuth resolution is

$$r_a = l_a \left[1 + \frac{R_{oT}}{R_{oRt}} - \frac{X^2 R_{oT}}{R_{oRt}^3}\right]^{-1}$$

$$r_a = l_a \left[1 + \frac{R_{oT} R_{oR}^2}{R_{oRt}^3}\right]^{-1} \tag{7.18}$$

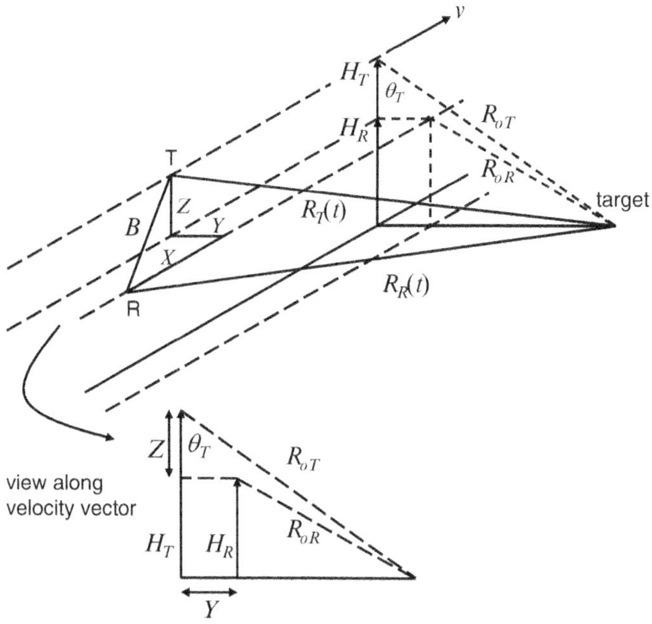

Fig. 7.12. Bistatic SAR using parallel orbits but with an arbitrarily inclined baseline: as in previous diagrams the along track dimensions are exaggerated for clarity; the beamwidth of the transmitting antenna, which has no squint, is illuminating the target

Now consider some special cases.

1. For monostatic SAR $R_{oRt}^2 = R_{oR}^2 = R_{oT}^2 = R_o^2$ so that $r_a = l_a/2$.
2. For cross track bistatic SAR X=0, Z=0 so that Y=B and R_{oRt}=R_{oR}. From (7.18) that gives

$$r_a = l_a \left[1 + \frac{R_{oT}}{R_{oR}}\right]^{-1}$$

which is the same as (7.6).

3. For along track bistatic SAR Y=0, Z=0 so that X=B and, from (7.16), R_{oR}=R_{oT}. Thus (7.18) becomes

$$r_a = l_a\left[1+\frac{R_{oT}^3}{R_{oRt}^3}\right]^{-1}$$

which is the same as (7.13).

7.4 The General Bistatic Configuration

In the analyses of Sect. 7.3 we have assumed the two platforms travel on parallel paths. While that will often approximate spacecraft bistatic radar systems, configurations based on aircraft or aircraft/spacecraft combinations often have the velocity vectors of the platforms inclined to each other[8]. Analysis sometimes uses a vector-based approach to avoid too much mathematical notational complexity. The processing of the SAR echoes to produce imagery via range and azimuth compression is also more complex, particularly the steps used to compensate for range migration[9].

Notwithstanding these complexities it is still possible to set up some general expressions for the resolution cell properties, although actual results may need computational solutions. To do so requires us to generalise the way we look at range and azimuth resolutions. Consider range resolution first. Recall that what a radar resolves is a difference between two targets (or pixels) by being able to separate their echoes in the slant direction; if the echoes are separated by more than the width of the (compressed) ranging pulse then the targets are resolvable. The limit of slant range resolution is given when the echo separation is no smaller than the pulse width. We thus write the slant range resolution as

$$r_r = c\tau = \frac{c}{\mathcal{B}_c}$$

in which τ is the compressed pulse width and $\mathcal{B}_c$ is the chirp bandwidth. Note that there is no "2" in the denominator here as there is in (3.1) and (3.5a). That is because we have said nothing about the slant path folding back on itself, as it does for monostatic radar. Instead the situation is now as depicted in Fig. 7.13. As radar users we are interested in resolving detail on the ground – shown as the x dimension in Fig. 7.13. Therefore we have to establish a general relationship between ground range resolution and slant range resolution – which means we have to understand how a change in ground coordinate x shows up as a change in slant range R. In other words we need to know $\frac{dR}{dx}$ for the general radar configuration. For the simple monostatic case of Fig. 3.6 we can see that

---

[8] See for example P. Dubois-Fernandez, H. Cantalloube, B. Vaizan, G. Krieger and A. Moreira, Chapter 5 Airborne Bistatic Synthetic Aperture Radar, in M. Cherniakov (ed), *Bistatic Radar Emerging Technology*, John Wiley and Sons, Chichester, 2008, and M. Antoniou, R. Saini and M. Cherniakov, Results of a space-surface bistatic SAR image formation algorithm, *IEEE Transactions on Geoscience and Remote Sensing*, vol. 45, no. 11, pt. 1, November 2007, pp. 3359-3371.

[9] See F.H. Wong, I.G. Cumming and Y.L. Neo, Focussing bistatic SAR data using the nonlinear chirp scaling algorithm, *IEEE Transactions on Geoscience and Remote Sensing*, vol. 46, no. 9, September 2008, pp. 2493-2505.

$\frac{dr_r}{dr_g} = 2\sin\theta$. In general, if we know $\frac{dR}{dx}$ we can determine the ground range resolution from

$$r_g = \frac{r_r}{dR/dx} = \frac{c}{B_c\, dR/dx} \tag{7.19}$$

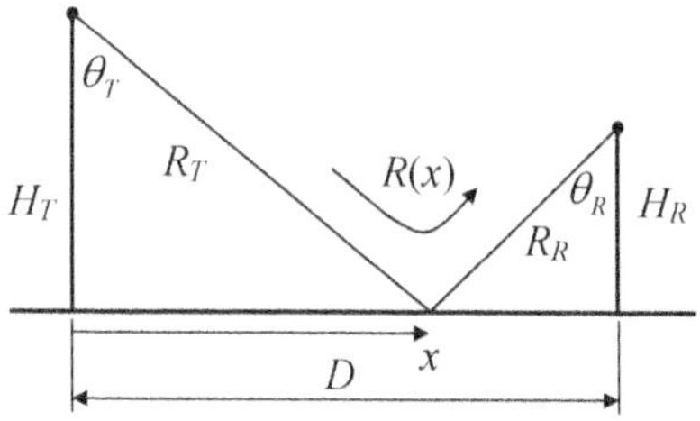

Fig. 7.13. Geometry for demonstrating the application of (7.19)

To apply this expression consider the simple case in Fig. 7.13. We have chosen the origin for the ground coordinate at the transmitter and assumed the received is at a distance D from the transmitter in the same plane as the target. This is just a two sided version of the cross track bistatic configuration considered earlier. We will generalise it shortly. The total slant range is given by

$$R = \sqrt{H_T^2 + x^2} + \sqrt{H_R^2 + (D-x)^2}$$

so that

$$\frac{dR}{dx} = \frac{x}{\sqrt{H_T^2 + x^2}} - \frac{D-x}{\sqrt{H_R^2 + (D-x)^2}} \tag{7.20}$$

which is just

$$\frac{dR}{dx} = \sin\theta_T - \sin\theta_R \tag{7.21}$$

When substituted into (7.19) that gives the required expression for ground range resolution. Note that if $H_T = H_R = H$ and $x = D/2$ then from (7.20) $dR/dx = 0$, so that there is no range resolution as we also noted in association with (7.2).

We are now in the position to consider the general bistatic case shown in Fig. 7.14 in which the transmitter and receiver are on arbitrarily inclined tracks at different altitudes and with different velocities. Again we choose a Cartesian rectangular coordinate system with the origin at the instantaneous transmitter position.

The total slant range from transmitter to target to receiver, as a function of target and receiver positions relative to the transmitter, is

$$R(x,y) = \sqrt{x^2 + y^2 + H_T^2} + \sqrt{(X_R - x)^2 + (Y_R - y)^2 + H_R^2} \tag{7.22}$$

We can apply (7.19) to find ground range resolution. The slant range, however, is now a function of both x and y so we need to find its incremental dependence on both – essentially we look for its differential with respect to both x and y together. Fortunately, there is a very helpful construct in the field of analytical geometry that we can use here – it is called the gradient operator. It allows us to differentiate a function of several independent variables to generate a vector for the gradient of the function. The gradient operator is represented by the symbol ∇ and defined, in three dimensions, as

$$\nabla f(x, y, z) = \frac{\partial f}{\partial x}\mathbf{i} + \frac{\partial f}{\partial y}\mathbf{j} + \frac{\partial f}{\partial z}\mathbf{k}$$

in which **i**, **j** and **k** are unit vectors that point in the x, y and z coordinate directions respectively; ∇ is usually pronounced "grad" or "del" although occasionally the older term "nabla" is used.

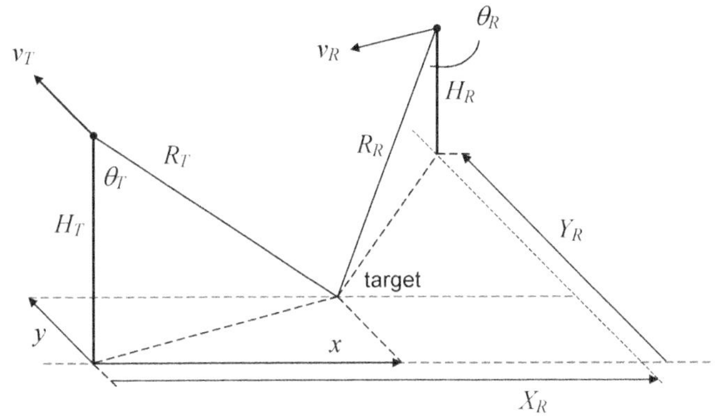

Fig. 7.14. General bistatic geometry

In our two dimensional case the gradient is simply

$$\nabla R(x, y) = \frac{\partial R}{\partial x}\mathbf{i} + \frac{\partial R}{\partial y}\mathbf{j}$$

The significance of having a vector expression for gradient is that its direction in space tells us the direction in which the change is greatest and the amount of change is the magnitude of the vector. In our analysis we are not so much concerned about the direction in which the change occurs as we are in its magnitude, which is given in the usual way for vectors as

$$|\nabla R(x, y)| = \sqrt{\left(\frac{\partial R}{\partial x}\right)^2 + \left(\frac{\partial R}{\partial y}\right)^2} \tag{7.23}$$

The ground range resolution expression of (7.19) then generalises[10] to

[10] G. Krieger and A. Moreira, Spaceborne bi- and multistatic SAR: potential and challenges, *IEE Proceedings on Radar, Sonar and Navigation*, vol. 153, no. 3, June 2006 give a more general form of this expression which incorporates the directional information from the gradient operator.

$$r_g = \frac{c}{\mathcal{B}_c \left|\nabla R(x,y)\right|} \tag{7.24}$$

Now from (7.22)

$$\frac{\partial R(x,y)}{\partial x} = \frac{x}{\sqrt{x^2+y^2+H_T^2}} - \frac{(X_R - x)}{\sqrt{(X_R-x)^2+(Y_R-y)^2+H_R^2}} \tag{7.25a}$$

$$\frac{\partial R(x,y)}{\partial y} = \frac{y}{\sqrt{x^2+y^2+H_T^2}} - \frac{(Y_R - y)}{\sqrt{(X_R-x)^2+(Y_R-y)^2+H_R^2}} \tag{7.25b}$$

Therefore, knowing the relative positions of the transmitter, receiver and target the ground range resolution can be calculated by substituting (7.25) into (7.23) and then into (7.24). Note that if $y=Y_R=0$ – i.e. the coplanar situation – this will give the same result as (7.21b). It will also reduce to the special cases of across track and tandem SAR with the appropriate choices of angles and baseline components.

We now turn to the azimuth resolution for the general bistatic configuration. As with range resolution we first determine a general expression. From (7.9) we see that the Doppler frequency component added to the radar centre frequency can be written

$$f_d(t) = \frac{v(t)}{\lambda} \tag{7.26}$$

in which $v(t)$ is the relative platform velocity *in the direction of signal travel*. It is written explicitly as a time-varying function for generality. Recall that we find azimuth resolution by noting that the Doppler shift component induces a chirp-like characteristic onto the radar signal. By correlating that received chirp against a replica, the resulting compressed pulse width determines the azimuth resolution in time. Multiplication of this result by the receiver platform velocity turns that into an expression in distance. The compressed pulse width is the reciprocal of the chirp bandwidth, which is the Doppler rate about $t=0$ (the rate at which the frequency changes in the middle of the induced chirp as the transmitter passes the target) multiplied by the duration of the chirp. Turning this word expression into a formula gives for azimuth resolution

$$r_a = \frac{v_R}{\left.\frac{\partial f_d(t)}{\partial t}\right|_{t=0} T_a} \tag{7.27}$$

In radar theory T_a, the chirp duration, is often called the receiver *coherent integration time*. We can develop an expression for T_a by assuming that the received azimuth chirp duration is set by the time that a point target is illuminated by the transmitted beam. In a sense this assumes that the receiver beamwidth on the ground is larger than that of the transmitter. By doing this the expression we derive for azimuth resolution can be reduced to special cases that we are familiar with. As in Sect. 3.6 T_a can be expressed in terms of the beamwidth of the transmitting antenna, the slant distance from the transmitter to the target and the velocity of the transmitter platform. The transmitting antenna's beamwidth is given by its real length in the direction of travel divided by the operating wavelength, so that

$$T_a = \frac{L_a}{v_T} = \frac{\lambda R_{oT}}{l_a v_T}$$

in which we assume that the transmitter slant range R_T does not vary greatly over the time T_a and can be represented by its value at t=0. The azimuth resolution expression of (7.27) then becomes

$$r_a = \frac{l_a v_R v_T}{\lambda R_{oT} \left. \frac{\partial f_d(t)}{\partial t} \right|_{t=0}}$$

Substituting from (7.26) this gives

$$r_a = \frac{l_a v_R v_T}{R_{oT} \left. \frac{\partial v(t)}{\partial t} \right|_{t=0}} \tag{7.28}$$

We now need to evaluate $v(t)$, the velocity component that gives rise to the Doppler shift of the radar carrier frequency. A little thought will show that it is the sum of the component of the transmitter platform velocity in the slant range direction and the receiver platform velocity component in its slant range direction. That is because the signal reaching and scattering from the ground is Doppler shifted by the transmitter platform motion. When it is received it then has an additional Doppler component added because of the receiver platform motion.

To find the components of velocity in the slant range directions we will concentrate just on the receiver situation since the results apply in general. Fig. 7.15 shows the receiver slant ray to the target and the receiver velocity vector; the orientation of the latter is shown described by three angles: the platform bearing with respect to the horizontal orientation of the slant ray ψ_R, the receiver platform orbital elevation angle ζ_R and the receiver instantaneous incidence angle θ_R. The last two could be replaced by a single angle but it suits our purposes here to keep them separate, as we will see. From Fig. 7.15 we note that the component of the velocity of the receiver platform in its slant range direction is

$$v_R^c = v_R \cos\zeta_R \cos\psi_R \sin\theta_R$$

Similarly the component of the transmitter platform velocity in its slant range direction will be

$$v_T^c = v_T \cos\zeta_T \cos\psi_T \sin\theta_T$$

Thus the velocity rate in (7.28) is

$$\frac{\partial v(t)}{\partial t} = \frac{\partial}{\partial t}\{v_R \cos\zeta_R \cos\psi_R \sin\theta_R + v_T \cos\zeta_T \cos\psi_T \sin\theta_T\} \tag{7.29}$$

The platform velocities are constants but each of the angles varies with platform-target relative motion and is thus a function of time. It is reasonable to assume that over the time of the synthetic aperture T_a, the angles only change slightly about their nominal values at a given instant of time, so that the trigonometric functions can be approximated, if required, by the linear terms of a Taylor series of the form

$$\cos\alpha = \cos\alpha_o - (\alpha - \alpha_o)\sin\alpha_o \tag{7.30a}$$

$$\sin\alpha = \sin\alpha_o + (\alpha - \alpha_o)\cos\alpha_o \tag{7.30b}$$

in which α_o is the nominal angle about which the trig functions are approximated. The derivatives in (7.29) then reduce to simple derivatives of the angles at t=0. They could be approximated from platform ephemeris data.

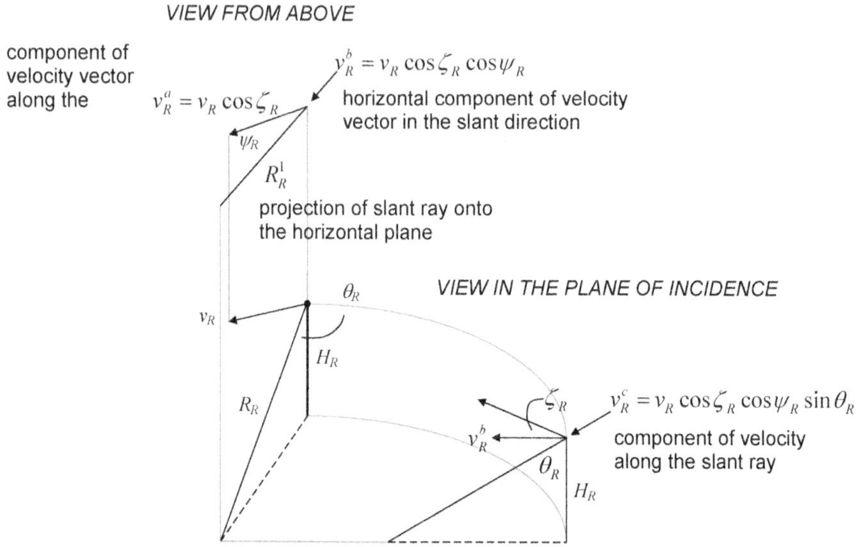

Fig. 7.15. Finding the component of platform velocity in the slant range direction

Consider now the special case of simple across track bistatic SAR shown in Fig. 7.8. For this we have $\zeta_T=\zeta_R=0$, $v_T=v_R=v$=constant. Also ψ_T and ψ_R are close to 90° over the length of the synthetic aperture (the coherent integration period). In addition, the incidence and scattering angles are constant for this arrangement. Fig. 7.16a shows the configuration with the relevant angles and distances defined, from which we see

$$\cos\psi_T = \frac{vt}{R_{oTg}} \text{ and } \cos\psi_R = \frac{vt}{R_{oRg}}$$

The subscripts on the distances refer to the slant ranges at t=0 (nominally transmitter broadside) projected onto the ground plane. With these expressions (7.29) for this configuration becomes

$$\frac{\partial v(t)}{\partial t} = v\frac{\partial}{\partial t}\{\cos\psi_R\sin\theta_R + \cos\psi_T\sin\theta_T\}$$

$$= v\frac{\partial}{\partial t}\{\frac{vt}{R_{oR}} + \frac{vt}{R_{oT}}\}$$

since v is constant and $R_{oRg} = R_{oR}\sin\theta_R$, $R_{oTg} = R_{oT}\sin\theta_T$.

Now
$$\frac{\partial}{\partial t}\frac{vt}{R_{oT}} = \frac{v}{R_{oT}} + vt\frac{\partial}{\partial t}R_{oT}^{-1}$$

so that
$$\left.\frac{\partial}{\partial t}\frac{vt}{R_{oT}}\right|_{t=0} = \frac{v}{R_{oT}}$$

Similarly
$$\left.\frac{\partial}{\partial t}\frac{vt}{R_{oR}}\right|_{t=0} = \frac{v}{R_{oR}}$$

Thus
$$\left.\frac{\partial v(t)}{\partial t}\right|_{t=0} = v^2\left(\frac{1}{R_{oR}} + \frac{1}{R_{oT}}\right)$$

which, on substituting into (7.28) gives the azimuth resolution as

$$r_a = \frac{l_a R_{oR}}{R_{oT} + R_{oR}}$$

This is the same as (7.6).

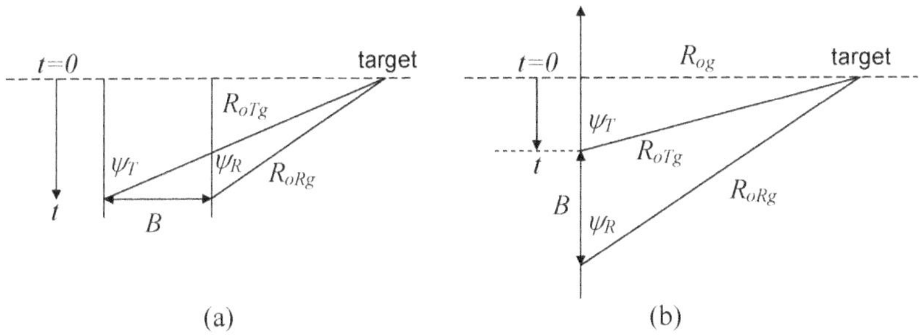

Fig. 7.16. Horizontal plane projections of (a) across track and (b) tandem bistatic SAR

Consider now the special case of the tandem (along track) configuration) shown in Fig. 7.9; however, instead of a slant plane view, Fig. 7.16b again shows the view from above – i.e. projected onto the horizontal plane. For this configuration $\zeta_T=\zeta_R=0$, $v_T=v_R=v=$constant as for the across track situation. From the figure we see

$$\cos\psi_T = \frac{vt}{R_{oTg}} \text{ and } \cos\psi_R = \frac{vt+B}{R_{oRg}}$$

As before
$$\frac{\partial v(t)}{\partial t} = v\frac{\partial}{\partial t}\{\cos\psi_R\sin\theta_R + \cos\psi_T\sin\theta_T\}$$

$$= v\frac{\partial}{\partial t}\{\frac{vt}{R_{oR}} + \frac{vt}{R_{oT}}\} \tag{7.31}$$

and
$$\frac{\partial}{\partial t}\cos\psi_T\bigg|_{t=0} = \frac{\partial}{\partial t}\frac{vt}{R_{oT}}\bigg|_{t=0} = \frac{v}{R_{oT}} \tag{7.32a}$$

However now
$$\frac{\partial}{\partial t}\cos\psi_R\bigg|_{t=0} = \frac{\partial}{\partial t}\frac{(vt+B)}{R_{oR}}\bigg|_{t=0} = \frac{v}{R_{oRt}} + (vt+B)\frac{\partial}{\partial t}R_{oR}^{-1}\bigg|_{t=0}$$

in which $R_{oRt} = \sqrt{R_{oT}^2 + B^2}$ is the slant range to the receiver at t=0 – i.e. transmitter broadside, in which R_{oT} is the transmitter slant range at broadside (see Fig. 7.10). Note also that the receiver slant range at general time t is given by $R_{oR}^2 = R_{oT}^2 + (vt+B)^2$. Thus

$$\frac{\partial}{\partial t}R_{oR}^{-1} = -v(vt+B)\left[R_o^2 + (vt+B)^2\right]^{-3/2}$$

so that
$$\frac{\partial}{\partial t}R_{oR}^{-1}\bigg|_{t=0} = -vB\left[R_o^2 + B^2\right]^{-3/2} = -vBR_{oRt}^{-3}$$

Thus
$$\frac{\partial}{\partial t}\cos\psi_R\bigg|_{t=0} = \frac{v}{R_{oRt}} - \frac{vB^2}{R_{oRt}^3} \tag{7.32b}$$

Substituting (7.32a,b) into (7.31) gives from (7.28)

$$r_a = l_a\left[1 + \frac{R_{oT}}{R_{oRt}} - \frac{R_{oT}B^2}{R_{oRt}^3}\right]^{-1} = l_a\left[1 + \frac{R_{oT}^3}{R_{oRt}^3}\right]^{-1}$$

which is the same as (7.13).

7.5 Other Bistatic Configurations

Bistatic SAR in remote sensing is still relatively new and no purpose-designed configurations are yet in operation. Some innovative arrangements have been proposed two of which are outlined here.

The TanDEM-X mission[11], consisting of two almost identical satellites based on the TerraSAR-X platform, both of which can transmit and receive, is primarily an interferometer designed to provide highly accurate global digital elevation models. The configuration is also intended to provide a versatile bistatic radar for general, polarimetric remote sensing studies. The satellites are planned for operation in several flight arrangements including (i) each working as a single monostatic radar platform, (ii) a fundamental bistatic configuration in which one platform transmits and both receive, and

[11] See G. Kreiger, A. Moreira, H. Fielder, I. Hajnsek, M. Werner, M. Younis and M. Zink, TanDEM-X: A satellite formation for high-resolution SAR interferometry, *IEEE Transactions on Geoscience and Remote Sensing*, vol. 45, no. 11, pt. 1, November 2007, pp. 3317-3341.

(iii) a so-called alternating bistatic mode in which the platforms alternate as transmitters with both receiving.

The interferometric cartwheel is another interesting arrangement[12], in which an existing radar satellite is used as the transmitter with reception taking place on a number of smaller receive-only satellites (microsatellites) arranged in a vertical rotating elliptical orbital configuration flying ahead of the transmitter in the same principal orbit. This is depicted in Fig. 7.17, using three receiving satellites. Interferograms are produced by interfering the images received from any two of the microsatellites. The elliptical microsatellite orbit has a semi-major axis twice the size of the semi-minor axis. With the three satellites equally spaced around the ellipse the vertical baseline formed between the two best placed microsatellites at any given time does not vary by more than 7.5% from its mean value even though they are orbiting. While the constellation (essentially a multistatic radar) is designed principally for topographic mapping (because of the vertical baseline) an effective horizontal baseline is created at the same time so that the cartwheel could also be applied to mapping change.

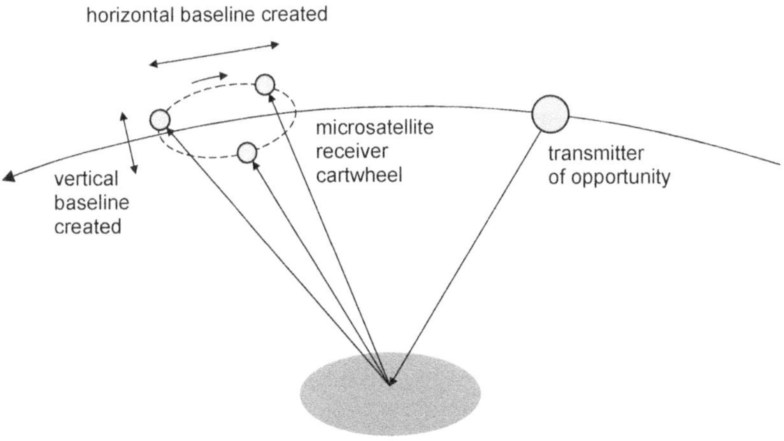

Fig. 7.17. The cartwheel concept in which a radar satellite designed for another purpose is used as a transmitter and there is a set of microsatellites in an elliptical sub-orbit acting as interferometric receivers

7.6 The Need for Transmitter-Receiver Synchronisation

With monostatic radar the transmitter and receiver, being on the same platform, can share signals, so that the time of transmission of the transmitted ranging chirps is known when needing to measure echo delay times. In bistatic radar there is usually a direct path from the transmitter to the receiver, as well as the signal scattered from the target, so that the receiver is aware of what has been transmitted. That requires the transmitter to be in line of sight to the receiver, unless they communicate via some other form of communications link.

[12] See D. Massonnet, Capabilities and limitations of the interferometric cartwheel, *IEEE Transactions on Geoscience and Remote Sensing*, vol. 39, no.3, March 2001, pp. 506-520.

7.7 Using Transmitters of Opportunity

An attractive form of bistatic radar is when the transmitter is some other pre-existing generator of electromagnetic energy at the wavelength of interest. Such a source could be a broadcast or communications satellite, one or several navigation satellites or even mobile phone towers and other forms of terrestrial transmitter[13]. Whether they are entirely suitable or not depends on the application. For remote sensing purposes navigation satellites are among the most suitable sources.

Optical remote sensing using the sun as an energy or illumination source is referred to as passive. Similarly, bistatic radar based on a source or illuminator of opportunity is called *passive bistatic radar*. In non-remote sensing applications a radar system using a source of opportunity is sometimes called *passive coherent location* or *hitchhiking*.

Using a transmitter of opportunity is a low cost option because the transmitter does not have to be provided explicitly. However, the power density produced at the earth's surface (for scattering) can be about 80dB below that produced by a typical remote sensing radar. That would suggest that rather larger resolution cells are required in order to gather sufficient signal for detection at the receiver; alternatively the system will not respond to weaker targets.

Synchronisation is a particular challenge when using a non-cooperative transmitter, especially if it has not been designed with radar-like purposes in mind. That is why GNSS navigation satellites are so attractive[14]. By their very design they transmit ranging signals that are used in trilateration to allow a target to locate itself. More than that, the radiated signals carry information on the time at which they were transmitted along with the satellite location, so that the receiver knows when the signal was sent and from where. Knowing when it is received allows the target to position itself on a spherical equidistant range line from the satellite. By receiving the signals from several satellites in this manner the target is able to locate itself in three dimensions and time, as illustrated in Fig. 7.18. Because time is such a crucial element in making a GNSS system such as GPS work effectively, three levels of clock are involved. The GPS receiver (target) contains a clock. There are more precise clocks on each of the satellites, and there is a highly precise clock at the GPS master station in Colorado Springs. The clocks on the satellites are regularly calibrated from the master clock. Any errors in the receiver clocks are compensated for in the algorithms used to process the received signals. At least four satellites are required for that operation.

There are nominally 24 satellites in the GPS constellation in a bird cage of orbits at an altitude of 20,200km. They are so arranged that at least six satellites are visible at any time from almost all parts of the earth's surface. They transmit their ranging signals on two L band frequencies ~1.575GHz and ~1.228GHz, right in the range of interest in radar remote sensing. Those signals are modulated onto a random set of binary digits called a pseudorandom sequence, unique to a particular satellite. In the GPS receiver the sequences are correlated against versions stored locally. Before correlation the sequences are 1ms in duration; after correlation they have been effectively compressed to 1μs. By

[13] See H.D Griffiths, From a different perspective: principles, practice and potential of bistatic radar, *Proc. International Conference on Radar 2003*, Adelaide, 3-5 Sept., 2003 or H.D. Griffiths, Bistatic and multistatic radar, *IEE Conference on Military Radar*, Shrivenham, 7 September 2004

[14] See T. Lindgren and D.M. Akos, A multistatic synthetic aperture radar for surface characterisation, *IEEE Transactions on Geoscience and Remote Sensing*, vol. 46, no. 8, August 2008, pp. 2249-2253.

using different binary sequences at each transmitter the receivers can distinguish among the satellites and thus the signals received from each.

Geosynchronous weather and telecommunications satellites can also be used as illuminators[15]. As with other passive systems the transmitter power is limited; moreover, since the transmitter does not move relative to the target, the azimuth resolution depends only of the Doppler rate established by the moving receiver. Although not passive, a multistatic radar network has been proposed using geostationary radar transmitters and orbiting receivers to improve imaging coverage[16].

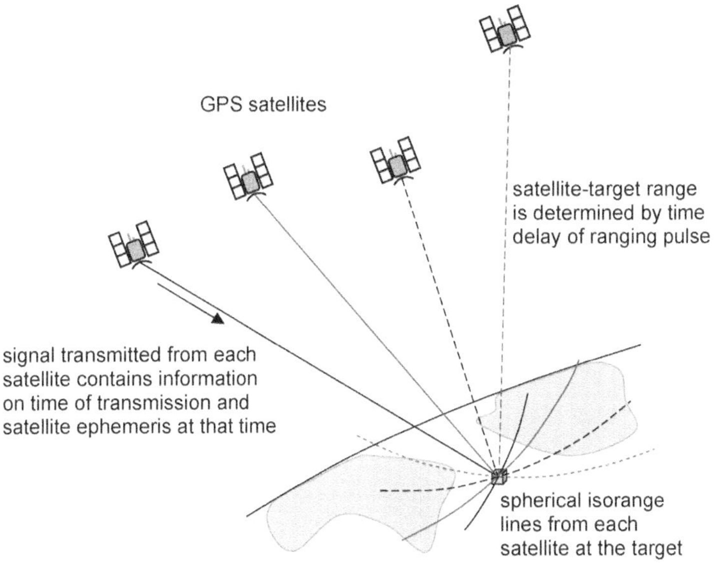

Fig. 7.18. The trilateration principle used by a receiver to determine its position from four GPS satellites

7.8 Geometric Distortion and Shadowing with Bistatic Radar

Because monostatic radar resolves in the slant plane, terrain altitude variations lead to geometric distortion particularly in the ground range direction as outlined in Sect. 4.1. The case is similar with bistatic radar, except the concept of the slant plane is a little more complex. To help visualise the situation it is useful to introduce the idea of isorange contours – they are the loci of points out from the radar in which the two way range from transmitter to target to receiver are constant. For monostatic radar they are circles, or strictly spheres in three dimensions. Based on the definition of an ellipse the isorange contours for bistatic radar (for which the sum of the transmitter-target and target-receiver

---

[15] See G. Krieger and A. Moreira, Spaceborne bi- and multistatic SAR: potential and challenges, *IEE Proceedings on Radar, Sonar and Navigation*, vol. 153, no. 3, June 2006

[16] See K. Sarabandi, J. Kellndorfer and L. Pierce, GLORIA: Geostationary/Low-Earth Orbiting Radar Image Acquisition System: A Multistatic GEO/LEO Synthetic Aperture Radar Satellite Constellation for Earth Observation, *Proceedings of the IEEE International Geoscience and Remote Sensing Symposium* (*IGARSS03*), Toulouse, 21-25 July 2003, vol. 2, pp. 733-775.

distances is constant) are ellipsoids in three dimensions, with the transmitter and receiver positions defining the foci.

For convenience we will concentrate on the across track baseline configuration so that we can consider the isorange ellipses in just the two dimensions of elevation and range as shown in Fig. 7.19. Clearly, if the baseline is small the ellipses are near circular so that the usual distortions of layover and relief displacement are similar to those for monostatic radar. For larger baselines the distortions will affected by the eccentricity of the isorange contours. Nevertheless it can be appreciated that the fundamental nature of layover and terrain relief distortions will be the same. Layover will also occur for along track bistatic radar, although to examine the likely situation the full isorange ellipsoid requires consideration.

Both the transmitter and receiver in bistatic radar will project shadows, as depicted in Fig. 7.20. The transmitter shadow is strictly the only true shadow; the shadow referred to for the receiver is actually a region of terrain from which the receiver cannot receive scattered radiation because the target occludes that area.

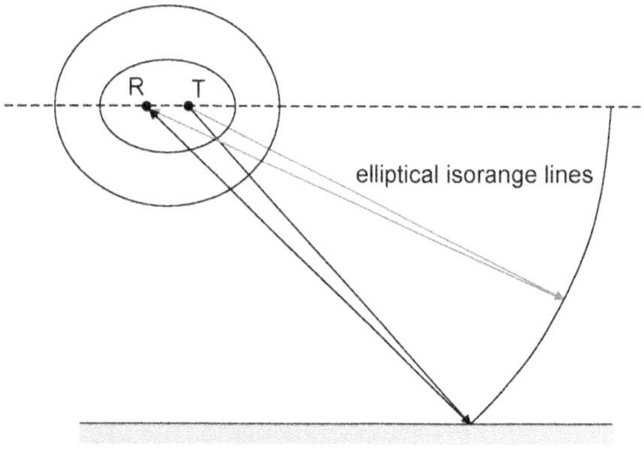

Fig. 7.19. Elliptical lines of constant range measured along the transmit and receive slant directions

7.9 Remote Sensing Benefits of Bistatic Radar

Although there is still the prospect of glint with bistatic radar – i.e. forward specular reflection of the transmitted signal in the direction of the receiver – there is much less opportunity for corner reflector like strong scattering. The dynamic range of an image is likely therefore to be smaller than for monostatic SAR. It has also been suggested that forward bistatic scattering from soils is less affected by surface roughness making it a better technology than monostatic SAR for assessing soil moisture[17].

Figure 7.21 shows X band monostatic and bistatic images of an urban region in which the more uniform dynamic range is evident. It was recorded in an experiment carried out by the Microwaves and Radar Institute of the German Aerospace Center (DLR) using an airborne receiver (F-SAR) with TerraSAR-X as the transmitter. Because of strong scattering the buildings in the monostatic image are very bright to the point that detail is

[17] See D. Masters, P. Axelrad and S. Katzberg, Initial results of land-reflected GPS bistatic radar measurements in SMEX02, *Remote Sensing of Environment*, vol, 92, 2004, pp. 507-520.

obscured. Similarly bright linear features, thought to be perimeter fences along the road running from the bottom centre to the right centre of the image and around one of the buildings, appear in the monostatic image but not the bistatic image. Likewise an array of solar panels on the bottom left hand part of the images is much brighter in the monostatic image than the bistatic one.

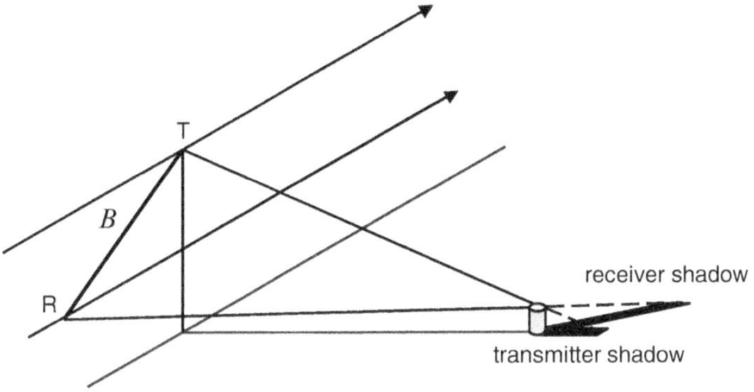

Fig. 7.20. Shadowing in bistatic radar

7.10 Bistatic Scattering

The definitions of radar cross section and scattering coefficient in Chapt. 3 involved only one system angle – the incidence angle. While recognising that scattering properties are also azimuthally dependent we usually ignore that in monostatic scattering. In bistatic scattering the situation is much more complex. First, the incidence and scattering angles are different. Also, if the transmitter and receiver platforms follow trajectories that are not parallel the illumination and scattering pathways and thus those angles generally change from pixel to pixel. Perhaps the best that can be said it that each situation will need to be examined afresh, especially since, almost by definition, there are no configuration or topological conventions established.

When we are interested in multipolarisation radar we need also to be careful about the coordinate system we choose to describe ray propagation and to define the scattering matrix of a target. We foreshadowed that concern in Sect. 3.17 and Appendix E by noting that two coordinate conventions are in use for radar – one better suited to monostatic situations and one for bistatic and multistatic arrangements. The former is the so-called backscatter alignment or antenna coordinate convention whereas the latter is the forward scatter alignment or wave coordinate convention.

Fig. 7.22 shows bistatic scattering from a target in the forward scattering alignment coordinates – in other words the orientations of the components of the field are consistent with the propagation direction both before and after scattering. Whereas in backscattering the scattered and incident field components are related by a scattering matrix called the Sinclair matrix, in the forward scattering convention the matrix is called the Jones matrix **T**:

$$\begin{bmatrix} E_H^s \\ E_V^s \end{bmatrix} = \begin{bmatrix} T_{HH} & T_{HV} \\ T_{VH} & T_{VV} \end{bmatrix} \begin{bmatrix} E_H^i \\ E_V^i \end{bmatrix} = \mathbf{T} \begin{bmatrix} E_H^i \\ E_V^i \end{bmatrix} \tag{7.33}$$

Strictly this expression only applies in the far field of the scatterer as we discussed also in the case of backscattering; it is more correct theoretically to write the Jones matrix in terms of the field received:

$$\begin{bmatrix} E_H^r \\ E_V^r \end{bmatrix} = \frac{e^{j\beta R}}{R}\mathbf{T}\begin{bmatrix} E_H^i \\ E_V^i \end{bmatrix}$$

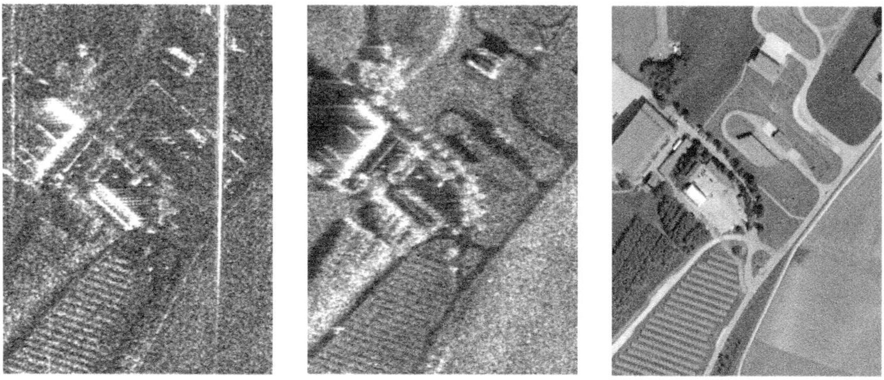

(a) (b) (c)

Fig. 7.21. (a) Monostatic (b) bistatic and air photo images demonstrating the more subdued dynamic range possible with bistatic imaging (from M. Rodr´ıguez-Cassol`a, S.V. Baumgartner, G. Krieger, A. Nottensteiner, R. Horn, U. Steinbrecher, R. Metzig, M. Limbach, P. Prats, J. Fischer, M. Schwerdt, A. Moreira, Bistatic spaceborne airborne experiment TerraSAR-X/F-SAR: data processing and results, *Proceedings of the International Geoscience and Remote Sensing Symposium 2008* (*IGARSS08*), vol. 3, Boston, 7-11 July 2008, pp. 451-454, ©2008 IEEE)

The Jones matrix for a scatterer is related to its Sinclair matrix by (E.1). Unlike the case for backscattering the Jones matrix is not symmetric – i.e. $T_{HV} \neq T_{VH}$ in general. That means that the target vector definitions of (3.45) and (3.48) apply and that the covariance and coherency matrices of Sect. 3.19 are four dimensional.

As with monostatic radar the Jones matrix is only good as a target descriptor when there is no unpolarised component of the scattered radiation. Again, it is better to describe the radiation in terms of its Stokes vector since then polarised, partially polarised and unpolarised situations can be handled. For backscattering the Stokes vectors were related by the Kennaugh matrix – see (3.60). In bistatic scattering they are related by the 4x4 Mueller matrix[18] $\mathcal{H}$:

[18] Note that there is confusion with the definitions of Mueller and Kennaugh matrices. See, for example, W-M Boerner, H. Mott, E. Luneberg, C. Livingstone, B. Brisco, R.J. Brown and J.S. Patterson, Polarimetry in Radar Remote Sensing: basic and applied concepts, in F.M. Henderson and A.J. Lewis, Principles and Applications of Imaging Radar, vol. 2, *Manual of Remote Sensing*, 3rd Ed., John Wiley and Sons, N.Y., 1998. In some treatments the Mueller matrix is applied to backscattering while the Kennaugh matrix is applied to forward scattering: see F.T. Ulaby and C. Elachi, *Radar Polarimetry for Geoscience Applications*, Artech House, Norwood Mass., 1990. Often the term Mueller matrix is used generically: see

$$\mathbf{s}^{s} = \mathcal{H}\mathbf{s}^{i} \text{ or } \mathbf{s}^{r} = \frac{1}{R^{2}}\mathcal{H}\mathbf{s}^{i} \tag{7.34}$$

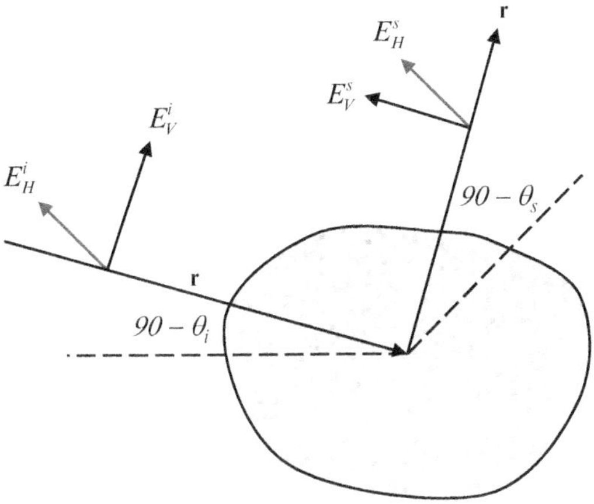

Fig. 7.22. Bistatic scattering in a forward scatter alignment coordinate (wave coordinate) system

I. Woodhouse, Introduction to Microwave Remote Sensing, CRC Taylor and Francis, Boca Raton, Florida, 2006.

Chapter 8

Radar Image Interpretation

8.1 Introduction

The principal goal of remote sensing is to interpret the data recorded in order to understand the region being imaged; interpretation can be based on qualitative or quantitative methods of analysis. Analyst expertise allows qualitative information extraction through photointerpretive methods in which visual clues around structure and contrast are used. With knowledge of the radar scattering behaviours of earth surface features, such as treated in Chapt. 5, the analyst can often make very good assessments of the types of land cover being imaged.

Visual interpretation can be complicated because the scattering mechanisms are very often composite; within an individual pixel, several distinct mechanisms can contribute to backscatter. That does not preclude visual interpretation, but the analyst needs to be critically aware of those complexities if successful results are to be obtained.

In quantitative radar image interpretation we seek to establish and map the most appropriate ground cover type for a resolution cell, or group of resolution cells, using computer-based labelling algorithms. Cells, or pixels, of a particular type can be counted to give quantitative estimates of ground covers, and symbols attached to pixels following interpretation allow thematic maps of the landscape to be generated.

Unlike ground cover type determination with optical image data, in which spectral responses largely characterise absorption and emission of materials on the earth's surface, in the case of radar imaging the properties that determine radar response are mainly related to the geometric nature of features and their moisture contents. Interpretation is therefore often focussed more on structural determination than on properties such as species, mineralogy, vegetation condition and stress such as we have come to associate with remote sensing imaging at optical wavelengths. That is not to say that we cannot differentiate species and condition with radar imaging. To do so, though, usually requires their association with geometric and moisture properties.

As a result of the phase of the radar signals being available it is possible to develop procedures for understanding something of the vertical structure of the landscape within a resolution cell (pixel). While that can assist in pixel labelling, it is not intended as a classification process. Rather it is an analytical technique similar to interferometry and tomography treated in Chapt. 6. In principle, it is a third line of analysis of radar imagery sitting alongside visual interpretation and quantitative thematic mapping.

This chapter reviews the these three approaches to the interpretation of radar imagery.

8.2 Analytical Complexity

With optical imagery recorded at a given time using a particular sensor there is generally only one type of data available for a pixel: that is the set of spectral measurements recorded at the wavelengths with which the sensor samples the landscape.

J.A. Richards, *Remote Sensing with Imaging Radar*, Signals and Communication Technology,
DOI: 10.1007/978-3-642-02020-9_8,

In contrast, with radar a complex set of measurements can be made, as depicted in Fig. 8.1. For each resolution cell backscatter measurements (in both amplitude and phase) can be produced for different polarisations, different wavelengths and potentially different incidence angles[1]. The last is not as common as the first two for a given mission, but sampling at a limited number of incidence angles is certainly feasible in many cases. In bistatic radar we may also have a range of scattering angles.

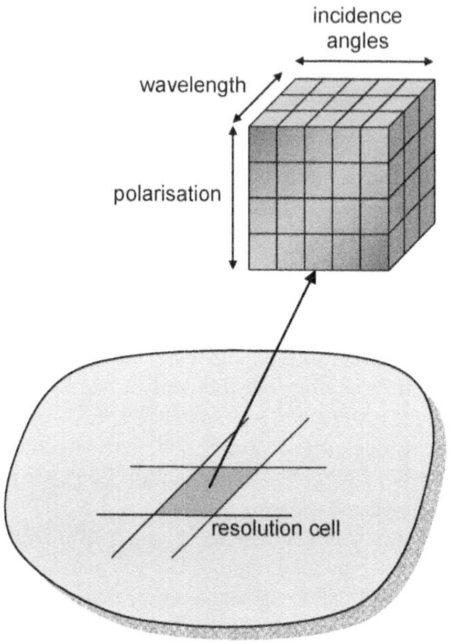

Fig. 8.1. Measurements available for a radar resolution cell

8.3 Visual Interpretation Through an Understanding of Scattering Behaviours

In forming an understanding of the landscape that has been imaged with a remote sensing platform an expert photointerpreter (human analyst) generally makes use of the spatial, temporal and brightness elements evident in the image. Spatial elements refer to shapes, sizes and textures and the recognition of elongate features that indicate roads and drainage systems. Image features that change in time between acquisitions constitute temporal clues. Apart from changes associated with topographic displacements detectable using interferometry, we will not pursue spatial and temporal keys in this chapter. Instead, we will concentrate on pixel brightness.

The brightness of a pixel can vary with any of the measurements indicated in Fig. 8.1. Often the photointerpreter will form an opinion about the region imaged by examining

[1] At the current time the dimensionality of the analysis problem with optical imagery is much greater than with radar data. Hyperspectral sensors generate several hundreds of bands, or features, whereas radar imagers tend to produce no more than about 12 channels of data. There is, however, a greater diversity of types among the features in radar imaging than for the bands in optical data. That can add to the complexity of analysis.

relative brightnesses across polarisations, incidence angles and wavelengths, as well as from position to position in an image. We proceed, therefore, by looking at the information available in each of the three dimensions in Fig. 8.1. We do so by considering the scattering coefficient as a function of incidence angle, wavelength and polarisation and, where possible and appropriate, we explore the scattering matrix.

8.3.1 The Role of Incidence Angle

In Fig. 8.2 we present a stylised set of co-polarised curves for the three principal scattering mechanisms generally encountered in practice – surface scattering (from both smooth and rough surfaces), volume scattering and hard target (dihedral corner reflector) scattering; the latter is reminiscent of tree trunk and urban scattering behaviours. Also shown are the samples of those response curves one would expect from two widely spaced incidence angles – one very low and one mid range. Those samples allow, at least in principle, discrimination among the cover types in a data space defined by the angles chosen.

A feature evident in Fig. 8.2 is that most contrast among cover types is given at the mid range incidence angles. The lower incidence angles are not as good. Fig. 5.5 also shows how poor the smaller incidence angles are for discriminating among surfaces with different degrees of roughness. Layover is also worse at smaller incidence angles as is topographic distortion, as seen in Chapt. 4.

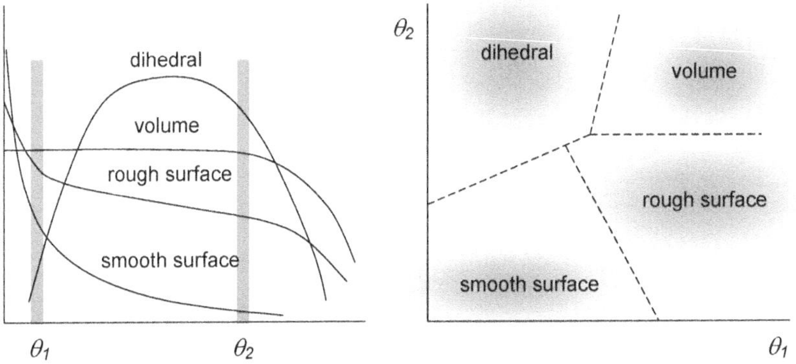

Fig. 8.2. Demonstration of the discrimination possible among the fundamental scattering types using the dependence of their co-polar behaviour on incidence angle; the dashed lines represent decision boundaries that could be used in classification (see Sect. 8.4.3)

Although mid-range angles of incidence are generally best for land based applications smaller incidence angles are to be preferred for sea surface observation, as noted in Sect. 5.7 and Fig. 5.34. An exception to this is if the ocean background has to be minimised to allow features such as ships to be more evident. Mid range angles are then better.

A general conclusion that might be drawn at this stage is that while smaller incidence angles are important for sea and oceanographic applications, mid range angles are more important for the land. Mid range angles also minimise the effect of topographic distortion.

Large angles of incidence are generally avoided because of the longer attenuating paths created through vegetation canopies and because of the greater chance of shadowing.

8.3.2 The Role of Wavelength

When examining the importance of wavelength, initial guidance is provided by the Rayleigh criterion of (5.4). For a surface with a given physical roughness that criterion allows us to assess whether a surface appears as smooth (specular) or rough (diffuse) for radar purposes. If, for illustration, we choose an incidence angle of 35° then a surface is specular for wavelengths in excess of about $7h$, where h is the variation in surface height. Conversely, the surface is more likely to appear rough for any wavelength less than about $7h$. For X band that means any surface will be rough if its variations exceed, say, 0.5cm. Practically, therefore, most natural surfaces will appear rough at X band. By comparison, at L band a surface will be rough if its vertical variations exceed about 3cm. It will behave as a specular surface otherwise. Considering the range of roughness variations encountered naturally it is highly likely that in L band imagery (and perhaps longer) considerably more variation in contrast will be evident for surfaces (soils etc) than for X band. That is also demonstrated in Fig. 5.10.

As demonstrated in Fig. 5.9 soil backscatter increases with increasing moisture content; the sensitivity of that change is higher at longer radar wavelengths than at shorter wavelengths[2].

Volumes will generally appear brighter and be more attenuating at shorter wavelengths. To get a feel for that we can consider the scattering characteristics of the elements that might compose a volume scatterer. If those elements are smaller than a wavelength then we know from the scattering of light in optical remote sensing that they will scatter incident energy according to λ^{-b} in which b has the value of 1 or smaller for Mie scattering from larger particles and 4 for Rayleigh scattering from very small particles[3]. Irrespective of the mechanism, the shorter the wavelength the greater the scattering, including backscattering. If the scattering elements are spherical we can use theoretical results from the scattering of radiation by dielectric spheres as a guide. The normalised backscattering cross section for a small non-magnetic dielectric sphere of radius a is approximately[4]

$$\frac{\sigma}{\pi a^2} = \left(\frac{2\pi a}{\lambda}\right)^4 \left|\frac{\varepsilon_r - 1}{\varepsilon_r + 2}\right|^2$$

in which ε_r is the dielectric constant of the material from which the sphere is composed. This shows that the radar cross section of the sphere increases with decreasing wavelength; it will be bigger at X band than L band meaning that backscattering from a volume composed of spheres will be greater at X band – in other words X band imagery will show greater volume scattering return than L band imagery. Even though this was done for spheres the lesson is the same for other shapes that might make up a volume,

[2] See D. A. Boyarskii, V. V. Tikhonov, and N. Yu. Komarova, Model of dielectric constant of bound water in soil for applications of microwave remote sensing, *Progress In Electromagnetics Research*, PIER vol 35, 2002, pp. 251–269.

[3] See J.R. Schott, *Remote Sensing; the Image Chain Approach*, Oxford University Press, New York, 1997.

[4] See G.T. Ruck, D.E. Barrick, W.D. Stuart and C.K. Krichbaum, *Radar Cross Section Handbook*, Plenum, N.Y., 1970, equation (3.3-7).

with the exception that spherical scatterers will not give any cross-polarisation. Scatterers that are in some way elongate, such as cylinders, ellipsoids or needles (sometimes referred to as dipoles) will generate a cross polarised response as discussed in Sect. 5.4.2.

In the range of wavelengths common to radar remote sensing the dielectric constant of vegetation changes from about 20 to 30 for a change in gravimetric moisture content[5] of 0.6 to 0.8 g/cm^3. Using these figures in the expression for the radar cross section of a dielectric sphere shows an RCS change of about 10%.

If a volume medium exhibits greater backscatter then clearly less energy propagates forward and the medium is seen to be higher in attenuation. Added to this forward propagating energy loss will be loss resulting from absorption in the material from which the scatterers are composed. That component is also wavelength dependent such that absorption is greater at shorter wavelengths. As a consequence volumetric media, such as forest canopies, will be quite opaque at short wavelengths, such as X band, but will be somewhat transparent at the longer wavelengths of L and P bands. Fig. 5.20 demonstrates the greater attenuation resulting from scattering at shorter wavelengths. Similar behaviours are to be expected for cross polarised returns.

The two-bounce dihedral model used to characterise trees, houses and ships at sea discussed in Sect. 5.5.2 is also a stronger scatterer at shorter wavelengths as can be assessed from the inverse wavelength dependence in (5.25-5.27). With forest canopies, however, the highly absorbing foliage at shorter wavelengths, such as X band, means that the double bounce trunk response is diminished in the overall backscatter response. Instead, longer wavelengths, such as L band, minimise canopy loss through absorption yet still provide a strong trunk-ground signal. Of course, if the foliage itself is of interest shorter wavelengths have the benefit that the response will be dominated by the canopy.

If monitoring structures at sea is of interest, including ships and oil platforms, short wavelengths would be preferred along with larger incidence angles. Not only is there no absorbing canopy but the sea response is minimised at the larger angles, as can be assessed from Fig. 5.34. This along with the fact that the radar cross section of the objects of interest is maximised, yields best contrast for visual analysis.

8.3.3 The Role of Polarisation

A very effective means by which to examine the polarisation domain is to use the polarisation synthesis process developed in Sect. 3.22 since through polarisation plots we have a complete summary of how various land covers behave as polarisation is altered.

For surfaces and volumes it is preferable to examine actual data because theoretical models are not completely adequate to explain what is observed in practice; examples are seen in Figs 3.27 and 5.36.

Rough surfaces and most volumes will generate an unpolarised component in their backscatter responses that adds to polarised returns. Those components often don't feature in models. Unpolarised scattering contributes a constant pedestal to polarisation plots as can be seen by returning to (3.74) which we could re-express as

$$\sigma = 4\pi \mathbf{s}^{ra}.\mathbf{s}^{b}$$

[5] See T.J. Schmugge and T.J. Jackson, A dielectric model of the vegetation effects on the microwave emission from soils, *IEEE Transactions on Geoscience and Remote Sensing*, vol. 30, no. 4, July 1992, pp. 757-760.

in which $\mathbf{s}^{ra}$ is the polarisation state of the receiving antenna, expressed as a Stokes vector, $\mathbf{s}^b = \mathbf{M}\mathbf{s}^t$ is the Stokes vector of the scattered wave at the receiving antenna and $\mathbf{M}$ is the Stokes scattering operator. If the scattered radiation is unpolarised then from (2.35) $\mathbf{s}^b=[\text{constant},0,0,0]^T$. The result of the dot product operation in the above expression is then just $4\pi \times$constant, which is independent of the polarisation and orientation (tilt) angles. It adds to any polarised returns so that they sit on a constant value pedestal, the height of which is determined by the relative level of unpolarised signal. As an illustration Fig. 8.3 shows a P band co-polarisation plot for a vegetated region in which there is a significant component of unpolarised radiation, showing as a pedestal.

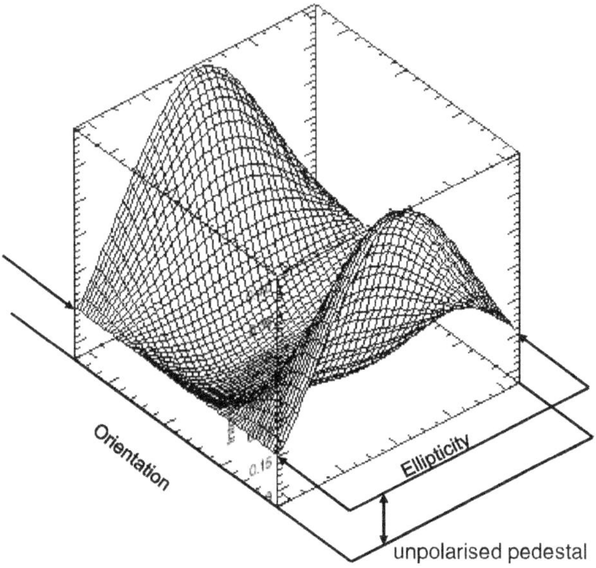

Fig. 8.3. Polarisation plot constructed from a P band AirSAR image of the Mt Gambier region in Australia; derived using ENVI™ (ITT Visual Information Solutions)

The polarisation dependent behaviour of the dihedral mechanism representing trunk scattering described in Sect. 5.5.2 can be found from the plots of Fig. 3.24 which apply to a dihedral corner reflector. For linear polarisation (ellipticity of zero) they show that for a vertically standing structure co-polarised responses are maximised for vertical and horizontal polarisation, as might be expected. There are no corresponding cross-polarised returns. In contrast, for orientation of the incoming linearly polarised field at 45° to the vertical there will be no co-polar response, but maximum cross-polar response.

Sometimes we are interested backscattering from a dihedral structure that is oriented away from the vertical. That can found most easily by rotating the polarisation of the incident waveform by the respective angle, scattering it from the normally oriented reflector and then rotating the scattered field back to align with the incident wave coordinate system. That saves the need to derive directly the scattering matrix for an inclined reflector. As an example, if we were interested in a dihedral at an orientation of 45° the result of those operations is

$$\begin{bmatrix} \cos 45^o & -\sin 45^o \\ \sin 45^o & \cos 45^o \end{bmatrix} \begin{bmatrix} 1 & 0 \\ 0 & -1 \end{bmatrix} \begin{bmatrix} \cos 45^o & \sin 45^o \\ -\sin 45^o & \cos 45^o \end{bmatrix} = \begin{bmatrix} 0 & 1 \\ 1 & 0 \end{bmatrix}$$

The resulting scattering matrix shows no co-polarised (HH and VV) scattering and maximum cross-polarised scattering. Fig. 8.4 shows the polarisation signatures, which should be compared to those of Fig. 3.24.

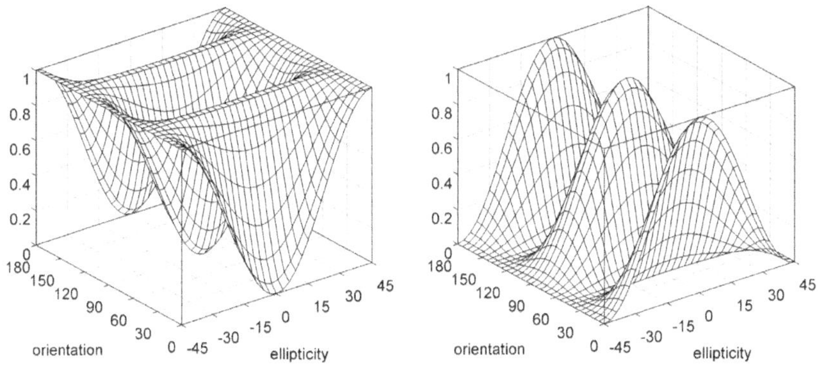

Fig. 8.4. Ideal co- and cross polarised signatures (left and right respectively) for a dihedral corner reflector angled at 45° to the polarisation of the incident wavefront

8.4 Quantitative Analysis of Radar Image Data for Thematic Mapping

8.4.1 Overview of Methods

The essential question in quantitative thematic mapping is: by analysing the recorded radar data how do we identify and label, at the pixel level, the region on the earth's surface being imaged? That, in turn, prompts us to ask what methods are available for such an analysis. Photointerpretive approaches are not practical to apply to the individual pixel because of the huge number of pixels that needs to be analysed and the difficulty a human interpreter has in handling the data at its full level of detail[6]. Computer assisted interpretation – called quantitative analysis in remote sensing – is therefore essential for wide scale thematic mapping.

At the heart of quantitative analysis is the mapping operation illustrated in Fig. 8.5. Mathematical or statistical models are developed that characterise the classes of interest to the user; those models are used to attach labels to each of the resolution cells. Usually the forms of the models are assumed and any necessary parameters are estimated by using previously labelled pixels – so-called training data. The most common model is that which assumes that the classes can be represented by multidimensional normal distributions. The training data is employed to estimate the mean and covariance parameters. That is called a supervised learning method. Unsupervised labelling processes, often based on clustering algorithms, are also used[7].

[6] See J.A. Richards and X. Jia, *Remote Sensing Digital Image Analysis*, 4th ed., Springer, Berlin, 2006.
[7] *ibid.*

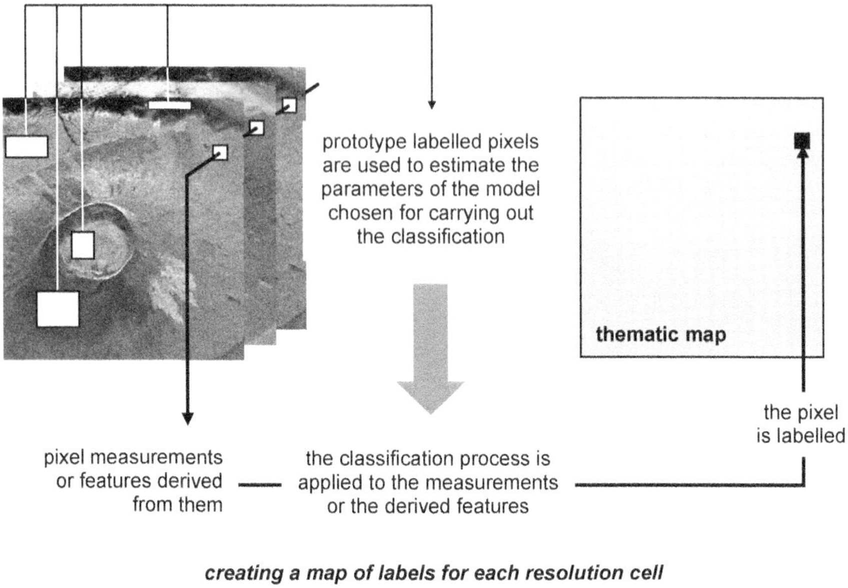

Fig. 8.5. Classification as a mapping from measurements to labels

Any of the classification procedures commonly used with optical remote sensing image data can be applied to radar imagery on the assumption that the classes of interest are able to be resolved well enough with the radar measurements available. As with optical data, it is not always the case that the classes or clusters able to be delineated in radar imagery will naturally map to the classes of interest to the user. The classes identifiable in the data are those which represent similarities of radar measurements – in optical remote sensing they are generally called *data classes*. It is unrealistic to expect that the classes of interest to the user – often called *information classes*, such as wheat, shallow water, clay, pine forest etc – will have a one-to-one association with the data classes and thus can be delivered directly from an analysis of the recorded image data. Part of the process of analysing the data is to form the link between data and information classes, a step often overlooked in simplistic classification exercises[8].

We could also devise classification methods more suited to the statistical nature of the radar data itself – in other words classifiers that are designed specifically with radar imagery in mind, rather than the more general purpose machine learning procedures used in a wide variety of scientific and engineering applications.

Another approach is to devise analytical and mapping procedures using an understanding the energy-matter interactions that take place when the landscape is irradiated with microwave energy. That can be based on devising backscatter models of more or less sophistication. Some models can be inverted to provide information on the region being imaged, as depicted in Fig. 8.6. Simple inversion models can be built empirically by curve fitting to experimental data but they are limited in scope. Inversion,

[8] See P.H. Swain and S.M. Davis, *Remote Sensing: The Quantitative Approach*, McGraw-Hill, N.Y., 1978 and J.A. Richards and X. Jia, *loc. cit.*

in general, is a non-trivial task and is usually not employed except in relatively simple circumstances. Several methods we explore in the following are tacitly inversion based, but they are set up in the first place with inversion in mind and thus seek to represent only dominant scattering behaviours rather than the full complexity of the landscape.

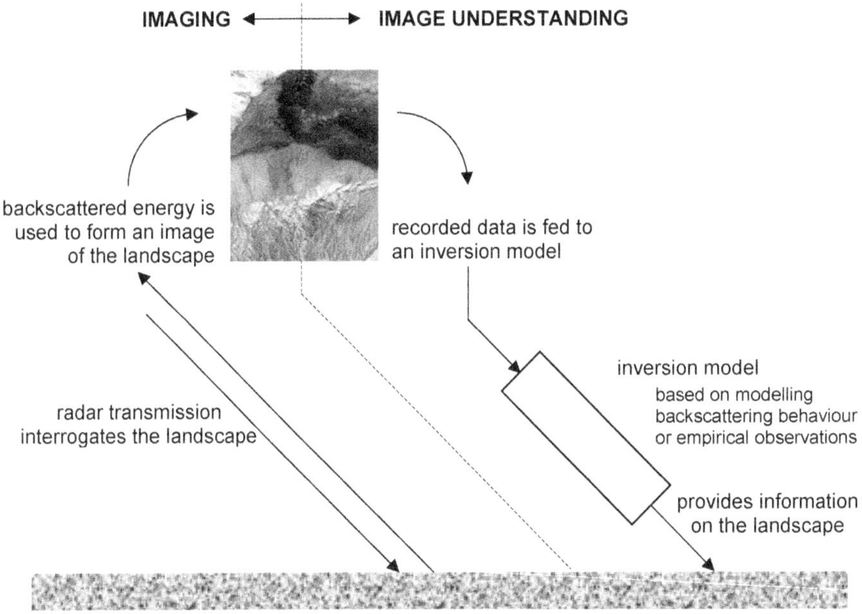

Fig. 8.6. Radar image interpretation as an inversion operation

So in summary there are three broad quantitative analytical approaches:

- application of standard remote sensing classification and labelling methods;
- derivation of mapping procedures that rely on the specific statistical nature of radar image data; and
- derivation of methods that depend on understanding the energy matter interaction in radar imaging.

Our treatment of quantitative radar analysis is subdivided into these three types. Before proceeding we need to be clear about the measurements or features available from radar data that provide the basis for quantitative analysis.

8.4.2 Features Available for Radar Quantitative Analysis

Before being able to appreciate fully the methods available for interpreting radar data at the level of the individual resolution cell it is important to see what types of feature are available for describing the scattering properties of a pixel. Those descriptors will be used as input data to analytical algorithms. The simplest descriptor is the scattering coefficient σ^o_{PQ}. Analogously, the components of the scattering matrix **S** can be used, as can derived features such as the target vector **k**. As shown in Fig. 8.7, tertiary level descriptors such as the covariance and coherency matrices, and even polarimetric complex coherence, are also valid measures that describe neighbourhoods of pixels as inputs to mapping and interpretation procedures.

Notwithstanding the pixel properties used as the basis for labelling it is important to re-emphasise that the response observed is dominated largely by dielectric constant and geometry. Examination of all of the expression for scattering coefficients and scattering matrices presented in Chapt. 5 will show that they are the target properties of importance. If we want to label a radar image into classes more identifiable by a user – information classes such as vegetation or soil type, forest or grassland, water or snow, for example – a bridge has to be established between those class types and their geometric and dielectric properties. Incidentally, the dielectric constant of most natural media is dominated by moisture content, as suggested in Fig. 5.3, so often we need to think about moisture content as a surrogate for dielectric properties.

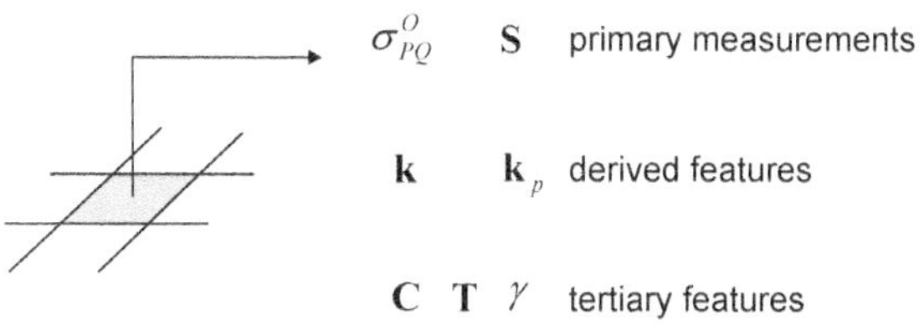

Fig. 8.7. Measurements and features for quantitative radar analysis

8.4.3 Application of Standard Classification Techniques

Traditional point classifiers such as Gaussian maximum likelihood classification, support vector machines and neural networks are adopted widely for thematic mapping with optical image data. Perhaps the biggest problem with applying them to radar imagery is the presence of speckle. Because it is multiplicative, the success of the classification will depend upon reducing the level of speckle, often through local averaging or through the application of speckle filters as treated in Sect. 4.3.3.

A very simple minimum distance classifier can be implemented by appropriately placing discriminating boundaries between the various scattering types shown in Fig. 8.2. It works because it summarises effectively the differing behaviours of surfaces, volumes and strong scatterers. The classifier could be unsupervised if we had some idea of where to place the boundaries because of prior knowledge of the dependence of scattering on incidence angle. More likely it would be supervised if we took labelled samples of each of the cover types and then found the class means (from which the boundaries are established).

The best that can be done for labels of course are those that correspond to the various scattering types. If we also take wavelength into account we could infer the physical cover types in some cases. For example, if the radar were L or P band one could speculate that the double bounce behaviour over land would be forest or urban, while if it were C band the double bounce might be associated with a crop.

8.4.4 Classification Based on Radar Image Statistics

8.4.4.1 A Maximum Likelihood Approach

This method uses the terrain properties summarised in the target vector of (3.47) as a feature for classification. We thus seek to classify the resolution cells of a radar image on the basis of the pixel measurement:

$$\mathbf{k} = \begin{bmatrix} S_{HH} \\ \sqrt{2}S_{HV} \\ S_{VV} \end{bmatrix}$$

which is a column list of the elements of the pixel's scattering matrix. Before proceeding to devise the classifier algorithm we need to understand something of the statistical properties of the target vector. We can do that by examining the statistics of the scattered signal, just as we did when investigating speckle in Sect. 4.3.1. In that section we observed that the signal received from a single pixel is composed of the sum of the signals from a large set of individual scatterers of the form

$$E_{rec} = \sum_k |E_k| e^{j\phi_k} \tag{8.1}$$

in which $|E_k|$ is the amplitude of the field received from the kth scatterer and ϕ_k is the corresponding phase angle. This assumes that there is no dominant scatterer in the pixel. We have removed the factors common to each return in (8.1), such as frequency and the overall phase delay between the pixel and the radar receiver, leaving only the amplitudes and relative phases among the scattering elements. Although not strictly correct we can assume, without loss of generality, that $|E_k|$ is proportional to the square root of the scattering coefficient for the pixel and is thus the same for each elemental scatterer so that (8.1) can be expressed, using $|E_k|=A$ for all k:

$$E_{rec} = A\sum_k e^{j\phi_k} = A\sum_k \cos\phi_k + jA\sum_k \sin\phi_k \tag{8.2}$$

We next assume, as we did for the analysis of speckle, that the phase angles ϕ_k are randomly distributed uniformly over the range $[0,2\pi]$. Consider now the expected value of the received signal:

$$\mathcal{E}\{E_{rec}\} = \mathcal{E}\{A\sum_k e^{j\phi_k}\} = A\mathcal{E}\{\sum_k \cos\phi_k\} + jA\mathcal{E}\{\sum_k \sin\phi_k\} \tag{8.3}$$

The expected value of a sum of random variables is the sum of their expected values. Also, the expected value of a trigonometric function with uniformly distributed arguments over a single cycle is zero. Thus the expected value of the received signal is zero. As an aside, this is not to be confused with the non-zero detected amplitude of the received signal. Writing (8.2) as

$$E_{rec} = A(I + jQ)$$

we see that its amplitude is

$$|E_{rec}| = A\sqrt{I^2 + Q^2}$$

which has a Rayleigh distribution. In radar we often square this to turn it into a measure of power received since that is directly related to the scattering coefficient of the pixel. The distribution function then becomes exponential, as we saw in Sect. 4.3.1. Both have non-zero means.

We return now to the fact that expected value of the received field in (8.3) is zero. This will be the case for any of the fields returning from the pixel irrespective of whether they result from like or cross polarised behaviour. Since the fields incident on the pixel are well-defined single sinusoids (complex exponentials), then the elements of the scattering matrix in (3.41), being the ratio of scattered to incident fields, will also have expected values of zero. That means that the target vector of (3.47) has an expected value of zero. Why is that important? It simplifies the description of class statistics.

We assume that the target vector of (3.47) comes from a class of such vectors (one for each pixel) that represent a given category of land cover or, perhaps more appropriately, category of scattering behaviour. We now make an assumption, common in remote sensing, that the classes can be described by a Gaussian distribution. In the case of optical multispectral data the class distribution models are assumed to be multivariate, with dimensionality the same as the number of spectral channels. Also, for optical data the measurement vectors are real (the reflectance in each band). For pixels in radar imagery described by the target vector of (3.47) the dimensionality is three and the vector components are complex. We thus assume we are working with radar classes that are described by three dimensional complex Gaussian[9] class conditional distribution functions.

A Gaussian distribution has two parameter sets: the multidimensional mean (expected value) and the covariance matrix, which describes the second order relationships among the components. We have shown above that the expected value of the target vector is zero. Therefore we need only work with the covariance matrix. The probability of finding a pixel from class ω_m in the radar image, with the three dimensional target (measurement) vector $\mathbf{k}$, is given by the Gaussian class conditional distribution function[10]

$$p(\mathbf{k} \mid \omega_m) = \frac{1}{\pi^q \mid \mathbf{C}_m \mid} \exp\{-\mathbf{k}^{*\mathrm{T}} \mathbf{C}_m^{-1} \mathbf{k}\} \text{ with } q=3 \tag{8.4a}$$

Note the conjugate transpose operation on the target vector in the exponent. For complex variables the covariance matrix is defined by (3.52), repeated here for convenience:

$$\mathbf{C}_m = \mathcal{E}\left(\mathbf{k}\mathbf{k}^{*\mathrm{T}}\right) = \begin{bmatrix} \langle S_{HH} S_{HH}^* \rangle & \langle \sqrt{2} S_{HH} S_{HV}^* \rangle & \langle S_{HH} S_{VV}^* \rangle \\ \langle \sqrt{2} S_{HV} S_{HH}^* \rangle & \langle 2 S_{HV} S_{HV}^* \rangle & \langle \sqrt{2} S_{HV} S_{VV}^* \rangle \\ \langle S_{VV} S_{HH}^* \rangle & \langle \sqrt{2} S_{VV} S_{HV}^* \rangle & \langle S_{VV} S_{VV}^* \rangle \end{bmatrix} \tag{8.4b}$$

---

[9] The complex variable $z=x+jy$ is distributed as a complex Gaussian with zero mean if each of x and y are Gaussian and are independent of each other.

[10] See N.R. Goodman, Statistical analysis based on a certain multi-variate complex Gaussian distribution (an introduction), *Annals of Mathematical Statistics*, vol. 34, 1963, pp. 152-177.

in which the angular brackets indicate averages over the samples available. Those samples will generally be the sets of class prototype labelled pixels with which to train the classification procedure.

As with the classification of optical multispectral data we assume we have available sufficient labelled samples for each of the classes of interest that we can obtain good estimates of their class covariance matrices. We then classify an unknown radar pixel, described by the target vector **k,** using the maximum *a posteriori* (MAP) decision rule:

$$\mathbf{k} \in \omega_m \quad \text{if } p(\omega_m \mid \mathbf{k}) > p(\omega_n \mid \mathbf{k}) \quad \text{for all } n \neq m \tag{8.5}$$

This says that the pixel, or radar resolution cell, described by the measurement vector **k** belongs to class ω_m because the probability that the correct class is ω_m is greater than the probability that the correct class is ω_n, for all n.

The rule in (8.5) cannot be applied directly because we can't estimate the posterior probabilities $p(\omega|\mathbf{k})$. Fortunately though we can use Bayes' theorem to express the posterior probabilities in terms of the class conditional density functions according to

$$p(\omega \mid \mathbf{k}) = \frac{p(\mathbf{k} \mid \omega)p(\omega)}{p(\mathbf{k})} \tag{8.6}$$

in which $p(\omega)$ is the so-called prior probability[11] that any pixel in the image will belong to class ω, and $p(\mathbf{k})$ is the probability that there are pixels in the image described by the measurement vector **k**; that turns out not to be important because when (8.6) is substituted into (8.5) it cancels out, leaving the decision rule as

$$\mathbf{k} \in \omega_m \quad \text{if } p(\mathbf{k} \mid \omega_m)p(\omega_m) > p(\mathbf{k} \mid \omega_n)p(\omega_n) \quad \text{for all } n \neq m \tag{8.7}$$

Classically, we call $g_m(\mathbf{k}) = p(\mathbf{k} \mid \omega_m)p(\omega_m)$ a *discriminant function* for class ω_m since it allows us to discriminate that class from the others. Because the class density function is assumed to be Gaussian it simplifies later expressions if we take the natural logarithm of the product of probabilities so that we obtain the more commonly used form of the discriminant function

$$\begin{aligned} g_m(\mathbf{k}) &= \ln\{p(\mathbf{k} \mid \omega_m)p(\omega_m)\} = \ln\{p(\mathbf{k} \mid \omega_m)\} + \ln\{p(\omega_m)\} \\ &= -\ln|\mathbf{C}_m| - \mathbf{k}^{*T}\mathbf{C}_m^{-1}\mathbf{k} + \ln\{p(\omega_m)\} \end{aligned} \tag{8.8}$$

In this last expression we have omitted -3lnπ since it doesn't add any discriminating information in a rule such as that in (8.7). When using the discriminant function of (8.8) we are looking for the class for which the function is largest. Sometimes we reverse the sign and look for the class that minimises the resulting expression. That is tantamount to minimising the distance measure:

$$d_m(\mathbf{k}) = -g_m(\mathbf{k}) = \ln|\mathbf{C}_m| + \mathbf{k}^{*T}\mathbf{C}_m^{-1}\mathbf{k} - \ln\{p(\omega_m)\} \tag{8.9a}$$

[11] Strictly this is called a non-informative prior to distinguish it from the conjugate prior, a subtlety that is often ignored in remote sensing: see C.M. Bishop, *Pattern Recognition and Machine Learning*, Springer, N.Y., 2006.

Often we don't know or cannot reasonably estimate the prior probability of class membership[12], so we assume all classes are equally likely and don't contribute any discriminating information in which case that term is omitted, leaving the distance as

$$d_m(\mathbf{k}) = -g_m(\mathbf{k}) = \ln|\mathbf{C}_m| + \{\mathbf{k}^{*T}\mathbf{C}_m^{-1}\mathbf{k}\} \tag{8.9b}$$

This is an interesting distance measure. If the covariance were the unit matrix it would reduce to the expression for Euclidean distance; in general though it is a distance measure that is different in the different dimensions of the target vector according to the entries in the covariance matrix. It is a special form of the Mahalanobis distance[13]. With (8.9) the classification decision rule of (8.7) becomes

$$\mathbf{k} \in \omega_m \quad \text{if } d_m(\mathbf{k}) < d_n(\mathbf{k}) \quad \text{for all } n \neq m \tag{8.10}$$

8.4.4.2 Handling Multi-look Data

To implement (8.9) requires the original data to be available in (single look complex) scattering matrix form so that the target vector of (3.47) can be created. Radar data is often provided in the form of the Stokes scattering operator of (3.75) or the single pixel covariance matrix given by $\mathbf{kk}^{*T}$. Following Lee et al[14] we form the multi-look average single pixel covariance, which we assume is the pixel description available in the image data provided:

$$\mathbf{Z} = \frac{1}{N}\sum_{n=1}^{N}\mathbf{k}_n\mathbf{k}_n^{*T} \tag{8.11}$$

Here N is the number of looks and $\mathbf{k}_n$ is the target vector for the nth look. We now define the matrix $\mathbf{A}=N\mathbf{Z}$ which has the complex Wishart distribution[15]

$$p(\mathbf{A}|\omega_m) = \frac{|\mathbf{A}|^{N-q}\exp\{-tr(\mathbf{C}_m^{-1}\mathbf{A})\}}{K|\mathbf{C}_m|^N} \tag{8.12}$$

in which the constant $K = \pi^{q(q-1)/2}\Gamma(n)\ldots\Gamma(n-q-1)$

is a class independent expression involving gamma functions; q is the dimensionality of the measurement space – in this case 3. The class conditional distribution of (8.12) shows the probability of finding a pixel from class ω_m with multi-look measurement $\mathbf{A}$. As in Sect. 8.4.4.1 we are really interested in the posterior probability that the class is ω_m given we have the measurement $\mathbf{A}$. Again we can use Bayes' theorem of (8.6) to express the

---

[12] The prior probability is usually taken to mean the probability with which class membership of the pixel can be guessed in the absence of the radar measurements, using any other available source of knowledge. For example, if there were four classes in a scene and we knew roughly their area proportions beforehand we could used those proportions to provide estimates of the priors.

[13] See J.A. Richards and X. Jia, *loc cit.*

[14] See J.S. Lee, M.R. Grunes and R. Kwok, Classification of multi-look polarimetric SAR imagery based on the complex Wishart distribution, *International Journal of Remote Sensing*, vol. 15, pp. 2299-2311, 1994.

[15] See Lee, Grunes and Kwok, *loc cit.*, and Goodman, *loc cit.*

posterior probability in terms of the class conditional distribution function of (8.12) so that, following a similar development to that in Sect. 8.4.4.1, we define the discriminant function for class ω_m as

$$g_m(\mathbf{A}) = (N-q)\ln|\mathbf{A}| - tr(\mathbf{C}_m^{-1}\mathbf{A}) - N\ln|\mathbf{C}_m| - \ln K + \ln p(\omega_m)$$

Only the terms involving $\mathbf{C}_m$ provide class discrimination so the others can be deleted. Reversing the result leads to a distance measure for use in a minimum distance classifier of the type developed in the previous section in terms of the target vector:

$$d_m(\mathbf{A}) = tr(\mathbf{C}_m^{-1}\mathbf{A}) + N\ln|\mathbf{C}_m| - \ln p(\omega_m)$$

Noting that $tr(\mathbf{A})=N\,tr(\mathbf{Z})$ we can express the distance rule in terms of the actual N look covariance matrix of (8.11)

$$d_m(\mathbf{Z}) = N\{tr(\mathbf{C}_m^{-1}\mathbf{Z}) + \ln|\mathbf{C}_m|\} - \ln p(\omega_m) \tag{8.13a}$$

If the priors are ignored, or regarded as equal, then the simpler form of distance results

$$d_m(\mathbf{Z}) = N tr(\mathbf{C}_m^{-1}\mathbf{Z}) + \ln|\mathbf{C}_m| \tag{8.13b}$$

As in Sect. 8.4.4.1 this distance measure is used in the decision rule of (8.10) after labelled training samples are used to estimate each of the class specific covariance matrices.

8.4.4.3 Relating the Scattering and Covariance Matrices, and the Stokes Scattering Operator

The classifiers of the previous two sections have used the scattering matrix (via the target vector), or the pixel-specific covariance matrix to describe the scattering properties of a pixel. Some data suppliers provide imagery in the form of the Stokes scattering operator for each pixel. Fortunately these matrices are easily related via the scattering matrix

$$\begin{bmatrix} S_{HH} & S_{HV} \\ S_{VH} & S_{VV} \end{bmatrix}$$

The single pixel covariance matrix derived from $\mathbf{kk}^{*T}$, in its general, non-reciprocal form, is

$$\mathbf{C} = \begin{bmatrix} \langle S_{HH}S_{HH}^*\rangle & \langle S_{HH}S_{HV}^*\rangle & \langle S_{HH}S_{VH}^*\rangle & \langle S_{HH}S_{VV}^*\rangle \\ \langle S_{HV}S_{HH}^*\rangle & \langle S_{HV}S_{HV}^*\rangle & \langle S_{HV}S_{VH}^*\rangle & \langle S_{HV}S_{VV}^*\rangle \\ \langle S_{VH}S_{HH}^*\rangle & \langle S_{VH}S_{HV}^*\rangle & \langle S_{VH}S_{VH}^*\rangle & \langle S_{VH}S_{VV}^*\rangle \\ \langle S_{VV}S_{HH}^*\rangle & \langle S_{VV}S_{HV}^*\rangle & \langle S_{VV}S_{VH}^*\rangle & \langle S_{VV}S_{VV}^*\rangle \end{bmatrix}$$

Equation (3.75) shows the relationship between the elements of the scattering matrix and those of the Stokes scattering operator, m_{ij}. Since the elements of the covariance matrix c_{ij} involve the products of pairs of scattering matrix elements, as do the elements of the Stokes scattering operator, it is possible to relate them. By inverting (3.72) we can show that the elements of the covariance matrix can be found from

$$\begin{aligned}
c_{11} &= \left\langle S_{HH}S_{HH}^*\right\rangle = m_{11}+m_{12}+m_{21}+m_{22}\\
c_{12} &= \left\langle S_{HH}S_{HV}^*\right\rangle = m_{13}+m_{23}-j(m_{14}+m_{24})\\
c_{13} &= \left\langle S_{HH}S_{VH}^*\right\rangle = m_{31}+m_{32}-j(m_{41}+m_{42})\\
c_{14} &= \left\langle S_{HH}S_{VV}^*\right\rangle = m_{33}-m_{44}-j(m_{34}+m_{43})\\
c_{21} &= \left\langle S_{HV}S_{HH}^*\right\rangle = m_{13}+m_{23}+j(m_{14}+m_{24})\\
c_{22} &= \left\langle S_{HV}S_{HV}^*\right\rangle = m_{11}-m_{12}+m_{21}-m_{22}\\
c_{23} &= \left\langle S_{HV}S_{VH}^*\right\rangle = m_{33}+m_{44}+j(m_{34}-m_{43})\\
c_{24} &= \left\langle S_{HV}S_{VV}^*\right\rangle = m_{31}-m_{32}+j(m_{42}-m_{41})\\
c_{31} &= \left\langle S_{VH}S_{HH}^*\right\rangle = m_{31}+m_{32}+j(m_{41}+m_{42})\\
c_{32} &= \left\langle S_{VH}S_{HV}^*\right\rangle = m_{33}+m_{44}+j(m_{43}-m_{34})\\
c_{33} &= \left\langle S_{VH}S_{VH}^*\right\rangle = m_{11}+m_{12}-m_{21}-m_{22}\\
c_{34} &= \left\langle S_{VH}S_{VV}^*\right\rangle = m_{13}-m_{23}+j(m_{24}-m_{14})\\
c_{41} &= \left\langle S_{VV}S_{HH}^*\right\rangle = m_{33}-m_{44}+j(m_{34}+m_{43})\\
c_{42} &= \left\langle S_{VV}S_{HV}^*\right\rangle = m_{31}-m_{32}+j(m_{41}-m_{42})\\
c_{43} &= \left\langle S_{VV}S_{VH}^*\right\rangle = m_{13}-m_{23}+j(m_{14}-m_{24})\\
c_{44} &= \left\langle S_{VV}S_{VV}^*\right\rangle = m_{11}-m_{12}-m_{21}+m_{22}
\end{aligned} \tag{8.14}$$

8.4.4.4 Adding Other Dimensionality

The classifiers considered above are based on the information contained in the polarisation dimension. If we can assume that the polarised returns are not strongly correlated across the wavelength ranges used in radar Lee[16] has suggested that distance discriminators of the types shown in (8.9) and (8.13) can be developed for each waveband and the results added. In principle, the same could be done for other incidence angles too, but in both cases the assumptions of independence need justification. It is also important that the dynamic ranges in all dimensions (apart from across polarisations) be comparable so that the measurements for one waveband or incidence angle do not bias the result.

[16] *ibid*

8.5 Interpretation Based on Structural Models

Several quite different analytical approaches are possible based on a knowledge of scattering behaviours. While some are inherently mathematical, others are similar to expert system methods since they exploit our understanding of how different structural cover types appear in radar image data.

8.5.1 Interpretation Using Polarisation Phase Difference

A very early classifier for radar data based on a knowledge of scattering behaviours used the changes in phase induced in the scattered signal at different polarisations by different scattering media. This allows segmentation into earth surface features that cause (a) non-coherent scattering, (b) one bounce coherent scattering or (c) double bounce scattering[17]. To appreciate how such an algorithm can be developed it is necessary to understand how scattering events affect polarisation phase difference. To see this consider the situations shown in Fig. 8.8 which involve scattering (reflection) from a conducting surface. Even though most surfaces we encounter in remote sensing will not be conductors, apart from some buildings and bridges, and calibrators such as corner reflectors, the principle of the results we derive here applies more generally, as can be appreciated by looking at scattering from dielectric interfaces[18].

In order to understand what is happening in Fig. 8.8 only one significant fact needs to be kept in mind: there can be no electric field tangential to a conductor. If we tried to create an electric field parallel to a conductor then the conductor would "short circuit" it, just as a piece of wire placed across the terminals of a battery will short circuit the battery. What does that mean for Fig. 8.8a? Since there is an electric field incident normally onto the conductor, and since there is a reflected field, they must oppose each other at the point of reflection so that their sum is zero. In other words, the polarity of the scattered field is opposite to that of the incident field. For the case of normal incidence shown in Fig. 8.8a that happens for both the vertically and horizontally polarised components; as a result the phase difference between them does not change.

Now examine the situation in Fig. 8.8b in which the ray is obliquely incident on the conducting interface. In the case of horizontal (perpendicular) polarisation there will be a change in polarity on reflection just as for the situation in Fig. 8.8a. For the vertically (parallel) polarised wave, the result will be as shown by the directional arrow. That can be appreciated by resolving the incoming vertical field into components parallel to and orthogonal to the interface as illustrated. The polarity of the orthogonal component is not affected by the reflection but that of the tangential component on reflection has to be reversed so that it cancels the tangential component of the incident wave.

In monostatic radar the situation in Fig. 8.8b is not encountered in isolation. Instead, it is part of a two or more bounce situation that causes the incident ray to be backscattered. Such a situation is shown in Fig. 8.9, using dihedral corner reflection for illustration. Tracking the changes we have just described for oblique incidence over the two reflections shows that the backscattered signal will have a 180° phase shift between its horizontal and vertical components compared with the incident ray (remembering that a reversal of sign or polarity is the same as adding a phase shift of 180° to a sinusoidally time varying signal).

[17] See J.J. van Zyl, Unsupervised classification of scattering behaviour using radar polarimetry data, *IEEE Trans Geoscience and Remote Sensing*, vol. 27, no. 1, January 1989, pp. 36-45.

[18] See J.D. Kraus and D.A. Fleisch, *Electromagnetics with Applications*, 5th ed., McGraw-Hill, N.Y., 2000.

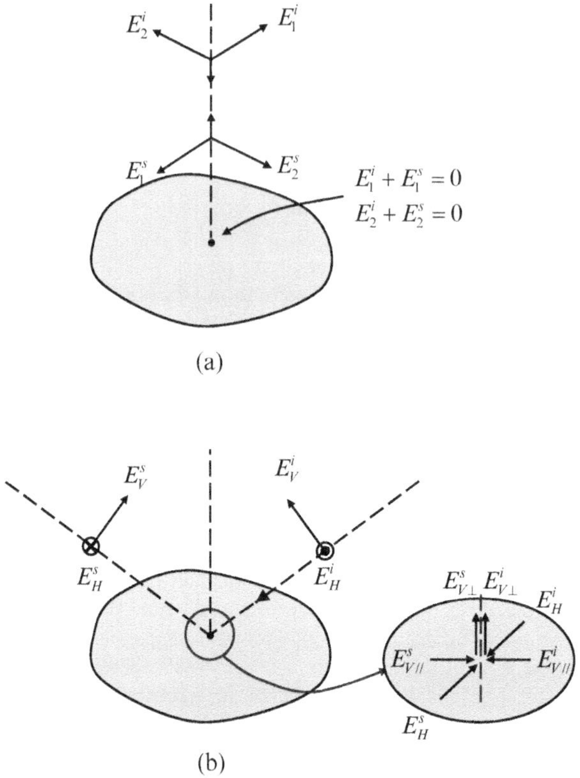

Fig. 8.8. (a) Scattering (reflection) from a conductor at vertical incidence with orthogonal polarisations (which in principle could be H and V) showing that there is no change in their relative phases after scattering (b) the effect on H and V polarised components of oblique scattering; the dot in the circle represents an arrowhead while the cross in the circle represents the tail of an arrow

We can now generalise: if the backscattered wave is the result of an odd number of reflections then there will be no change in the phase difference between its horizontally and vertically polarised components; if it undergoes an even number of reflections there will be a 180° phase shift between the components.

If the scattering medium were not a perfect conductor there can be tangential components of electric field at the interface and the situation will be a little different from that just described. Nevertheless, there will in most cases be phase differences in the vicinity of 0° and 180° respectively allowing that knowledge to be used to construct a classifier in the following manner.

The product

$$S_{HH}^* S_{VV} = |S_{HH}||S_{VV}|e^{j(\phi_{VV} - \phi_{HH})}$$

identifies the phase difference between the two linear polarisations. Usually it is averaged over a small group of resolution cells to reduce variability resulting from speckle. Then

if $\arg\langle S_{HH}^{*}S_{VV}\rangle$ is in the vicinity of zero
then we have an odd number of bounces
or
if $\arg\langle S_{HH}^{*}S_{VV}\rangle$ is in the vicinity of 180°
then we have an even number of bounces

We can associate an even number of bounces with dihedral corner reflector behaviour. That can indicate urban regions, or forests at longer wavelengths. Odd numbers of bounces can be associated with relatively smooth surfaces or even direct scattering from foliage at shorter wavelengths. For very diffuse scattering media there will be little correlation between the like polarised terms so that $\langle S_{HH}^{*}S_{VV}\rangle \approx 0$. By setting up rule sets such as these it is possible to form a simple unsupervised classifier of multi-polarisation radar image data.

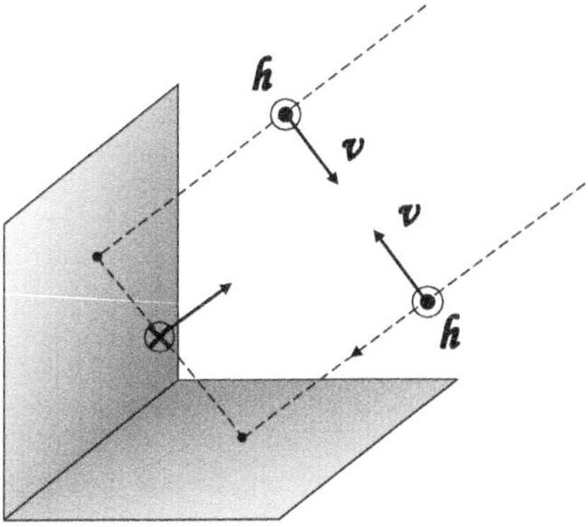

Fig. 8.9. Demonstrating how 180° relative phase shift happens between the H and V polarised components with two reflections

8.5.2 Interpretation Through Structural Decomposition

End member analysis, often employed with optical remote sensing imagery, seeks to understand the class composition of a pixel in terms of a number of pure classes, or end members. It is assumed that the spectral response of the pixel is a weighted sum of the responses of the end members and the task is to find the weighting coefficients. Maps of those coefficients can then be produced to show the abundances of the end members, by pixel. Usually there are far fewer end members than the dimensionality of the measurements space so that least squares estimates of the weighting coefficients are employed.

A similar approach is can be followed with multi-polarisation radar although the end members as such are structural types. Three different measurement dimensions are available for a pixel – each of HH, HV and VV – so it is possible to decompose the recorded data for each resolution cell into a weighted sum of three fundamental structural types. The responses could be those characteristic of surface scattering, volume scattering and dihedral corner reflector double bounce scattering, rather than cover types as such, but a limited range of ground cover types can often be induced from the weighted sum. Although not a classification procedure, decomposition of the measured scattering data in this manner does allow interpretation and thus a description of recorded radar pixels.

8.5.2.1 Decomposing the Scattering Matrix

It is logical to commence by examining the scattering matrix since it contains the target response by polarisation. We assume that it is possible to represent the matrix for a given pixel in the form

$$\mathbf{S} = \sum_{i=1}^{3} p_i \mathbf{S}_i \tag{8.15}$$

in which the $\mathbf{S}_i$ are the scattering matrices of the fundamental scatterers that compose the composite response, and the p_i are weighting or abundance coefficients. Scattering matrices can be added because the fields backscattered from the individual scattering components can be added provided we know their amplitudes and phases. Those properties are incorporated in the complex elements of the scattering matrix. If the pixel were composed of a specular background, a dihedral corner reflector and a trihedral corner reflector (at the same absolute distance from the radar) then the composite scattering matrix would be

$$S = p_1 \begin{bmatrix} a & 0 \\ 0 & -a \end{bmatrix} + p_2 \begin{bmatrix} b & 0 \\ 0 & b \end{bmatrix}$$

where the a and b are the amplitudes derived from Table 4.1.

While theoretically appealing, this approach has one significant limitation – the components being summed must be expressible as scattering matrices. That is not readily done if the significant scattering mechanisms in the resolution cell are distributed (such as a volume scatterer) and have substantial components of unpolarised returns. In such cases, which of course occur often in remote sensing, it is better to look at decomposing features that can handle unpolarised behaviour. Because the covariance and coherency matrices are based on the expected values of the elements of the scattering matrices via the respective target vectors (effectively through ensemble averaging) they can incorporate unpolarised components of radar returns and thus can form the basis of decomposition models. They are also tantamount to the scattering coefficients used often in radar imagery to describe backscattered levels of power density.

8.5.2.2 Decomposing the Covariance Matrix: the Freeman-Durden Approach[19]

The Freeman-Durden decomposition was developed principally for interpreting forest backscattering by seeking to resolve the covariance matrix into three component covariances: one associated with volume scattering, one with double bounce dihedral scattering representing the effect of a tree trunk, and one associated with surface (or single bounce) scattering. The model assumes that the three components are statistically

[19] A. Freeman and S.L. Durden, A three-component scattering model for polarimetric SAR Data, *IEEE Transactions on Geoscience and Remote Sensing*, vol. 35, no.3, May 1998, pp. 963-973.

independent allowing the component covariances to be added. The recorded covariance matrix is therefore expressed

$$\mathbf{C} = f_v\mathbf{C}_{volume} + f_d\mathbf{C}_{dihedral} + f_s\mathbf{C}_{surface} \tag{8.16}$$

where f_v, f_d and f_s are weighting coefficients. It is a non-coherent model since by adding the covariances we are effectively adding powers as against electric fields

Before proceeding further it is important to recognise that this model was derived by Freeman and Durden with the ordering of the elements of the scattering matrix (used in constructing the covariance matrix) in the opposite sense to that used here; they used the convention

$$\begin{bmatrix} S_{VV} & S_{VH} \\ S_{VH} & S_{HH} \end{bmatrix} \text{ rather than } \begin{bmatrix} S_{HH} & S_{HV} \\ S_{HV} & S_{VV} \end{bmatrix}$$

We will stay with the our convention of writing the horizontal transmit polarisation first.

The surface model used in the Freeman-Durden decomposition is based on the Bragg small roughness model which has vertical and horizontal co-polarised responses but no cross-polarised behaviour. Its (normalised) scattering matrix is

$$\mathbf{S}_{surface} = \begin{bmatrix} \rho_H & 0 \\ 0 & \rho_V \end{bmatrix} \tag{8.17}$$

where ρ_V and ρ_H are the Fresnel reflection coefficients of the surface given by (5.3). Applying (3.52) the corresponding covariance matrix is

$$\mathbf{C}_{surface} = \begin{bmatrix} \rho_H\rho_H^* & 0 & \rho_H\rho_V^* \\ 0 & 0 & 0 \\ \rho_V\rho_H^* & 0 & \rho_V\rho_V^* \end{bmatrix}$$

Dividing throughout by $\rho_V\rho_V^*$ gives

$$\mathbf{C}_{surface} = \begin{bmatrix} |\beta|^2 & 0 & \beta \\ 0 & 0 & 0 \\ \beta^* & 0 & 1 \end{bmatrix} \tag{8.18}$$

in which $\beta=\rho_H/\rho_V$.

Although there are other factors that should have been included in these matrices to reflect the full detail of the Bragg model, they are essentially incorporated in the relevant scaling factor in (8.16). It is the structure of (8.18) that is important.

The dihedral trunk-ground model used in the Freeman-Durden decomposition is of the form (see Sect. 5.5.2)

$$\mathbf{S}_{dihedral} = \begin{bmatrix} \rho_{gH}\rho_{tH} & 0 \\ 0 & -\rho_{gV}\rho_{tV} \end{bmatrix} \tag{8.19}$$

in which the reflection coefficients are for the trunk or ground (t, g) for vertical and horizontal polarisation as appropriate. There is no cross-polarised response, implying that the trunks are vertical. Since there can be a significant canopy over the trunks an exponential two way propagation term can be incorporated into each of the elements of the scattering matrix:

$$\mathbf{S}_{dihedral} = \begin{bmatrix} e^{2\gamma_H r}\rho_{gH}\rho_{tH} & 0 \\ 0 & e^{2\gamma_V r}\rho_{gV}\rho_{tV} \end{bmatrix}$$

in which r is path length (the slant path) through the canopy and γ_H and γ_V respectively are propagation constants for H and V polarisation. The exponent on the vertically polarised term has also been used take up the negative sign so that the vertically polarised entry is shown as positive for convenience. The corresponding covariance matrix is

$$\mathbf{C}_{dihedral} = \begin{bmatrix} |\alpha|^2 & 0 & \alpha \\ 0 & 0 & 0 \\ \alpha^* & 0 & 1 \end{bmatrix} \tag{8.20}$$

with $\alpha = e^{2(\rho_H - \rho_V)r}\dfrac{\rho_{gH}\rho_{tH}}{\rho_{gV}\rho_{tV}}$. Again, other factors such as the sizes of the trunks get picked up the scaling factor in (8.16).

The third component, involving canopy volume scattering, is developed by using a random distribution of thin cylinders to represent branches and twigs. This is made simple by starting with the scattering matrix of a single cylinder at an angle ϕ with respect to the vertical and then determining its response to an arbitrarily inclined incoming ray. We can do that by rotating the coordinate system of the wave to align its vertical component to the cylinder axis, applying the scattering matrix to find the cylinder response, and then rotating the response back to the original orientation of the incident wave vector. If the cylinder is inclined ϕ anti-clockwise from the vertical then using (2.20)

$$\mathbf{S}(\phi) = \begin{bmatrix} \cos\phi & -\sin\phi \\ \sin\phi & \cos\phi \end{bmatrix} \mathbf{S} \begin{bmatrix} \cos\phi & \sin\phi \\ -\sin\phi & \cos\phi \end{bmatrix} \tag{8.21}$$

in which

$$\mathbf{S} = \begin{bmatrix} 0 & 0 \\ 0 & 1 \end{bmatrix} \tag{8.22}$$

is the normalised scattering matrix of a thin cylinder irradiated with vertically polarised radiation. This is a very convenient method that does not require any complex expressions for aligned cylinders. Expanding (8.21) gives

$$\mathbf{S}(\phi) = \begin{bmatrix} \sin^2\phi & -\sin\phi\cos\phi \\ -\sin\phi\cos\phi & \cos^2\phi \end{bmatrix}$$

If ϕ is distributed uniformly then it can be shown that the corresponding covariance matrix is

$$\mathbf{C}_{volume}=\pi\begin{bmatrix}1 & 0 & 1/3\\ 0 & 2/3 & 0\\ 1/3 & 0 & 1\end{bmatrix} \tag{8.23}$$

The common factor π can be absorbed into the weighting coefficient f_v in (8.16). With (8.18), (8.20) and (8.23) the Freeman-Durden decomposition of (8.16) is

$$\mathbf{C}=f_v\begin{bmatrix}1 & 0 & 1/3\\ 0 & 2/3 & 0\\ 1/3 & 0 & 1\end{bmatrix}+f_d\begin{bmatrix}|\alpha|^2 & 0 & \alpha\\ 0 & 0 & 0\\ \alpha^* & 0 & 1\end{bmatrix}+f_s\begin{bmatrix}|\beta|^2 & 0 & \beta\\ 0 & 0 & 0\\ \beta^* & 0 & 1\end{bmatrix} \tag{8.24}$$

What we do now is to assume that the recorded covariance matrix for a resolution cell can be approximated by (8.24). We need to determine the unknown proportions f_v, f_d and f_s to find the relative abundances of each scattering type in the resolution cell. However, we don't know α and β, since we generally don't have knowledge of the respective dielectric constants needed for computing the reflection coefficients. Thus there are five unknowns needing to be found to make this decomposition work.

From (8.4b) and (8.24) we can see that the measured and modelled covariance matrix elements are related by

$$c_{11}=<|S_{HH}|^2>=f_v+|\alpha|^2 f_d+|\beta|^2 f_s \tag{8.25a}$$

$$c_{22}=2<|S_{HV}|^2>=\frac{2}{3}f_v \tag{8.25b}$$

$$c_{33}=<|S_{VV}|^2>=f_v+f_d+f_s \tag{8.25c}$$

$$c_{13}=<S_{HH}S_{VV}^*>=\frac{1}{3}f_v+\alpha f_d+\beta f_s \tag{8.25d}$$

These are just four equations in the five unknowns and thus the problem is under-specified. Interestingly, though, (8.25b) shows $f_v=3<|S_{vh}|^2>$ not only giving an abundance value for the volume term but allowing (8.25a,c,d) to be reduced to

$$<|S_{HH}|^2>-3<|S_{HV}|^2>=|\alpha|^2 f_d+|\beta|^2 f_s \tag{8.26a}$$

$$<|S_{VV}|^2>-3<|S_{HV}|^2>=f_d+f_s \tag{8.26b}$$

$$<S_{HH}S_{VV}^*>-<|S_{HV}|^2>=\alpha^* f_d+\beta^* f_s \tag{8.26c}$$

We now have three equations in four unknowns.

The total power carried by the response for a given resolution cell is called the *span* for the pixel and is given by the collection of the squares of the like and cross polarised responses:

$$\text{Total power} = \text{span}=<|S_{HH}|^2>+<|S_{VV}|^2>+2<|S_{VH}|^2>$$

The cross polarised term is doubled since it is the result of an assumption of reciprocity for monostatic backscattering. From (8.25a,b,c) this is

$$\text{Total power}=\frac{8}{3}f_v+(1+|\alpha|^2)f_d+(1+|\beta|^2)f_s \qquad (8.27)$$

The total power must also be equal to that from the assumed three backscattering mechanisms, for each of which we sum the diagonal elements of their covariance matrices (i.e. the traces of those matrices) multiplied by the relevant weighting coefficient. That gives

$$\text{Total power}=P_v+P_d+P_s$$

in which

$$P_v=\frac{8}{3}f_v$$

$$P_d=(1+|\alpha|^2)f_d$$

$$P_s=(1+|\beta|^2)f_s$$

leading again to (8.27). Unfortunately, therefore, the calculation of span does not provide another independent equation in the required unknowns, so we are still left with needing to determine four unknowns from three equations. In their solution Freeman and Durden chose $\alpha=-1$ if they assess, through an examination of the measurement in the left hand side of (8.26c), that surface scatter is dominant after the volume scattering effect has been removed. Otherwise, if dihedral behaviour is seen to be dominant, they fix $\beta=1$. Once they have determined either of those parameters they can then find the remaining three from (8.26a,b,c).

It is important to recognise that the Freeman-Durden decomposition is not unique nor theoretically determined. It is however practical in a forest context since it picks up the most important scattering mechanisms, apart perhaps from the weaker volume-ground component. The same form of model could be devised for other ground cover communities by choosing the most appropriate scattering mechanisms; those mechanisms then have to be modelled and means for finding their parameters need to be developed.

8.5.2.3 Decomposing the Coherency Matrix: the Cloude-Pottier Approach[20]

We now present an approach based upon diagonalising the coherency matrix. For backscattering from reciprocal media the matrix is three dimensional so that, again in principle, at most only three fundamental structural components can be determined. Nevertheless the method provides a useful basis for unsupervised and supervised procedures since many cover types reflect in structural components or their combinations. Ideally, we would like to identify the dominant scattering mechanism for each resolution cell if one exists.

Typically, the scattering mechanisms encountered in practice might be one of those listed in Table 8.1 along with their idealised (normalised) scattering matrices. It is important to re-emphasise before proceeding that simple scattering matrix descriptions are available only for pure, simple surfaces or targets such as those shown. Pixels which have a significant component of unpolarised return cannot effectively be described by a

[20] See S.R. Cloude and E. Pottier, An entropy based classification scheme for land based applications of polarimetric SAR, *IEEE Transactions on Geoscience and Remote Sensing*, vol. 35, no. 1, January 1997, pp. 68-78.

scattering matrix. Provided we pursue the analysis based not on the scattering matrices, but on covariance or coherency matrices derived from ensemble averages, we can still identify pixels that are dominated by the responses typical of the elements in Table 8.1.

We adopt as a starting point the Pauli basis form of the target vector in (3.49). The expected value of its outer product – see (3.55) – over an ensemble of measurements leads to the coherency matrix **T** of (3.57) in the case of backscattering. As with the covariance matrix, it is easy to see that the coherency matrix is Hermitian. That means that its eigenvalues are real and that the matrix of eigenvectors used to find its diagonal form is unitary[21]. That simplifies analysis and leads directly to the decomposition being sought.

From (B.14) we can express **T** in its diagonal form

$$\mathbf{T} = \mathbf{G}\boldsymbol{\Lambda}\mathbf{G}^{-1} \tag{8.28}$$

in which $\boldsymbol{\Lambda}$ is the diagonal matrix of eigenvalues of **T** and **G** is a unitary matrix of the eigenvectors of **T**, arranged by column. Since **G** is unitary its inverse is equal to its conjugate transpose so that (8.28) can be written

$$\mathbf{T} = \mathbf{G}\boldsymbol{\Lambda}\mathbf{G}^{*T} \tag{8.29}$$

Expanding this we have

$$\mathbf{T} = [\mathbf{g}_1 \quad \mathbf{g}_2 \quad \mathbf{g}_3]\begin{bmatrix} \lambda_1 & 0 & 0 \\ 0 & \lambda_2 & 0 \\ 0 & 0 & \lambda_3 \end{bmatrix}[\mathbf{g}_1^* \quad \mathbf{g}_2^* \quad \mathbf{g}_3^*]^T \tag{8.30}$$

i.e.

$$\mathbf{T} = \lambda_1\mathbf{g}_1\mathbf{g}_1^{*T} + \lambda_2\mathbf{g}_2\mathbf{g}_2^{*T} + \lambda_3\mathbf{g}_3\mathbf{g}_3^{*T} \tag{8.31}$$

which shows that the coherency matrix can be resolved into three independent components, weighted by the eigenvalues λ_i[22]. This is reminiscent of the principal components transformation used with optical multispectral data which generates as many orthogonal and uncorrelated components as there are original bands in the data. Here there are only three separate polarisation measurements so we can, at most, only generate three elements in this particular, alternative description of the properties of the scattering medium.

We would hope that in most remote sensing radar studies there would only be one dominant scattering mechanism per pixel. If that pixel is part of a particular cover type then the cover type response on the average would be dominated by that mechanism – in other words we hope that the chance of mixed pixels is minimised. Candidate mechanisms would include surface scattering, volume scattering and dihedral reflections as discussed earlier. Of course natural media are not always that simple and some pixels will exhibit composite responses – surface, canopy and trunk scattering together in forest stands is an example.

---

[21] See Appendix B.

[22] Unfortunately in mathematics the symbol λ is used for eigenvalues. In radar studies that has the potential to be confused with the symbol for wavelength; usually the context identifies which one is meant.

Table 8.1 Fundamental pure scatterers and their scattering matrices

Scatterer		Scattering Matrix	Comments
single bounce volume scattering from a medium composed of spherical scatterers	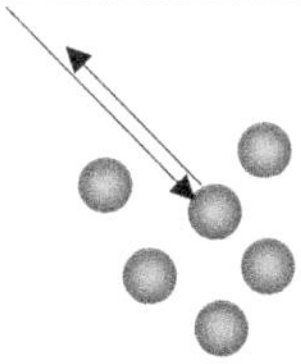	$\mathbf{S}=\begin{bmatrix}1 & 0\\ 0 & 1\end{bmatrix}$	both the horizontal and vertical responses are the same, with no opportunity to generate cross polarised responses
single bounce volume scattering from anisotropic scatterers	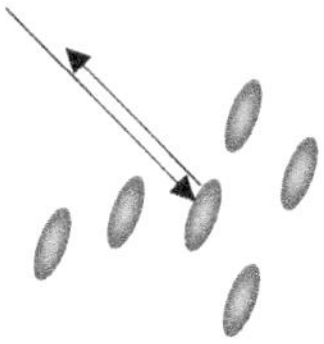	$\mathbf{S}=\begin{bmatrix}a & 0\\ 0 & b\end{bmatrix}$	*a* and *b* are complex elements reflecting the shape anisotropy of the scatterer
single bounce volume scattering from a medium composed of thin needle like scatterers	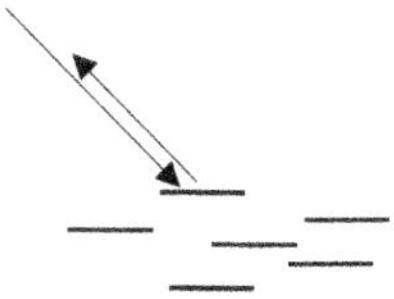	$\mathbf{S}=\begin{bmatrix}1 & 0\\ 0 & 0\end{bmatrix}$	this matrix assumes they are horizontally aligned so that the is no vertical response; scattering matrices for other orientations can be derived by rotating the coordinate system (see Sect. 2.11)
dihedral corner reflector	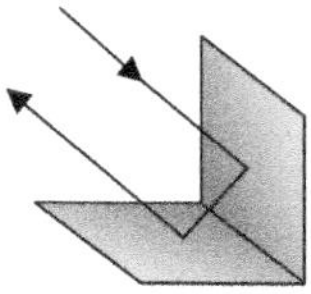	$\mathbf{S}=\begin{bmatrix}1 & 0\\ 0 & -1\end{bmatrix}$	this can also represent trunk-ground interaction for a tree, in which the reflector is oriented for maximum response
trihedral corner reflector	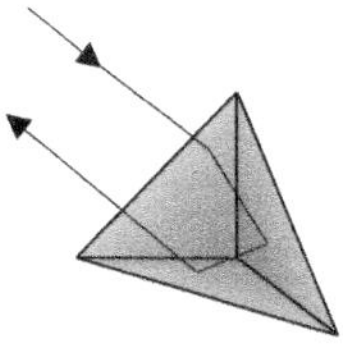	$\mathbf{S}=\begin{bmatrix}1 & 0\\ 0 & 1\end{bmatrix}$	
surface scattering	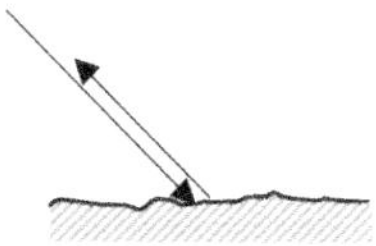	$\mathbf{S}=\begin{bmatrix}a & 0\\ 0 & b\end{bmatrix}$	based on the Bragg model, in which the elements *a* and *b* are related to the reflection coefficients of the surface

To determine how the response is composed we can examine the weighting factors – i.e. the eigenvalues – in (8.31). If two are zero then there is only one fundamental response type, whereas if all three are of comparable magnitudes then the response is a mixture.

Under what conditions will there be only one non-zero eigenvalue of a 3x3 coherency matrix? That happens when the coherency matrix has unit rank[23] which means, in turn, that it has no sub-matrices (more properly called principal minors) larger than 1x1 with

[23] See Appendix B

non-zero determinants. We cannot determine that uniquely by examining the coherency matrix but we can develop some valuable guidance.

Consider the most general form of the matrix for backscattering, repeated here for convenience:

$$\mathbf{T} = \frac{1}{2}\begin{bmatrix} <(S_{HH}+S_{VV})(S_{HH}^*+S_{VV}^*)> & <(S_{HH}+S_{VV})(S_{HH}^*-S_{VV}^*)> & 2<(S_{HH}+S_{VV})S_{HV}^*> \\ <(S_{HH}-S_{VV})(S_{HH}^*+S_{VV}^*)> & <(S_{HH}-S_{VV})(S_{HH}^*-S_{VV}^*)> & 2<(S_{HH}-S_{VV})S_{HV}^*> \\ 2<(S_{HH}^*+S_{VV}^*)S_{HV}> & <2(S_{HH}^*-S_{VV}^*)S_{HV}> & 4<S_{HV}S_{HV}^*> \end{bmatrix} \tag{8.32}$$

Suppose now the resolution cell contained a single trihedral corner reflector with the scattering matrix shown in Table 8.1. Substituting this into (8.32) gives

$$\mathbf{T} = \begin{bmatrix} 2 & 0 & 0 \\ 0 & 0 & 0 \\ 0 & 0 & 0 \end{bmatrix}$$

which is of rank 1 since the largest sub matrix with a non-zero determinant is of size 1x1. It only has one eigenvalue, $\lambda=2$, which is easily shown (see Appendix B), signifying that there is a dominant scatterer (the trihedral reflector).

Suppose now that the resolution cell contains, instead, a dihedral corner reflector with the scattering matrix in Table 8.1. For this the coherency matrix is

$$\mathbf{T} = \begin{bmatrix} 0 & 0 & 0 \\ 0 & 2 & 0 \\ 0 & 0 & 0 \end{bmatrix}$$

which again is of rank 1 and which has only one non-zero eigenvalue.

Therefore if the measured coherency matrix yielded a single non-zero eigenvalue it is *possible* that the pixel is composed of either a dihedral or trihedral corner reflector. As an aside, note that if a target is non-depolarising so that $S_{HV}=0$ then at most the coherency matrix can be of rank 2 since (8.32) then has one column and one row full of zeros and cannot have a non-zero determinant of size 3x3.

If an analysis of the coherency matrix led to rank 1, then we could conclude that the measured backscatter from the resolution cell was the result of a single pure scatterer or a scattering type that did not lead to significant cross-polarisation. Often that will not be the case and the rank is more likely to be 2 or 3, signifying a mixture of fundamental scatterers or even a random scatterer in the resolution cell.

Thus the relative magnitudes of the eigenvalues is an important measure. It is helpful to turn them into a set of proportions or probabilities because that opens up some other measures we may wish to consider. We therefore normalise them by defining

$$p_i = \frac{\lambda_i}{\sum_{i=1}^{3}\lambda_i} \tag{8.33}$$

A useful measure of the distribution of probabilities is called *entropy*. It was devised for understanding the information carried by messages in telecommunications systems, but finds wider applications in coding theory and image processing. For our three element system it is defined as

$$H = \sum_{i=1}^{3} p_i \log_3 \frac{1}{p_i} = -\sum_{i=1}^{3} p_i \log_3 p_i \tag{8.34}$$

The basis (radix) for the logarithm is chosen as 3 so that when all the probabilities are equal (1/3) the entropy has 1 as its maximum value. Thus the entropy will be high if the measured radar scattering is made up of several comparably important scatterers. At the other extreme, if there is a dominant scatterer, then the entropy approaches 0; it will be exactly zero if there is only one non-zero probability, which will be 1. That can be shown by expanding the logarithm in its power series and noting that $\lim_{x\to 0}(x\log x) = 0$.

Entropy is a useful feature to use in radar classification because it tells us something about the likely mixture of scattering types in a region. Another helpful feature relates to the eigenvectors $\mathbf{g}_i$ of $\mathbf{T}$ in (8.31) since they tell us something about the *types* of the individual scattering mechanisms. The first element of an eigenvector can be written as $\cos\alpha_i$ with the angle α_i, different for each eigenvector[24]. We can find an average value for α across all three eigenvectors by computing

$$\alpha = \sum_{i=1}^{3} p_i \alpha_i = \sum_{i=1}^{3} p_i \cos^{-1} g_{i1} \tag{8.35}$$

The pair of H and α can now be used as a feature set for classification because they seem, *prima facie*, to provide some form of discrimination among differing scattering mechanisms. If they could be related specifically to actual scattering types then they could form the basis of unsupervised classification[25].

It is instructive at this stage to take an example from Cloude and Pottier. Consider the shaped anisotropic scatterer in the second row of Table 8.1. Rather than just a single scatterer at a given orientation with respect to the incoming radar beam imagine we have many scatterers with random orientations to the ray. We can derive the scattering matrix for an arbitrarily oriented scatter by using the device of (8.21): we rotate the beam to align with the scatterer, so that the form of the scattering matrix in the Table is applicable, and then rotate the result back again. Thus the scattering matrix for an anisotropic scatterer at angle ϕ with respect to the horizontal plane is

$$\mathbf{S}(\theta) = \begin{bmatrix} \cos\phi & -\sin\phi \\ \sin\phi & \cos\phi \end{bmatrix} \begin{bmatrix} a & 0 \\ 0 & b \end{bmatrix} \begin{bmatrix} \cos\phi & \sin\phi \\ -\sin\phi & \cos\phi \end{bmatrix}$$

$$= \begin{bmatrix} a\cos^2\phi + b\sin^2\phi & a\sin\phi\cos\phi - b\sin\phi\cos\phi \\ a\sin\phi\cos\phi - b\sin\phi\cos\phi & a\sin^2\phi + b\cos^2\phi \end{bmatrix}$$

[24] Cloude and Pottier, 1997, *loc. cit.* This is a very convenient construct because, as we will see, the angle is a more sensitive discriminator than the eigenvector element it is derived from.
[25] See Cloude and Pottier, *ibid.*

As an aside, note that the reciprocity condition has been preserved on rotation of the scatterer – i.e. $S_{HV}(\phi)=S_{VH}(\phi)$. From the last expression we can determine the target vector in the Pauli basis, $\mathbf{k}_p$, as defined in (3.49):

$$\mathbf{k}_p = \frac{1}{\sqrt{2}}\begin{bmatrix} S_{HH}+S_{VV} \\ S_{HH}-S_{VV} \\ 2S_{HV} \end{bmatrix} = \frac{1}{\sqrt{2}}\begin{bmatrix} a+b \\ (a-b)\cos 2\phi \\ (a-b)\sin 2\phi \end{bmatrix}$$

so that the coherency matrix

$$\mathbf{T} = \mathcal{E}(k_p k_p^{*T})$$

is the expected value with respect to ϕ of

$$\mathbf{k}_p\mathbf{k}_p^{*T} = \frac{1}{2}\begin{bmatrix} |a+b|^2 & (a+b)(a-b)^*\cos 2\phi & (a+b)(a-b)^*\sin 2\phi \\ (a-b)(a+b)^*\cos 2\phi & (a-b)(a-b)^*\cos^2 2\phi & (a-b)(a-b)^*\sin 2\phi\cos 2\phi \\ (a-b)(a+b)^*\sin 2\phi & (a-b)(a-b)^*\sin 2\phi\cos 2\phi & (a-b)(a-b)^*\sin^2 2\phi \end{bmatrix}$$

When the expectation is taken as an average over all orientations we get the coherency matrix[26]

$$\mathbf{T} = \begin{bmatrix} 0.5|a+b|^2 & 0 & 0 \\ 0 & 0.25(a-b)(a-b)^* & 0 \\ 0 & 0 & 0.25(a-b)(a-b)^* \end{bmatrix} \quad (8.36)$$

We can analyse this coherency matrix for various combinations of a and b to understand the dominant behaviours of the ensemble of scatterers. First note that, in principle, it has three eigenvalues[27]. Its rank can therefore be 3 and there may be no dominant scatterer as such. Interestingly the second and third eigenvalues are equal; they are therefore called *degenerate* minor eigenvalues.

For the special case of $a=b$ so that vertically and horizontally polarised radiation scatter in the same manner– in other words the scatterers are no longer anisotropic – the two minor eigenvalues go to zero, leaving a rank 1 matrix. The entropy is zero, implying a dominant scattering mechanism.

If b= –a, we see from Table 8.1 that we have the situation of a random distribution of dihedral corner reflectors in which case the coherency matrix becomes

$$\mathbf{T} = \begin{bmatrix} 0 & 0 & 0 \\ 0 & bb^* & 0 \\ 0 & 0 & bb^* \end{bmatrix} \quad (8.37)$$

Thus we have two equal eigenvalues, so that the entropy is $2\text{x}0.5\log_3 2=0.62$. The dominance of a single mechanism is thus not strongly indicated for this randomly

[26] The definite integral of a trig function over a full period is zero, while the integral of the square of a trig function over a full period is 0.5.

[27] The eigenvalues of a diagonal matrix are the diagonal entries, by definition. See Appendix B.

oriented ensemble of dihedral reflectors, as is to be expected. In contrast if there were a single dihedral reflector in a resolution cell then from Table 8.1

$$\mathbf{k}_p = \frac{1}{\sqrt{2}}\begin{bmatrix}0\\2\\0\end{bmatrix}$$

so that

$$\mathbf{T} = \begin{bmatrix}0 & 0 & 0\\0 & 2 & 0\\0 & 0 & 0\end{bmatrix} \tag{8.38}$$

which has only one non-zero eigenvalue, and is of rank 1, signifying as noted earlier a single dominant scatterer – the single corner reflector.

If b=0 we have, by reference to Table 8.1, a distribution of needle-like scatterers. From (8.36) we see

$$\mathbf{T} = \begin{bmatrix}0.5|a|^2 & 0 & 0\\0 & 0.25|a|^2 & 0\\0 & 0 & 0.25|a|^2\end{bmatrix}$$

Here we have three eigenvalues in the proportions 2,1,1 so that the entropy is 0.91, showing a random, non-dominant scattering event. In contrast, for a single needle

$$\mathbf{k}_p = \frac{1}{\sqrt{2}}\begin{bmatrix}1\\1\\0\end{bmatrix}$$

and

$$\mathbf{T} = \frac{1}{2}\begin{bmatrix}1 & 1 & 0\\1 & 1 & 0\\0 & 0 & 0\end{bmatrix}$$

This is of rank 1 (since the 2x2 determinant is zero) and will have only one non-zero eigenvalue equal to 2 and thus will have zero entropy.

If we take the case of a general anisotropic scatterer ($a \neq b$) and assume a and b to be real with a=nb, then we find entropies of 0.35, 0.58 and 0.75 for n=2,3 and 5 respectively.

Cloude and Pottier also examined the case of multiple scatterings from a volume of identical particles and found that if all but single scatterings are ignored then the entropy is low, signifying that the fundamental mechanism can be dominant. As the order of scattering increases entropy steadily rises towards 1.0 indicating increased randomness in the backscattered signal.

Consider now a Bragg surface with the scattering matrix shown in Table 8.1; its parameters depend on the angle of incidence of the radar system, but if that is within the usual range of, say, less than 50°, the above analysis for anisotropic volume scatterers applies also to scattering from slightly rough surfaces.

Observations such as these can be used to form associations between entropy and scattering types as summarised in Table 8.2. Unfortunately, as noted, entropy on its own is not enough to allow a good separation of differing scattering types and at least one further feature is needed. This is where the actual eigenvectors themselves are important. In particular the alpha angle seen in (8.35) is a sufficient measure of the nature of the dominant and other eigenvectors for helping to separate the scattering types, as we will now demonstrate.

Table 8.2 Summary of the entropies of scattering types

Entropy range	Scattering types
low (dominant scatterer)	single dihedral corner reflector single needle scatterer (little practical interest unless dipolar) volume of isotropic scatterers slightly rough surfaces
medium	random orientation of corner reflectors (little practical interest) random orientation of mildly anisotropic volume scatterers
high (no dominant scatterer)	random orientation of needles random orientation of strongly isotropic particles

Consider two low entropy scatterers: a slightly rough surface from Table 8.1 with b=1.4a (corresponding to an incidence angle of about 35°)[28] and the single dihedral corner reflector from Table 8.1. From (8.36) the surface has the eigenvalues 2.88, 0.04 and 0.04. That gives an entropy of 0.133. For the singe dihedral (which we assume dominates its pixel's backscatter) there is only a single eigenvalue as determined above, so that the entropy is 0. How can we separate these two low entropy cases? Consider the eigenvector average angle from (8.35).

Take the surface first. With b=1.4a, (8.36) shows that the coherency matrix is

$$\mathbf{T} = \begin{bmatrix} 2.88 & 0 & 0 \\ 0 & 0.04 & 0 \\ 0 & 0 & 0.04 \end{bmatrix}$$

Since it is diagonal its entries are its eigenvalues. The eigenvectors are found by using the procedure of Sect. B.10 and, specifically, solving (B.8) with the particular value of λ inserted, i.e.

$$\begin{bmatrix} 2.88-\lambda_i & 0 & 0 \\ 0 & 0.04-\lambda_i & 0 \\ 0 & 0 & 0.04-\lambda_i \end{bmatrix} \begin{bmatrix} g_{1i} \\ g_{2i} \\ g_{3i} \end{bmatrix} = 0 \text{ with i=1,2,3}$$

[28] See Fig. 9 of S.R. Cloude and E. Pottier, A review of target decomposition theorems in radar polarimetry, *IEEE Transaction on Geoscience and Remote Sensing*, vol. 34, no. 2, March 1996, pp. 498-518.

to get the three distinct eigenvectors. For λ_1=2.88 this gives

$$0.g_{11}=0$$
$$-2.84g_{21}=0$$
$$-2.84g_{31}=0$$

Clearly $g_{21}=g_{31}=0$, but what about g_{11}? It seems indeterminate. Fortunately, there is a constraint we haven't used; that is that the eigenvectors are of unit magnitude[29], so that $g_{11}=1$. Therefore the first eigenvector is

$$\mathbf{g}_1=\begin{bmatrix}1\\0\\0\end{bmatrix}$$

By the same analysis, the other two are

$$\mathbf{g}_2=\begin{bmatrix}0\\1\\0\end{bmatrix}\quad \mathbf{g}_3=\begin{bmatrix}0\\0\\1\end{bmatrix}$$

Recall that the first element of each eigenvalue in the Cloude and Pottier decomposition is expressed

$$g_{1i}=\cos\alpha_i$$

so that for the slightly rough surface we see $\alpha_1=0°$, $\alpha_2=\alpha_3=90°$. Using these in (8.35) with the probabilities computed from (8.33) gives α=2.4°.

Now consider the dihedral corner reflector. Equation (8.38) shows that there is only one eigenvalue of value λ=2; therefore there will also only be one eigenvector, found by solving

$$\begin{bmatrix}-2 & 0 & 0\\ 0 & 2-2 & 0\\ 0 & 0 & -2\end{bmatrix}\begin{bmatrix}g_{11}\\ g_{21}\\ g_{31}\end{bmatrix}=0$$

which, again using the unit magnitude of the eigenvector, gives

$$\mathbf{g}_1=\begin{bmatrix}0\\1\\0\end{bmatrix}$$

so that α=90°. Note that there is no averaging here since there is only the one eigenvector. Using the alpha angle we can separate the surface and the dihedral reflector,

[29] A requirement of the matrix of eigenvectors being unitary.

even though their entropies are both close to zero. Note that for the case of a random collection of dihedral reflectors we find α=45°.

If we take the case of a random orientation of needle-like scatterers with eigenvalues in the proportions 2,1,1 as identified above then the average α is 45°. Also α=45° for a single needle scatterer, even though the entropy is zero.

Table 8.3 summarises these results and other observations in Cloude and Pottier.

Table 8.3 Summary of scattering types by alpha angle

Alpha angle range	**Scattering types**
near 0°	slightly rough surfaces
near 45°	random orientation of strongly anisotropic volume scatterers random orientation of needles a single needle scatterer random collection of dihedral reflectors
near 90°	single corner reflector behaviour

Although simple, that analysis suggests that the combination of entropy and alpha angle can be used to provide a form of target discrimination. Cloude and Pottier summarise the association of those measures with particular target types in an H-α diagram, shown in Fig. 8.10. Their descriptions of the various sectors is based upon the observations that the above style of analysis reveals. Fig. 8.11 shows the sectors described in terms of likely scattering types.

Two further comments are important. First, Cloude and Pottier attribute mid range alpha angles to "dipole" behaviour. That can be appreciated by looking at the analysis we carried out earlier for the random orientation of needles for which the eigenvalues are 2,1,1. The corresponding eigenvectors are

$$\begin{bmatrix}1\\0\\0\end{bmatrix} \quad \begin{bmatrix}0\\1\\0\end{bmatrix} \quad \begin{bmatrix}0\\0\\1\end{bmatrix}$$

Thus α_1=0° and α_2=α_3=90°. Noting that the probabilities are 0.5, 0.25 and 0.25, then the average alpha angle is α=45° as noted earlier. Also, the entropy was seen to be high (0.91). In contrast for a single needle the entropy is zero while the alpha angle will be seen to be 45°.

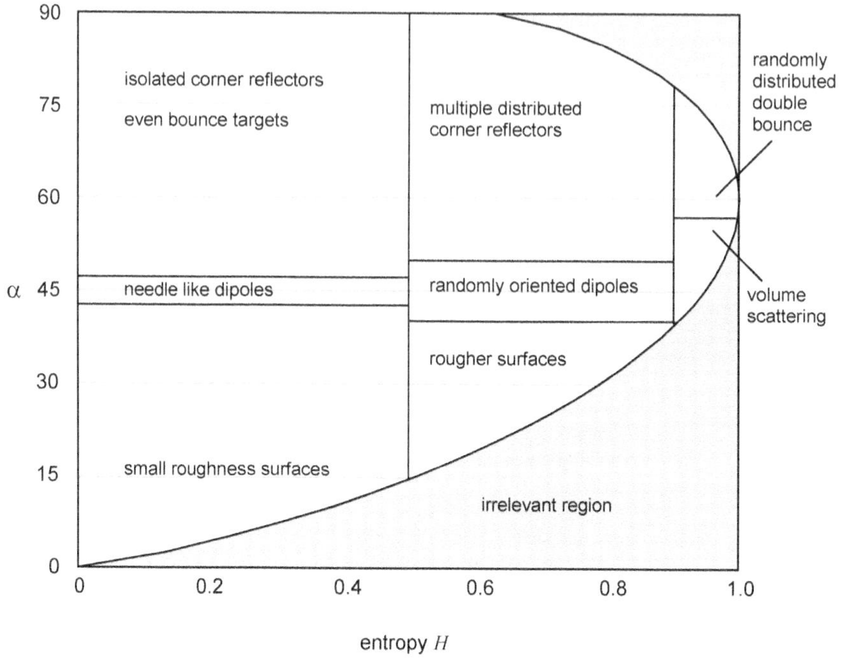

Fig. 8.10. The Cloude-Pottier H-α diagram expressed in terms of scattering types

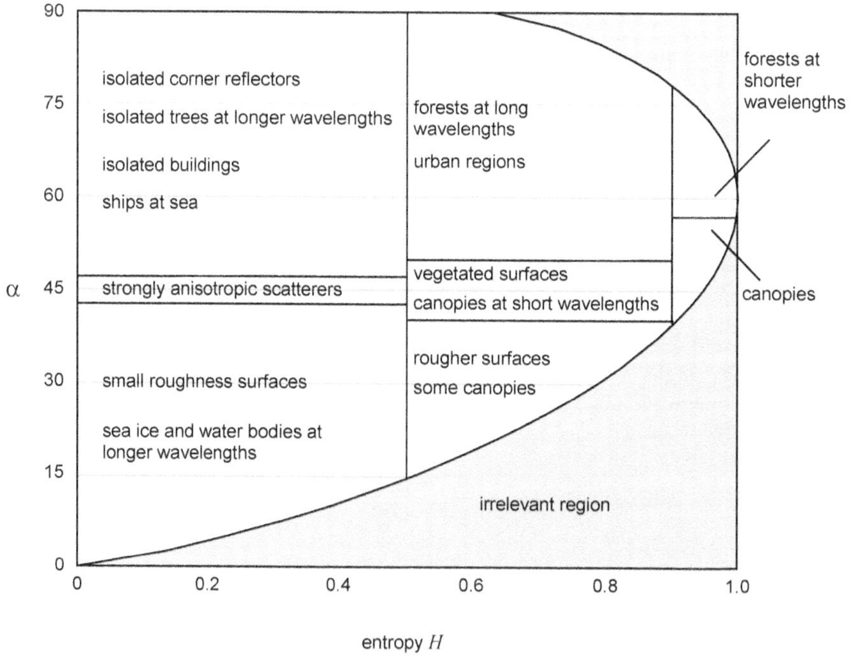

Fig. 8.11. The Cloude-Pottier H-α diagram expressed in terms of likely cover types

Secondly, there are regions of entropy and alpha that cannot co-exist, simply because there are certain maximum values for entropy for each alpha angle. The entropy maximum as a function of α is seen as a limiting curve in Fig. 8.10. We can determine that limit in the following manner, at least for the case of a diagonal coherency matrix which, as we have seen, covers targets without strongly cross polarising behaviour or those for which we average over distributions about the line of sight of the radar. Suppose the coherency matrix is

$$\begin{bmatrix} a & 0 & 0 \\ 0 & b & 0 \\ 0 & 0 & c \end{bmatrix}$$

Because it is diagonal, its eigenvalues will be *a, b* and *c*. The respective eigenvectors are

$$\begin{bmatrix} 1 \\ 0 \\ 0 \end{bmatrix}, \begin{bmatrix} 0 \\ 1 \\ 0 \end{bmatrix} \text{ and } \begin{bmatrix} 0 \\ 0 \\ 1 \end{bmatrix}$$

so that the individual alpha angles are 0°, 90° and 90°. Thus the average alpha angle is

$$\alpha = 90(p_2 + p_3) = 90(1 - p_1) \tag{8.39}$$

because the probabilities sum to unity. Entropy is given by

$$H = p_1 \log_3 p_1^{-1} + p_2 \log_3 p_2^{-1} + p_3 \log_3 p_3^{-1} \tag{8.40}$$

Because there is a one to one relationship between α and p_1 in (8.39), specifying α sets p_1 in our quest to find an expression for the limiting H-α curve. Once p_1 has been specified we need to think about what values to give p_2 and p_3. Recall from the discussion above when introducing entropy, that entropy is maximised when the probabilities are equal. That will not necessarily be the case here since we have already set a value for p_1 and have thus accounted for its contribution to the overall entropy. Continuing with the same reasoning though, the residual entropy will be maximised if the other two probabilities are the same. That is easily demonstrated numerically if necessary. Thus the maximum entropy, noting the unity sum of the probabilities, is given by

$$H_{\max} = p_1 \log_3 p_1^{-1} + 2\frac{(1-p_1)}{2}\log_3 \frac{2}{(1-p_1)}$$

Substituting from (8.39), gives

$$H_{\max} = \frac{90-\alpha}{90}\log_3 \frac{90}{90-\alpha} + \frac{\alpha}{90}\log_3 \frac{180}{\alpha} \tag{8.41}$$

which is plotted in Fig. 8.10. Cloude and Pottier[30] show that (8.41) applies even for a non-diagonal coherency matrix.

---

[30] See Cloude and Pottier, *loc cit.*

Two other measures that can be used with H and α are anisotropy A and span, defined respectively as

$$A = \frac{\lambda_2 - \lambda_3}{\lambda_2 + \lambda_3} \tag{8.42}$$

$$span = \lambda_1 + \lambda_2 + \lambda_3 = tr\mathbf{T} \tag{8.43}$$

Anisotropy is particularly interesting since it shows how different the two minor eigenvalues are. Recall that if they are equal, giving zero anisotropy, then the location of the relevant scatterer is on the limiting curves of the graphs of Figs. 8.10 and 8.11 when the coherence matrix is diagonal, signifying that the situation for the secondary scattering mechanisms is least clear. If one is zero the anisotropy will have unit magnitude which suggests there is an identifiable secondary scattering mechanism.

Figure 8.12 shows an AirSAR image of a part of the city of Brisbane, Australia with four individual cover types picked out. The H-α diagrams for each cover type in each of C, L and P bands are shown, which can be seen broadly to fall into the respective sectors identified in Fig. 8.11. The differentiation is perhaps best at L band and poorest at P, most likely because at P band most of the cover types look like slightly rough surfaces. This is demonstrated further in Fig. 8.13 which shows H-α plots for the full image.

Because the H-α plots are segmented by possible cover type as shown in Fig. 8.11 it is possible to use the boundaries in that diagram as the basis of an unsupervised classification based on entropy and alpha angle.

8.5.2.4 Coherency Shape Parameters as Features for PolInSAR Classification

The complex coherence for PolInSAR in (6.29) and (6.37) can be used in SAR image segmentation as a basis for thematic mapping. As noted in Sect. 6.14.1 it incorporates information on the scattering properties of the pixel viewed from the perspectives of each of the radars in an interferometer – i.e. from each end of the baseline with the polarisations chosen. A different polarisation is chosen for each radar in (6.37) but if the filter vectors are chosen to be the same then we have

$$\gamma = |\gamma| e^{j\psi} = \frac{\mathbf{w}^{*T}\mathbf{\Omega}_{12}\mathbf{w}}{\sqrt{\mathbf{w}^{*T}\mathbf{T}_{11}\mathbf{w}}\sqrt{\mathbf{w}^{*T}\mathbf{T}_{22}\mathbf{w}}} \quad \text{with } \mathbf{\Omega}_{12} = \langle \mathbf{k}_1 \mathbf{k}_2^{*T} \rangle$$

The coherency matrices $\mathbf{T}_{11}$, $\mathbf{T}_{22}$ and $\mathbf{\Omega}_{12}$ are properties of the resolution cell, or group of similar resolution cells, being imaged by the interferometer; they are derived from the Pauli form[31] of the target vector $\mathbf{k}_P$. The filter vector $\mathbf{w}$ allows the effect of any polarisation configuration to be incorporated into the computation of complex coherency. By varying $\mathbf{w}$ over its full range, while keeping its magnitude at unity, γ takes values appropriate to each polarisation configuration determined by the particular value of $\mathbf{w}$. The range of γ so generated, sometimes called the *coherence region*[32], tends to cluster in

[31] See (3.48, 3.49)

[32] See T. Flynn, M. Tabb and R. Carande, Coherence region shape extraction for vegetation parameter estimation in polarimetric SAR interferometry, *Proceedings of the International Geoscience and Remote Sensing Symposium 2002* (*IGARSS02*), vol. 5, June 2002, pp. 2596-2598.

the complex coherence diagram of Fig. 6.20. It is to be hoped that different scattering types will yield distinct clusters, both in position in the complex space, and in shape. The position, shape and orientation of these clusters can be used as features for segmentation and classification purposes[33].

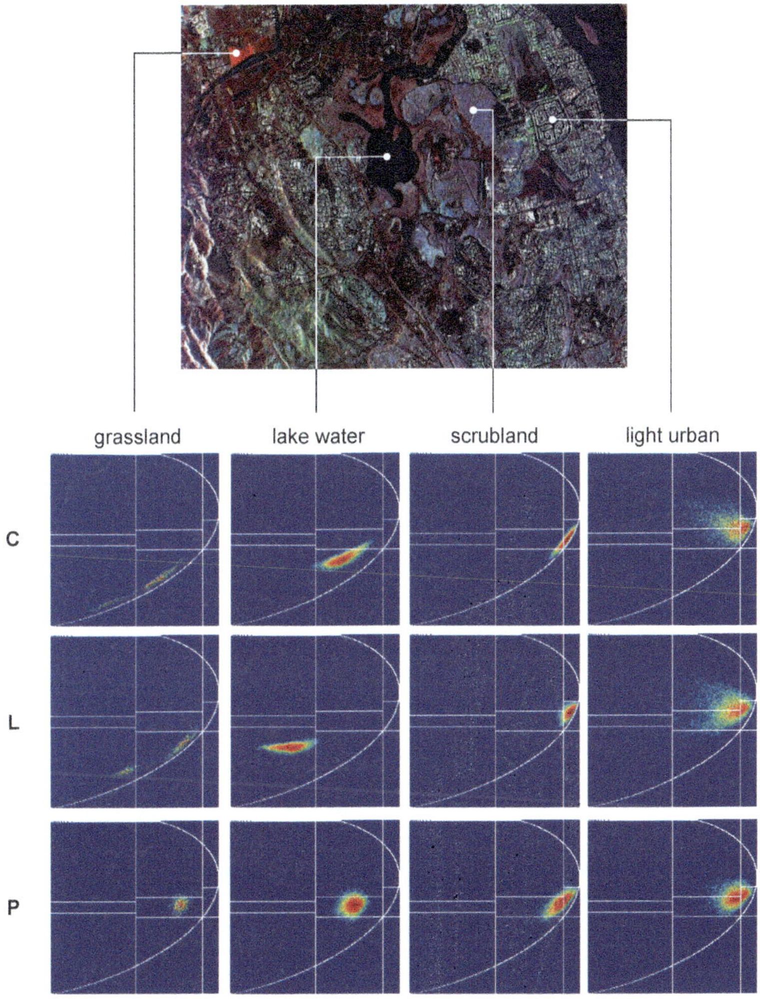

Fig. 8.12. Portion of a quad polarised AirSAR scene of the city of Brisbane, Australia with C band total power shown as red, L band total power as green and P band total power as blue, along with entropy alpha angle plots at three different wavelengths for the cover types shown: the image is in ground range format and was processed using ENVI™ (ITT Visual Information Solutions); the entropy alpha angle plots were produced using POLSARPRO V3.0

[33] See M. Neumann, A. Reigber and L. Ferro–Famil, Data classification based on PolInSAR coherence shapes, *Proceedings of the International Geoscience and Remote Sensing Symposium 2005* (*IGARSS05*), Seoul, vol. 7, 2005, pp. 4582-4585.

Fig. 8.14 illustrates a typical coherence cluster in the complex domain. The shape is shown as elliptical because that is a good approximation to the clusters seen in practice. Its features are the distance to the centroid of the cluster and the associated angle (which are the mean absolute coherence and mean phase), the major and minor axes of the cluster ellipse determined from the eigenvalues (principal components) of the covariance matrix of the cluster itself, and the orientation of the cluster. They could be used, as is, or they can form the basis of derived features[34].

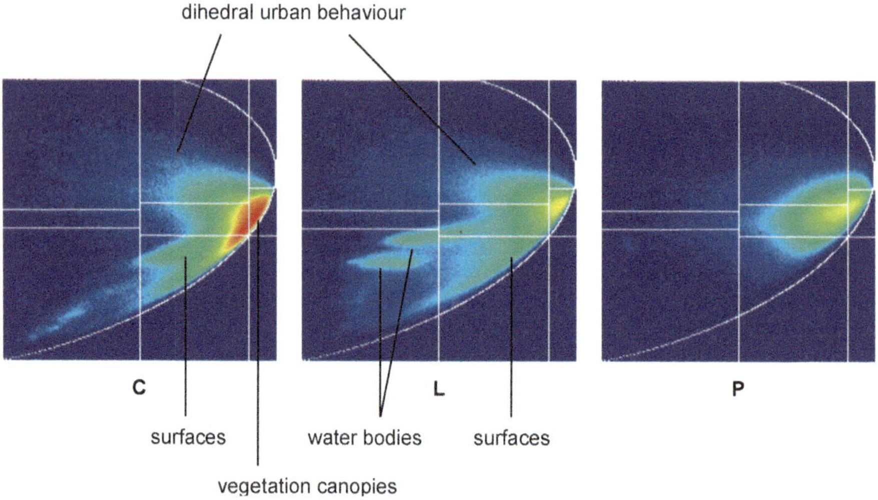

Fig. 8.13. Entropy alpha angle plots for the full Brisbane AirSAR scene of Fig. 8.12, produced using POLSARPRO V3.0

8.6 Interferometric Coherence as a Discriminator

Interferometric coherence in (6.29) is a measure of the correlation between the two different measurements taken of the same resolution cell. They could be measurements from different times, different ends of a baseline (InSAR) and/or with different polarisations (PolSAR or PolInSAR). It would be expected to be low for cover types that have changed with time and high for those that remain fairly constant. Forest canopies (and the sea surface as an extreme example) will demonstrate low coherence whereas for soil surfaces, urban regions and grasslands the coherence might be expected to be high. Coherence is therefore often a convenient feature to include in a classification owing to its ability to provide that coarse level of discrimination.

More generally, if the complex coherence associated with polarimetric radar is examined it is clear that it contains cross-correlations among all of the measurements of a pixel in all polarisation configurations. We can see that by inspecting the numerator of (6.29) which is just the joint image coherency matrix of (6.38) and which can be written

[34] *ibid.*

$$\mathbf{\Omega} = \frac{1}{2}\begin{bmatrix} <(S_{1HH}+S_{1VV})(S_{2HH}^{*}+S_{2VV}^{*})> & <(S_{1HH}+S_{1VV})(S_{2HH}^{*}-S_{2VV}^{*})> & 2<(S_{1HH}+S_{1VV})S_{2HV}^{*}> \\ <(S_{1HH}-S_{1VV})(S_{2HH}^{*}+S_{2VV}^{*})> & <(S_{1HH}-S_{1VV})(S_{2HH}^{*}-S_{2VV}^{*})> & 2<(S_{1HH}-S_{1VV})S_{2HV}^{*}> \\ 2<S_{1HV}(S_{2HH}^{*}+S_{2VV}^{*})> & 2<S_{1HV}(S_{2HH}^{*}-S_{2VV}^{*})> & 4<S_{1HV}S_{2HV}^{*}> \end{bmatrix}$$

The subscripts 1 or 2 have been added to indicate the two different images used in forming the polarimetric interferometric pair. $\mathbf{\Omega}$ should therefore be a good source of information to use for interpretation, especially when based on physical models of the scattering medium from which we can estimate likely coherences. The electromagnetically important properties of surfaces, for example, can be estimated by using simple surface models that incorporate vertical roughness, correlation length and soil moisture to estimate their effects on coherence[35].

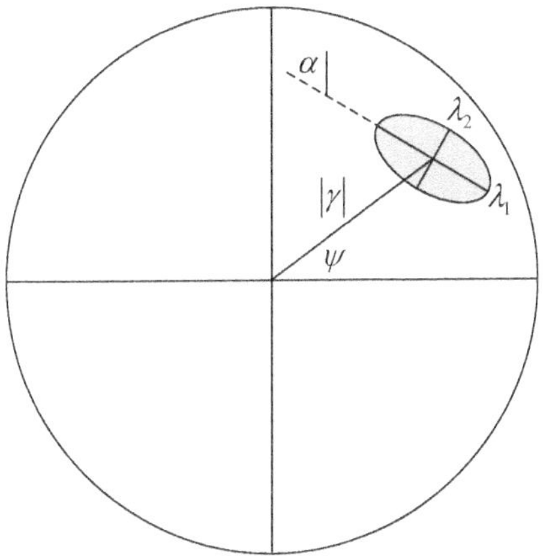

Fig. 8.14. Defining a coherence cluster by its location, shape and orientation; λ_1 and λ_2 are eigenvalues of the cluster covariance matrix indicating the principal axes of the coherence ellipse

8.7 Some Comparative Classification Results

With so many candidate approaches to thematic mapping using radar image data it is reasonable to ask whether any stand out above the rest in terms of performance. It is important to recognise however, as with the classification of optical data, success often depends on the methodology that surrounds the use of a particular algorithm and the skill of the analyst, particularly during the training phase of classification.

In Table 8.4 we have summarised a number of investigations to give a comparative indication of performance. For the reasons just given it is important not to place too much emphasis on this material, but just to use it for guidance. There is an interesting

[35] See I. Hajnsek, K.P. Papathanassiou, A. Moreira and S. R. Cloude, Surface parameter estimation using interferometric and polarimetric SAR, *Proceedings of the International Geoscience and Remote Sensing Symposium 2002 (IGARSS02)*, vol. 1, 24-28 June 2002, pp. 420-422, and I. Hajnsek and P Prats Soil moisture estimation in time with airborne D-InSAR, *Proceedings of the International Geoscience and Remote Sensing Symposium 2008 (IGARSS08)*, vol. 3, 7-11 July 2008, pp. 546-549.

precautionary tale from a recent study of classification results with remote sensing imagery[36]: over the fifteen year period 1989-2004, notwithstanding the development of new techniques, average classification performance did not improve.

We have only chosen a representative set of results to use in Table 8.4, and not all the results for a given study have necessarily been included. Studies for which no quantitative results are given have not been used. Regrettably with many radar studies it has become common to cite qualitative segmentation results rather than comparisons against ground truth or reference data; that limits the values of those types of study.

Interestingly, the results summarised in Table 8.4 don't suggest that one approach is naturally superior to another, at least at this high level of comparison. In reality, it is important to match the data to the application. SAR classification performs best when the classes have good structural or dielectric constant differentiation, such as with the examples in the table involving forests and sea ice mapping. For crop thematic mapping and when there are a number of classes quite different from forest included in a forest mapping study the results are generally not as good. It is important to consider the combination of optical and radar imagery in more general thematic mapping since classes that present difficulties for one data type may be readily resolved in the other. Moreover, it is important to be aware that some of the classes of interest to the user may not be reachable with any data type on its own and may need inferences form both optical and radar imagery to be revealed in thematic mapping[37].

Table 8.4 Some radar classification studies

Technique and source	**Features**	**Class types and results**
Supervised labelling using the Wishart classifier *Lee et al*[38]	P,L and C four look data.	Ice data set with four classes: open water, first year ice, multi-year ice and ice ridges. Overall accuracies achieved on the training data were 79% at P band, 86% at L band and 81% at C band. When all bands were used together 94% was achieved. Theoretical simulations to improve training estimates gave higher values.
Knowledge based classification built on simulated and experiential scattering behaviours. Includes post classification modal filtering to improve homogeneity. This work also summarises SAR classification to 1996. *Dobson et al*[39]	ERS-1 and JERS-1 backscattering coefficients and expert rules in the form of linear discriminants.	Five classes: surface, short vegetation, upland conifers, lowland conifers, decurrent broadleaf. Testing set classification accuracies of about 94% were achieved.

[36] See G.G. Wilkinson, Results and implications of a study of fifteen years of satellite image classification experiments, *IEEE Transactions on Geoscience and Remote Sensing*, vol. 43, no. 3, March 2005, pp. 433-440.

[37] See, for example, J.A. Richards, Analysis of remotely sensed data: the formative decades and the future, *IEEE Transactions on Geoscience and Remote Sensing*, vol. 43, no. 3, March 2005, pp. 422-432.

[38] J-S Lee, M.R. Grunes and R. Kwok, Classification of multi-look polarimetric SAR imagery based on complex Wishart distribution, *Int. J. Remote Sensing*, vol. 15, 1994, pp. 2299-2311.

Isodata unsupervised classification using the Wishart distance measure, initialised with *H-α* categories similar to Fig. 8.11. *Ferro-famil et al*[40]	P and L fully polarimetric single look complex data.	Two soil and four forest age classes. Results typically in the range 84-98% with the exception of poor performance on soil.
Supervised Wishart classification, with speckle reduction on both fully polarimetric and individual combinations of polarisation and intensity measurements. *J-S Lee et al*[41]	P, L and C fully polarimetric single look complex and intensity data.	Crop exercise: six crop classes, water, forest, lucerne, bare soil, grass. Overall accuracy 71% for P band, 82% for L band and 67% for C band, but 91% when all used. Forest exercise: six age classes and bare soil. Overall accuracy 79% for P band, 65% for L band and 43% for C band.
Support vector machine applied to intensity data, with speckle filtering. *Fukuda and Hirosawa*[42]	P, L and C fully polarimetric intensity data.	Nine crop classes, forest, water and two bare soil classes. Accuracies in separating class pairs are in the range of 90%.
Supervised Bayesian hierarchical classifier (decision tree), maximum likelihood classification and Isodata unsupervised clustering. *Kouskoulas et al*[43]	L and C band backscatter coefficients at HH, HV and VV plus VV/HH average complex coherence	Four crop classes: wheat, alfalfa, corn and soybeans. Overall results on a testing set were 74% with clustering, 84% with standard maximum likelihood classification and 93% with the Bayesian hierarchical classification technique.
Supervised maximum likelihood classification. *Karathanassi and Dabboor*[44]	Absolute values of the Pauli components (S_{HH}-S_{VV}), S_{VH} and (S_{HH}+S_{VV}) from E-SAR data.	Four classes: urban, forest, vegetation and runways. Overall accuracy without speckle filtering was 80%, rising to 91% when the Lee filter with a 21x21 window used.
Support vector machines and random forests of decision trees, on individual and fused SAR and TM data sets; pixels were aggregated to various sizes of object for classification. *Waske and van der Linden*[45]	Thematic mapper bands, ASAR and ERS-2 backscatter (intensity) data.	Five crop, one soil, one forest and one urban class. Individual class results in the range 64-96% for SAR data alone and 63-98% for TM data alone. Fused results in the range of 76-97%.

[39] M.C. Dobson, L.E. Pierce and F.T Ulaby, Knowledge-based land-cover classification using ERS-1/JERS-1 SAR composites, *IEEE Transactions on Geoscience and Remote Sensing*, vol. 34, no. 1, January 1996, pp. 83-99.

[40] L. Ferro-Famil, E. Pottier and J-S Lee, Unsupervised classification of multifrequency and fully polarimetric SAR images based on the H/A/Alpha-Wishart classifier, *IEEE Transactions on Geoscience and Remote Sensing*, vol. 39, no. 11, November 2001, pp. 2332-2342.

[41] J-S Lee, M.R. Grunes and E. Pottier, Quantitative comparison of classification capability: fully polarimetric versus dual and single-polarization SAR, *IEEE Transactions on Geoscience and Remote Sensing*, vol. 39, no. 11, November 2001, pp. 2343-2351.

[42] S. Fukuda and H. Hirosawa, Polarimetric SAR image classification using support vector machines, *IEICE Transactions on Electronics*, vol. E84-C, 2001, pp. 1939-1945.

[43] Y. Kouskoulas, F.T. Ulaby and L.E. Pierce, The Bayesian hierarchical classifier (BHC) and its application to short vegetation using multifrequency polarimetric SAR, *IEEE Transactions on Geoscience and Remote Sensing*, vol. 42, no. 2, February 2004, pp. 469-477.

[44] V. Karathanassi and M. Dabboor, Land cover classification using E-SAR polarimetric data, *Proc. Commission VII, ISPRS Congress*, Istanbul, 2004, pp. 280-285.

8.8 Finding Pixel Vertical Detail Using Interferometric Coherence

We now turn to interpretation of the vertical profile of a pixel. This is not a pixel labelling process, but is an analytical procedure that reveals information made possible because of the coherent nature of radar imagery.

A simple model for interpreting forest structure information makes use of complex coherence with multi-polarisation data; this is an alternate to polarisation coherence tomography treated in Sect. 6.15.5. Called the random volume over ground (RVOG) model[46], it uses the composite complex coherence

$$\gamma = e^{j\Delta\phi_{topo}}\frac{\gamma_v + m}{1+m} = e^{j\Delta\phi_{topo}}[\gamma_v + \frac{m}{1+m}(1-\gamma_v)] \tag{8.44}$$

in which γ_v is the complex coherence of the vegetation layer and m is the ratio of the ground to the volume power received. Varying this parameter essentially allows the effect of vegetation density to be examined. It assumes that range spectral filtering has been performed (Sect. 6.16) and that the scattering from the surface under the canopy is direct backscattering and not specular surface scattering followed by a subsequent vegetation scatter back to the radar (i.e. there is no double bounce mechanism).

The vegetation coherence is given by

$$\gamma_v = \frac{\int_0^{h_v}\exp(2\kappa_e h\sec\theta + jk_h h)dh}{\int_0^{h_v}\exp(2\kappa_e h\sec\theta)dh} \tag{8.45}$$

in which κ_e is the power extinction coefficient of the volume layer which extends from the surface at h=0 to a height h=h_v, θ is the incidence angle and k_h is the vertical wave number, given by

$$k_h = \frac{2k\Delta\theta}{\sin\theta} \tag{8.46}$$

where $\Delta\theta$ is the change in incidence angle associated with the different viewing positions of the radars at either end of the interferometric baseline, and $k=2\pi/\lambda$.

Note that the denominator of (8.45) is

$$\int_0^{h_v}\exp(2\kappa_e h\sec\theta)dh = \cos\theta\frac{e^{2\kappa_e h_v\sec\theta}-1}{2\kappa_e}$$

[45] B. Waske and S. van der Linden, Classifying multilevel imagery from SAR and optical sensors by decision fusion, *IEEE Transactions on Geoscience and Remote Sensing*, vol. 46, no. 5, May 2008, pp. 1457-1466.

[46] See R.N. Treuhaft and P.R. Siqueira, Vertical structure of vegetated land surfaces from interferometric and polarimetric radar, *Radio Science*, vol 35, 2000, pp. 141-177, S.R. Cloude and K.P. Papathanassiou, Polarimetric SAR Interferometry, *IEEE Transactions on Geoscience and Remote Sensing*, vol. 36, no. 5 part 1, September 1998, pp. 1551-1565, and K.P. Papathanassiou and S.R. Cloude, Single-Baseline Polarimetric SAR Interferometry, *IEEE Transactions on Geoscience and Remote Sensing*, vol. 39, no.11, November 2001, pp. 2352-2363

and its numerator can be expressed

$$\int_0^{h_v} \exp(2\kappa_e h \sec\theta + jk_h h)dh = \frac{e^{2\kappa_e h_v \sec\theta + jk_h h_v} - 1}{2\kappa_e \sec\theta + jk_h}$$

so that the vegetation coherence can be written

$$\gamma_v = 2\kappa_e h_v \sec\theta \frac{e^{2\kappa_e h_v \sec\theta} e^{jk_z h_v/2} \text{sinc}(k_h h_v/2 - j\kappa_e h_v \sec\theta)}{e^{2\kappa_e h_v \sec\theta} - 1} \qquad (8.47)$$

which is the same as (6.68), derived for polarisation coherence tomography, noting that the topographic phase term is incorporated in (8.44). Before proceeding, note that if the canopy were lossless, so that in the limit $\kappa_e \to 0$, then (8.47) reduces to

$$\gamma_v = e^{jk_h h_v/2} \text{sinc}(k_h h_v/2)$$

which is the same as (6.66) given again that the topographic phase term has been taken care of in (8.44).

We now return to the task of structural identification of the vegetation canopy from (8.44). The topographic phase term is not polarisation sensitive, and from (8.47) neither is volume coherence on the assumption that the extinction coefficient is independent of polarisation. Given the volume model is based on a totally random volume of scatterers that is an acceptable assumption.

The complex coherence of (8.44) varies along a straight line in the complex plane with variations in the ground to volume power ratio *m*, as shown in Fig. 8.15. Interestingly, the line meets the unit circle at the angle that corresponds to the topographic phase. That can be seen by letting *m*=0 (no surface contribution) so that the complex coherence is

$$\gamma = |\gamma_v| e^{j\Delta\phi_{topo} + \angle\gamma_v}$$

That point is shown in Fig. 8.15. If the vegetation coherence were unity (i.e. no volume decorrelation, such as might happen with a very lossy canopy as seen in Fig. 6.29) then the net complex coherence is just $\gamma = e^{j\Delta\phi_{topo}}$, which is on the unit circle. As the vegetation coherence falls from unity then the complex coherence moves away from that boundary point.

The only term in (8.44) that could depend on polarisation is the ground to volume power ratio *m*, which is therefore sometimes written as a function of the filter vector **w**. That is not essential if we know what polarisations we are using or interested in. We normally append a subscript to *m* to signify polarisation.

In (8.44), with (8.47), there are 3 unknowns: the ratio of ground to volume contributions *m*, the canopy extinction coefficient κ_e and the canopy height h_v. There are four if we also have to estimate the topographic phase $\Delta\phi_{topo}$. But in the measurement of complex coherence there are only two pieces of information – its magnitude and phase. However, by using three different polarisation configurations – say HH, VV and HV – we can set up the following set of equations that have six unknowns and six measured quantities.

$$\gamma_{HH} = e^{j\Delta\phi_{topo}}[\gamma_v + \frac{m_{HH}}{1+m_{HH}}(1-\gamma_v)] \quad (8.48a)$$

$$\gamma_{VV} = e^{j\Delta\phi_{topo}}[\gamma_v + \frac{m_{VV}}{1+m_{VV}}(1-\gamma_v)] \quad (8.48b)$$

$$\gamma_{HV} = e^{j\Delta\phi_{topo}}[\gamma_v + \frac{m_{HV}}{1+m_{HV}}(1-\gamma_v)] \quad (8.48c)$$

In order to incorporate a ground term it is tacitly assumed that there is sufficient penetration that some energy reaches the ground and is backscattered to contribute to the coherence seen by the interferometer. That would suggest that for most forest-like applications the wavelength used is long – say L band or even P band. This model has also been applied at X band[47], although reasonably the cross polar contribution from the ground is assumed to be negligible, simplifying ((8.48c) to $\gamma_{HV} = e^{j\Delta\phi_{topo}}\gamma_v$.

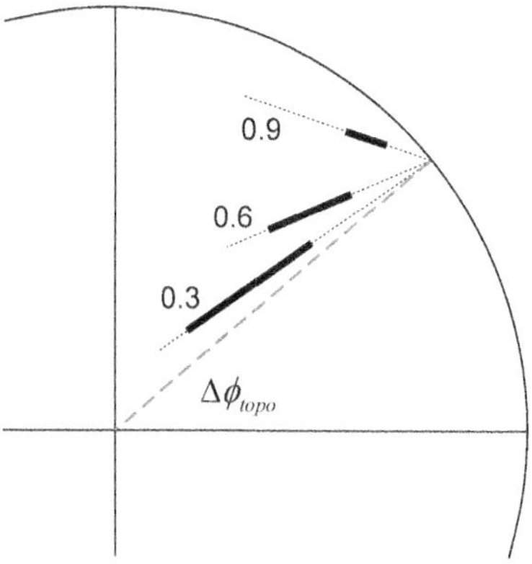

Fig. 8.15. Plots (solid lines) the complex coherence of (8.44) for a topographic phase angle of 0.7rad, and a volume complex coherence of $|\gamma|\exp(j0.2)$ with $|\gamma|$=0.3, 0.6, 0.9 and for m varying over the full range of 0 to 1; the dotted trend lines converge to the point on the unit circle corresponding to the topographic phase

[47] See F. Garestier, P.C. Dubois-Fernandez and K.P. Papathanassiou, Pine forest height inversion using single-pass X band PolInSAR data, *IEEE Transactions on Geoscience and Remote Sensing*, vol. 46, no. 1, January 2008, pp. 59-68.

CHAPTER 9

PASSIVE MICROWAVE IMAGING

9.1 Introduction

Although the theme of this book is radar remote sensing, imaging with passive microwave is a complementary technology that warrants an introduction to identify its role alongside radar. This chapter lays the framework for passive microwave imaging, drawing in part from the scattering treatment developed for imaging radar in Chapt. 5

Notwithstanding the very small natural power density levels available from the earth as seen in Chapt. 2, passive microwave remote sensing is possible provided sufficiently large resolution cells are used so that measurable power levels can be obtained. It is an important remote sensing technology, particularly for sea, ice and snow mapping and in the assessment of soil moisture.

In principle, passive microwave imaging is similar to image data gathering at optical wavelengths: upwelling radiation is detected (using a *radiometer*) and converted to a brightness value from which an image is formed. The source of energy may not always be just the earth's surface. The atmosphere can also generate measureable energy at certain wavelengths as can sub-surface features. Moreover, there is a finite level of solar microwave radiation scattered from the earth's surface which can contribute to the total power level detected. Those components are depicted in Fig. 9.1.

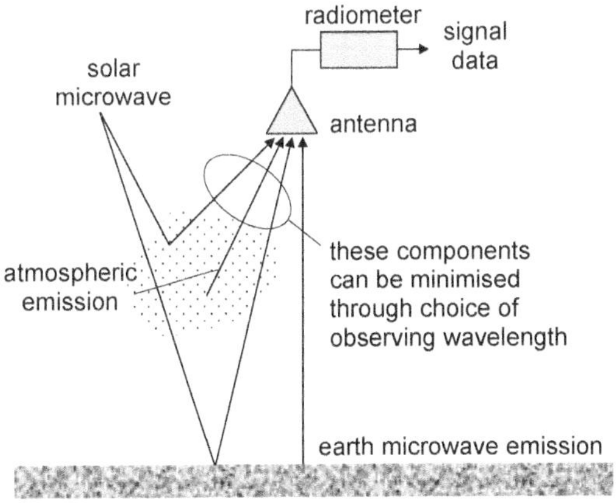

Fig. 9.1. The components of passive microwave energy theoretically available for measurement

Because it is a passive technology the synthetic aperture techniques used with radar are not available for generating fine spatial resolutions with microwave radiometry.

J.A. Richards, *Remote Sensing with Imaging Radar*, Signals and Communication Technology,
DOI: 10.1007/978-3-642-02020-9_9,

Consequently, pixels sizes are generally or the order of 10km or so. The terms "aperture synthesis" and "synthetic aperture" are nevertheless still found in connection with passive imaging. However they refer to the use of arrays of small antennas to synthesise the large aperture needed to gather the weak radiometric signal, rather than as techniques to enhance spatial resolution[1].

9.2 Radiometric Brightness Temperature

As with active radar techniques we use the received power level to build up the microwave image of a scene. Received power itself, though, is not a good indicator of the intrinsic properties of the material being imaged since it will vary with the bandwidth over which the measurements are made and with the pixel size used. Instead, we need a quantity that can be derived from the received power but which is invariant with system parameters like spatial resolution and measurement bandwidth. Equation (2.6) shows that, in the microwave range of wavelengths, the spectral power density emitted by an object is directly proportional to the temperature of its surface. Thus the power density detected by a radiometer will also be directly proportional to the temperature of the object being observed. The actual power in watts received by an antenna when it is irradiated by a black body can be expressed[2]

$$P = kT\mathcal{B} \tag{9.1}$$

In which T is the surface temperature of the body (degrees K) and $\mathcal{B}$ is the bandwidth (Hz) over which the microwave emission is observed; k is Boltzmann's constant, which has the value $1.38065\text{x}10^{-23}\text{JK}^{-1}$. As a consequence of (9.1) we could infer the temperature of the region being observed from

$$T = \frac{P}{k\mathcal{B}} \tag{9.2}$$

As discussed in Sect. 2.1 a real scene does not behave as an ideal black body but emits a lower level of energy, described by its emissivity ε, with $0 \le \varepsilon \le 1$. If P_r were the actual power received from a real surface then (9.1) is modified to

$$P_r = \varepsilon kT\mathcal{B}$$

which can be re-arranged

$$P_r = k(\varepsilon T)\mathcal{B} = kT_B\mathcal{B}$$

in which we have introduced the *radiometric brightness temperature* in degrees Kelvin

$$T_B = \varepsilon T = \frac{P_r}{k\mathcal{B}} \tag{9.3}$$

that characterises the material being imaged. It is determined by the real (i.e. physical) temperature of the material (sometimes written as T_O) and its emissivity. Whereas we talk

---

[1] See D.M. LeVine, Synthetic aperture radiometer systems, *IEEE Transactions on Microwave Theory and Techniques*, vol. 47, no. 12, December 1999, pp. 2228-2236.

[2] Strictly, this requires the antenna to be completely surrounded by the black body to be true. Fortunately, in our treatment of passive imaging we do not need to observe that theoretical requirement.

of scattering coefficient for a radar imaging system, we will talk of brightness temperature for passive microwave imaging.

The radiometric brightness temperature can be polarisation dependent. In other words the upwelling microwave energy (or power in 9.3) can be a function of the polarisation of observation. It is therefore convenient to summarise the polarisation dependence of the brightness temperature using a Stokes vector; normally the modified form of (2.34) is used in which we write

$$\mathbf{s}=\begin{bmatrix}|E_V|^2\\|E_H|^2\\2\,\mathrm{Re}\langle E_H E_V^*\rangle\\2\,\mathrm{Im}\langle E_H E_V^*\rangle\end{bmatrix}=\frac{\lambda^2}{k\eta}\begin{bmatrix}T_V\\T_H\\T_U\\T_V\end{bmatrix}=\frac{\lambda^2}{k\eta}\mathbf{T}_B \tag{9.4}$$

in which the T_U and T_V components relate to the ellipticity of the polarisation and thus the correlation between the horizontal and vertical components. They are generally small compared with the first two elements and, in the past, had been considered negligible. They are now known to provide important discriminating information when the region being image is anisotropic or asymmetric in its emissive properties. $\mathbf{T}_B$ is the brightness temperature vector. Using this we can generalise (9.3)

$$\mathbf{T}_B = T\boldsymbol{\varepsilon} \tag{9.5}$$

in which $\boldsymbol{\varepsilon}$ is an emissivity vector. Since emissivity is unity for an ideal black body, the emissivity vector for a black body is $[1\ 1\ 0\ 0]^T$; in other words it radiates horizontally and vertically and there is no relation between the horizontal and vertical radiation.

The constant of proportionality in (9.4) arises from equating the received power expressed in (9.3) with that captured by an antenna for which the aperture is proportional to λ^2; the power density incident on that antenna is given by $|E||H|=\frac{|E|^2}{\eta}$ in which η is the impedance of free space (377Ω) – see (2.7). Sometimes (9.4) is written in the form of (2.30)

$$\mathbf{s}=\begin{bmatrix}|E_H|^2+|E_V|^2\\|E_H|^2-|E_V|^2\\2\,\mathrm{Re}\langle E_H E_V^*\rangle\\2\,\mathrm{Im}\langle E_H E_V^*\rangle\end{bmatrix}=\frac{\lambda^2}{k\eta}\begin{bmatrix}T_H+T_V\\T_H-T_V\\T_{+45^o}-T_{-45^o}\\T_L-T_R\end{bmatrix}$$

in which T_{+45°} and T_{-45°} are temperatures at the two linear polarisations inclined at 45°, and T_L and T_R are temperatures at left and right circular polarisations.

9.3 Relating Microwave Emission to Surface Characteristics

Just as it is important to know how scattering coefficients and scattering matrices depend upon the nature of the earth's surface (dielectric constant and roughness in particular) and system parameters (wavelength, incidence angle and polarisation), it is also important to understand how radiometric brightness temperature depends on those quantities since that is the type of information needed for interpreting passive microwave images. Fortunately, we don't have to start anew to discover the relationships of importance. Rather, we can make use of the material on surface and volume scattering with radar that was derived in Chapt. 5.

Before proceeding, there is one definition we need to be clear about. In radar we talk explicitly about incidence angle as the angle that the transmitted ray makes with the earth's surface. In passive imaging there is no transmitted signal; rather we observe the radiation emitted from the earth's surface in a given angular direction. We should therefore talk about observation angle; in practice however the term incidence angle is still sometimes used, even though it is inappropriate.

We now come to an important principle that underpins passive microwave remote sensing. When an object is in thermal (or thermodynamic) equilibrium – i.e. its temperature is constant – the amount of energy it is able to emit is the same as the amount of energy it is capable of absorbing. If the energy it is emitting is different from the energy it is absorbing then it will either be warming up or cooling down. It can only be at a stable temperature when the two are in balance. From a passive imaging viewpoint we are interested in the emitted energy (or power) from which we can infer radiometric brightness temperature. To find that we use the principle of thermal equilibrium and search instead for the power absorbed, since we have a pathway for finding that quantity.

In radar the power absorbed is that fraction which is not reflected or scattered. It is determined by what we might call the *absorptivity* of the medium. Fig. 9.2 shows the relationship between the incident, reflected and absorbed components of power. If the reflectivity is represented by Γ, then the absorptivity is $1-\Gamma$. With the assumption of thermal equilibrium that is also equal to the emissivity of the surface, so we have the important relationship

$$\varepsilon_P = 1 - \Gamma_P \tag{9.6}$$

where the subscript P refers to polarisation.

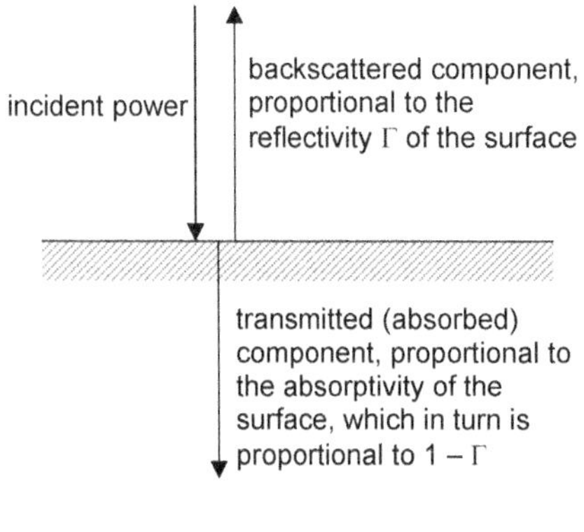

Fig. 9.2. Relationship between the absorptivity and reflectivity of a surface

For a specular surface Γ_P will be the power reflection coefficient of (5.1). For a rough surface it is related to the surface scattering coefficient, but it must account for the totality of scattering from the surface and not just that in the "backscattered" direction since we really want to know how much travels across the boundary and, in principle, is absorbed. We can only do that if we take into account all the scattered power as shown in Fig. 9.3.

Just as with radar imaging, radiated emissions depend on the angle of view. This is seen easily in the specular case of Fig. 5.2, for which

$$\varepsilon_p(\theta) = 1 - \Gamma(\theta) = 1 - |\rho_p(\theta)|^2 \tag{9.7}$$

where $\rho_P(\theta)$ is the Fresnel reflection coefficient for the surface discussed in Sect. 5.3.1, and which is explicitly dependent on incidence angle.

As an illustration consider the case of a surface which is a good electrical conductor – such as a metallic plate. This is unlikely in remote sensing, but provides some useful guidance for what is to come. It can be shown that $\rho = -1$ for such a surface, irrespective of polarisation or incidence (observation) angle. Thus the emissivity of the surface, from (9.6), is zero and its radiometric brightness temperature, from (9.3), is also zero. In a passive image, which has been calibrated to make brightness increase with emissivity or brightness temperature, a metallic surface will therefore show as black.

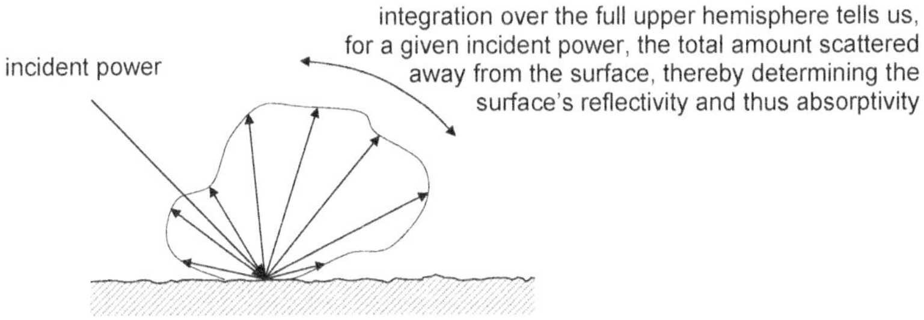

Fig. 9.3. All the power scattered into the upper hemisphere has to be found when determining reflectivity for a rough surface

Now consider the more realistic case of a still water body viewed directly from above (i.e. vertical "incidence"). For convenience we can assume that the dielectric constant of water is about 81 so that from (5.2) the Fresnel reflection coefficient is about 0.8. Therefore the emissivity will be 1–0.64=0.32. The brightness temperature of the water is then $0.32T_o$, where T_o is the physical temperature. For a water surface temperature of 293K, this gives a brightness temperature of 105K.

Now examine the case of a still water body viewed from any angle. Equation (5.3) gives the Fresnel reflection coefficients as a function of angle of incidence and polarisation. Substituting those expressions into (9.7) with the assumption, again, that the dielectric constant of the water is 81 yields the curves of Fig. 9.4. The strong peak in the vertically polarised curve is the result of an effect know as the Brewster angle, that does not occur for horizontally polarised radiation.[3] For lossless dielectric media the Fresnel reflection coefficient is zero at the Brewster angle. Note that both curves converge to 105K when the surface is viewed from directly above and that there is a strong polarisation dependence as observation angle increases. At the peak in the curve for

[3] See J.A. Richards, *Radio Wave Propagation: An Introduction for the Non-Specialist*, Springer, Berlin, 2008.

vertical (parallel) polarisation, corresponding to the Brewster angle, the brightness temperature is equal to the physical temperature.

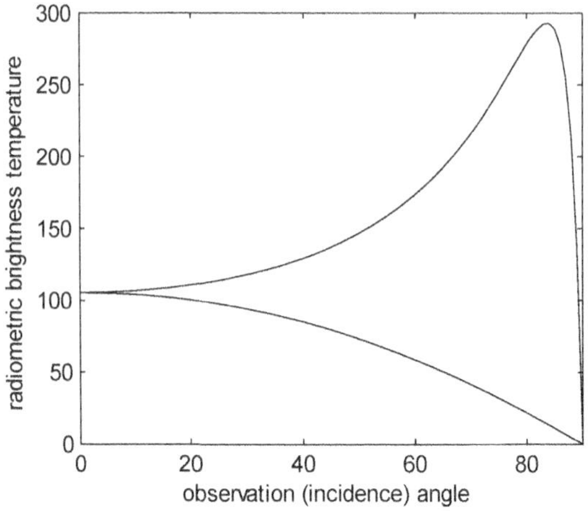

Fig. 9.4. Radiometric brightness temperature for still water as a function of observation angle and polarisation

9.4 Emission from Rough Surfaces

The previous section looked at emission from ideally smooth surfaces. Consider now the other extreme of an ideally rough surface, characterised by Lambertian scattering described by (5.5a) and illustrated in Fig. 9.5. In order to find the emissivity of this surface in terms of its absorptivity and thus reflectivity it is necessary to integrate the scattered energy over the whole upper hemisphere. We also have to take into account any depolarisation as part of the scattering process. When (5.5a) is integrated in this manner[4] the reflectivity of the Lambertian surface is seen to be

$$\Gamma(\theta) = \frac{\sigma^o(0)}{4}$$

so that

$$\varepsilon_p(\theta) = 1 - \frac{\sigma^o(0)}{4} \tag{9.8}$$

Thus the emissivity, and therefore the image tone for a very rough surface, is independent of observation angle, polarisation and frequency, provided the Rayleigh roughness criterion (5.4) holds.

---

[4] See F.T. Ulaby, R.K. Moore and A.K. Fung, *Microwave Remote Sensing: Active and Passive, Vol 1*, Addison-Wesley, Reading Mass., 1982, p 251.

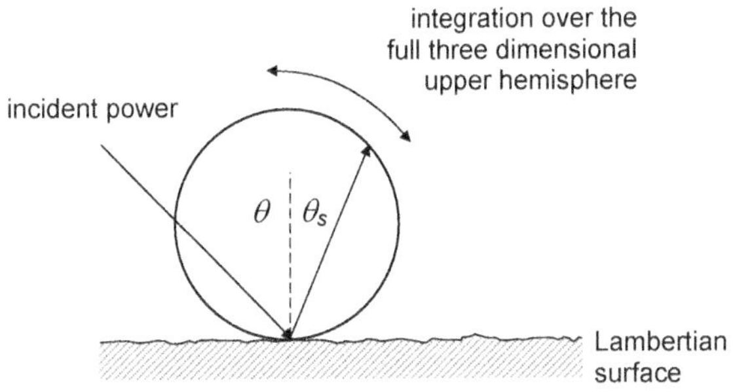

Fig. 9.5. A Lambertian surface

We have now examined the two extreme cases of surface roughness. Other surfaces will have emission characteristics between the behaviours of those extremes as illustrated in the curves of Fig. 9.6. As roughness increases, emissivity and thus radiometric brightness temperature increases and is less dependent on observation angle; it also exhibits less variation with polarisation. The reason that rougher surfaces absorb and thus emit more can be appreciated by noting the greater likelihood of multiple interactions (and thus chances for absorption) with local surface variations as illustrated in Fig. 9.7. For smoother surfaces the radiometric brightness temperature is lower, more sensitive to observation angle (as seen in Fig. 9.4) and shows more variation with polarisation.

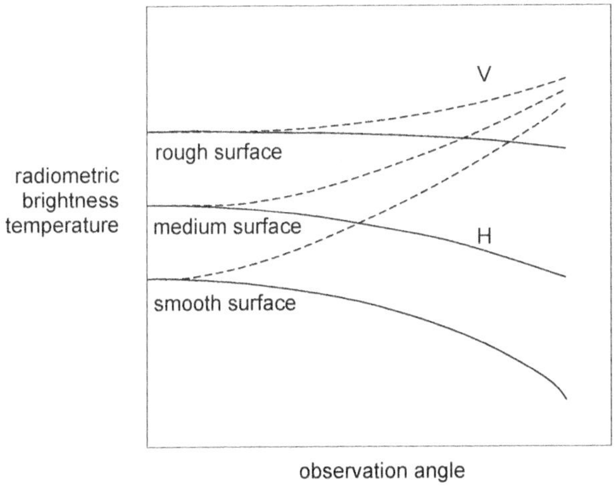

Fig. 9.6. Dependence of radiometric brightness temperature on surface roughness and polarisation

9.5 Dependence on Surface Dielectric Constant

Since the reflectivity of a surface, both specular and diffuse, increases with dielectric constant and thus water content, absorptivity and thus emissivity will decrease. Radiometric brightness temperature will therefore decrease with increasing moisture, as illustrated in Fig. 9.8 for the case of a smooth sandy surface at 1.4GHz. Those curves have been produced using (5.3) in (9.7) and with the values of dielectric constant in Table 9.1 for the moisture contents shown. The sensitivity to moisture evident in Fig. 9.8 demonstrates the value of passive microwave imaging for soil moisture studies. For a real situation there will also be a small imaginary part of the dielectric constant associated with energy loss; that has been ignored in constructing Fig. 9.8.

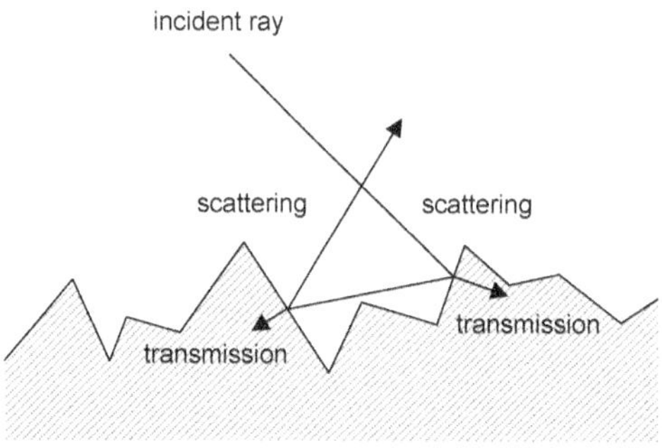

Fig. 9.7. Demonstrating the enhanced possibility of multiple surface interactions, and thus opportunities for transmission and absorption, with increased surface roughness

Table 9.1 Dielectric constants of sand with varying moisture contents (From Fig. 7.4 of J.A. Richards, *Radio Wave Propagation: An Introduction for the Non-Specialist*, Springer, Berlin, 2008.)

Volumetric moisture content	Approximate dielectric constant (real part only)
10%	6.3
20%	11.4
30%	18.2

9.6 Sea Surface Emission

Figure 9.4 shows the emission from a still water body. The sea surface however is most often roughened by waves. From the discussion in Sect. 9.4 concerned with soils we can induce that the rougher the sea surface the higher its brightness temperature. Since the roughness of the sea surface depends on wind speed, brightness temperature can be used as an indicator of wind speed.

Apart from the level of wind speed, knowledge of its direction over the ocean is also important. Measuring fully polarimetric brightness temperature data represented by the Stokes parameters in (9.4), allows wind speed direction (the vector wind field) to be estimated; there is an ambiguity when relying on the first two Stokes parameters alone which can be resolved using the third and fourth parameters. Consequently, T_U and T_V have become important indicators of asymmetry (and thus ocean anisotropy). The dependence on wind speed comes through its effect on the vector emissivity of (9.5) provided the contributions of the atmosphere and sky to the recorded brightness temperature have been removed. To a good approximation the components of the emissivity vector can be expressed

$$\varepsilon_V = a_{0V} + a_{1V}\cos\phi + a_{2V}\cos 2\phi$$
$$\varepsilon_H = a_{0H} + a_{1H}\cos\phi + a_{2H}\cos 2\phi$$
$$\varepsilon_U = b_U\sin\phi + b_U\sin 2\phi$$
$$\varepsilon_V = b_V\sin\phi + b_V\sin 2\phi$$

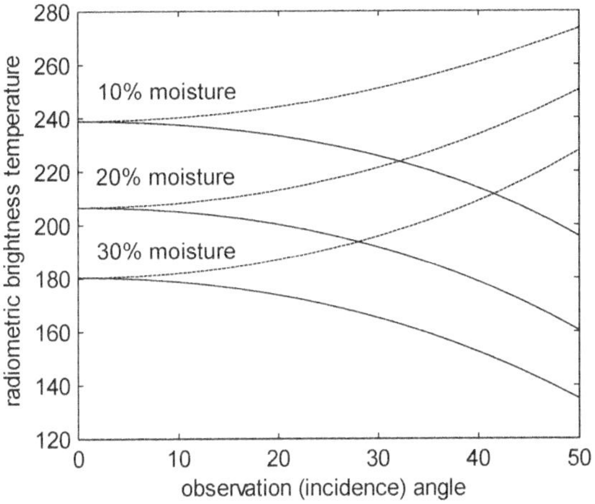

Fig. 9.8. Computed radiometric brightness temperature of smooth sand at 1.4GHz as a function of moisture content using the dielectric constants of Table 9.1, ignoring the effect of the (small) imaginary component of the dielectric constant and any surface roughness; the full lines represent horizontal polarisation and the dotted (upper) lines vertical polarisation

The coefficients in these expansions, often referred to as the harmonic coefficients because of they weight the trig functions of ϕ, all have a roughly linear dependence on wind speed and are weakly dependent on other physical parameters[5]; ϕ is the angle between the direction of the wind and the look direction of the radiometer. Interestingly the first two emissivities have an even dependence on relative wind direction and the third and fourth an odd dependence. Wind vector algorithms can be derived based on these properties[6]. Moreover, the third and fourth Stokes parameters are less affected by geophysical noise[7]. More recently, the third and fourth Stokes parameters have been

[5] See S.H. Brown, C.S. Ruf and D.R. Lyzenga and S. Cox, A nonlinear optimisation algorithm for WindSat wind vector retrievals, *IEEE Transactions on Geoscience and Remote Sensing*, vol. 44, no. 3, March 2006, pp. 611-621.

[6] See S.H. Brown, C.S. Ruf and D.R. Lyzenga and S. Cox, *loc cit.*, and M.H. Bettenhausen, C. Smith, R.. Bevilacqua, N-Y Wang, P.W. Gaiser and S. Cox, A nonlinear optimisation algorithm for WindSat wind vector retrievals, *IEEE Transactions on Geoscience and Remote Sensing*, vol. 44, no. 3, March 2006, pp. 597-610.

[7] J.R. Piepmeier and A.J. Gasiewski, High resolution passive polarimetric mapping of ocean surface wind vector fields, *IEEE Transactions on Geoscience and Remote Sensing*, vol. 39, no.3, March 2001, pp. 606-622.

shown also to be good indicators of asymmetry in the surface structures of polar ice sheets[8].

9.7 Brightness Temperature of Volume Media

Volume scattering media such as vegetation canopies and sea ice have radar scattering coefficients that are moderately independent of observation angle and are generally fairly large because of multiple scattering. Consequently, unless the medium is particularly lossy, the emissivity of inhomogeneous, volumetric materials will be low and not strongly dependent on the angle of observation. If the canopy is weakly absorbing it may be difficult to measure those characteristics because of interfering emission from an underlying surface. Such a composite situation is treated in Sect. 9.8 following. If the canopy is strongly absorbing then it will also be a strong emitter with a high brightness temperature and the effect of any emission from an understory will be minimised. Fig. 9.9 shows the indicative dependence of brightness temperature on observation angle for a volume medium at the extremes of absorption.

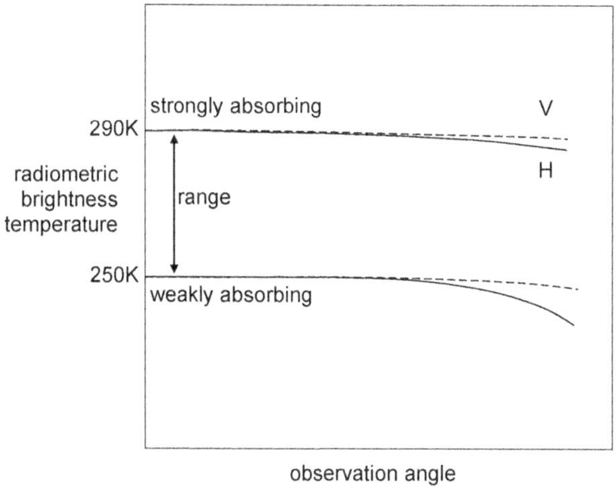

Fig. 9.9. Likely range of radiometric brightness temperatures for a volume medium

9.8 Layered Media: Vegetation over Soil

It is difficult to treat vegetation in isolation unless it is so strongly absorbing that no energy passes completely through it. More typically the observed brightness temperature will include a significant component from the underlying surface. We now consider that situation. Assume the canopy is weakly scattering but is composed of elements that are absorbing. We will also restrict ourselves to the case of a specular underlying surface as shown in Fig. 9.10, although the results to be derived apply in general.

[8] See L. Li, P. Gaiser, M.R. Albert, D. Long and E.M. Twarog, WindSat passive microwave polarimetric signatures of the Greenland ice sheet, *IEEE Transactions on Geoscience and Remote Sensing*, vol. 46, no. 9, September 2008, pp. 2622-2631.

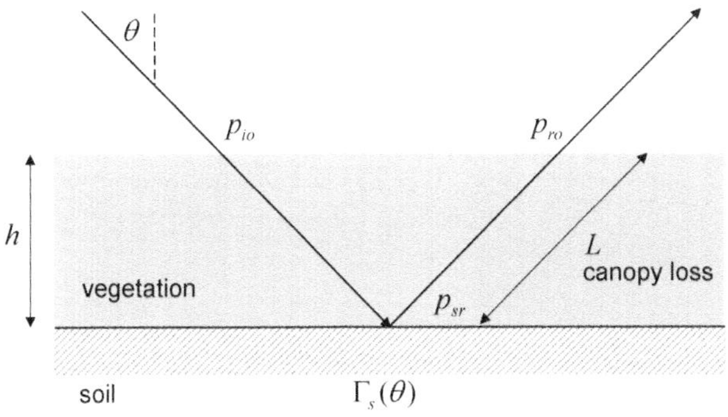

Fig. 9.10. Vegetation layer over a smooth surface

If the power reflection coefficient at the soil surface is $\Gamma_s(\theta)$, then the power density emerging from the canopy after two way transmission though the canopy and reflection from the surface is

$$p_{ro} = \frac{p_{sr}}{L} = \Gamma_s(\theta)\frac{p_{io}}{L^2} \tag{9.9}$$

where L is the canopy loss, given by

$$L = \exp(\kappa_e h \sec\theta) \tag{9.10}$$

in which κ_e is the extinction coefficient of the canopy, attributed entirely to material absorption since scattering loss is assumed negligible. From (9.9) the effective reflectivity of the canopy/ground combination is

$$\Gamma_c(\theta) = \frac{p_{ro}}{p_{io}} = \frac{\Gamma_s(\theta)}{L^2} \qquad E(r) = E_o \qquad \gamma R)$$

so that the effective emissivity of the canopy/ground is

$$\varepsilon_c(\theta) = 1 - \Gamma_c(\theta) = 1 - \frac{\Gamma_s(\theta)}{L^2} \tag{9.11a}$$

If the soil emissivity is $\varepsilon_s(\theta)$ then

$$\varepsilon_c(\theta) = 1 - \frac{1-\varepsilon_s(\theta)}{L^2} \tag{9.11b}$$

For a lossless canopy $\kappa_e = 0$ so that L=1, giving $\varepsilon_c(\theta) = \varepsilon_s(\theta)$, whereas for a high loss canopy $\kappa_e \to \infty$ so that $L \to \infty$, giving $\varepsilon_c(\theta) = 1$; in other words a very high loss canopy will exhibit total emission and a brightness temperature equal to the physical temperature.

Equation (9.11b) can be converted to radiometric brightness temperature through multiplication by the physical temperature T_o:

$$T_c(\theta) = T_o\left\{1 - \frac{1}{L^2}\right\} + \frac{T_s(\theta)}{L^2} \tag{9.11c}$$

in which $T_s(\theta)$ is the soil brightness temperature. This tells us that the sensitivity of radiometric brightness temperature to soil moisture content is reduced by the square of the loss imposed by an overlying vegetation layer. Fig. 9.11 demonstrates that effect with a 0.5m deep vegetation layer over sand. The canopy extinction coefficient has been chosen as $6dBm^{-1}$ and the angle of view is 30°.

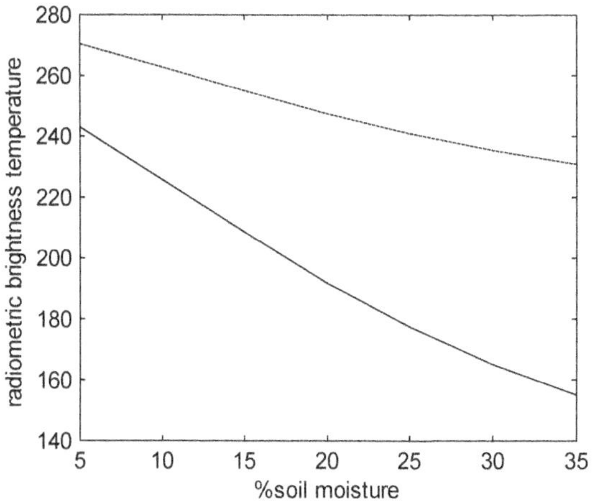

Fig. 9.11. Effect of an overlying canopy on the measurement of soil brightness temperature; the lower curve is just sand brightness temperature and the upper curve is the reduced sensitivity resulting from the vegetation layer

9.9. Passive Microwave Remote Sensing of the Atmosphere

Figure 2.7 shows that the earth's atmosphere absorbs incident electromagnetic radiation at very high microwave frequencies and is very strongly absorbing in certain wavebands. As a consequence of thermal equilibrium it will also be a strong emitter in those bands. Therefore, if we wanted to detect atmospheric constituents we would do so at those frequencies for which atmospheric emission is strongest for the constituents of interest. We would use 22GHz for water vapour measurement and 60GHz for oxygen detection. At 60GHz and 120GHz none of the earth's own emission succeeds in travelling up through the atmosphere; at 22 GHz, however, it is necessary to image over a cold surface such as the sea to minimise the earth's contribution.

ERRATUM TO: REMOTE SENSING WITH IMAGING RADAR

John A. Richards

J.A. Richards, *Remote Sensing with Imaging Radar*, Signals and Communication Technology,
DOI: 10.1007/978-3-642-02020-9, © Springer-Verlag Berlin Heidelberg 2009

DOI: 10.1007/978-3-642-02020-9_10

The publisher regrets the following information was in error when the book was produced. The correction list for this book is provided below.

Location	Correction
Pages 24, 82, eqn. (2.7b), (2.7c) and (3.39)	"η" should be "η^{-1}"
Page39, eqn. (2.33)	Interchange rows 3 and 4 in each column vector
Page 39, first equation	exp missing in E_v expression
Page 49	In the two equations following the second paragraph α should be αt
Page 51, line following eqn. (2.50)	1000kms^{-1} should be 300ms^{-1}
Page 59, eqn. (3.3)	Add 2π before α
Page 64, end of para following eqn. (3.11b)	5760Hzs^{-1} should be 560 Hz^{-1}
Page 65, third last line of second last paragraph	Change 5s to 3s
Page 69, Fig. 3.12	Denominator of r_g should include $\sin\theta$
Page 71, eqn. (3.26) and next four lines	v in numerator should be c 90m should be 16m azimuth resolution 15m should be 5.3m antenna length 4 should be 3.5 safety factor

Page 91, eqn. (3.63)	Interchange rows 3 and 4 in each column vector
p.175, last line	C band should be L band
p. 176, Fig 5.35 caption	Reference should be L. Fu and B. Holt, Seasat Views Oceans and Sea Ice with Synthetic Aperture Radar, JPL Publication 81-120, 15 February 1982.
pp 181, 182, 185, eqns. (6.1), (6.3a) and (6.12)	(j should be j(
P 191, eqn. (6.17)	$\cos(\theta-\alpha)$ should be $\cos\theta$
P 197, eqn. (6.21)	Delete conjugate from e_2 in denominator
P 244, first para after (7.7)	Fourth sentence should read "That is when the receiver is on the side of the transmitter closest to the target". The next sentence should commence "Because of the correspondingly longer path from the transmitter to the receiver with a negative baseline, such an arrangement …"
Page 279, eqn. (8.13b)	Eqn (8.13b) should be $d_m(\mathbf{Z}) = N\{tr(\mathbf{C}_m^{-1}\mathbf{Z}) + \ln\lvert\mathbf{C}_m\rvert\}$
Page 335, eqn. (B.12)	$\mathbf{\Lambda G}$ should be $\mathbf{G\Lambda}$
Page 337, Table C2	Insert new third line p pico $\times 10^{-12}$ Correct previous two lines a atto $\times 10^{-18}$ f femto $\times 10^{-15}$

The online version of the original book can be found at
http://dx.doi.org/10.1007/978-3-642-02020-9

APPENDIX A

COMPLEX NUMBERS

Despite their name complex numbers are not complicated; nor are they difficult to handle. It is not even necessary to have a feeling for what they mean theoretically. Rather, they are convenient tools with which to manipulate some of the quantities we encounter in radar imaging, particularly concerning the electromagnetic energy that is used to irradiate the landscape and which is received, after scattering, to form an image.

The basis for complex number theory rests in describing the square root of a negative number; that in itself is not so important as the properties that flow from it. We describe the square root of minus one by the symbol *j*:

$$j = \sqrt{-1}$$

In mathematics and physics instead of *j* the symbol *i* is used; *j* however is commonplace in electrical engineering to avoid confusion with the symbol for current. Note $jxj = -1$.

We can express the square root of any negative number in terms of the symbol *j*. For example

$$\sqrt{-9} = \sqrt{-1x9} = \sqrt{-1}x\sqrt{9} = j3$$

The number that multiplies the *j* is called an imaginary part or an *imaginary number* to distinguish it from the real numbers with which we are familiar in everyday life (for counting and describing real things). If we add a real and an imaginary number we then have a *complex number*:

$$z = a + jb$$

in which a is called the real part of the complex number z and b is called its imaginary part, written respectively as

$$a = \mathcal{Re}\{z\} \text{ and } b = Im\{z\}$$

We can add or subtract two complex numbers by adding or subtracting their components:

$$z_1 \pm z_2 = a_1 \pm a_2 + j(b_1 \pm b_2)$$

Complex numbers can also be multiplied and divided using the normal rules of algebra:

$$z_1 z_2 = (a_1 + jb_1)(a_2 + jb_2) = a_1 a_2 - b_1 b_2 + j(a_1 b_2 + a_2 b_1)$$

$$\frac{z_1}{z_2} = \frac{a_1 + jb_1}{a_2 + jb_2} = \frac{(a_1 + jb_1)(a_2 - jb_2)}{(a_2 + jb_2)(a_2 - jb_2)} = \frac{1}{a_2^2 + b_2^2}[a_1 a_2 + b_1 b_2 + j(a_2 b_1 - a_1 b_2)]$$

We will demonstrate shortly a more convenient way to carry out multiplication and division.

It is helpful for later developments to plot the components of a complex number on a graph with Cartesian coordinates representing the real part horizontally and the imaginary part vertically, as shown in Fig. A.1. That *complex plane* is called an *Argand diagram.* Note that we can now describe the complex number in polar coordinate form, by the length of the vector from the origin R and the angle measured up from the positive real axis ϕ. R is often called the *modulus* or *magnitude* of the complex number and ϕ is its *argument.* Using geometry and trigonometry the polar coordinates are related to the Cartesian components by

$$R = \sqrt{a^2 + b^2} \text{ and } \phi = \tan^{-1}\frac{b}{a} \tag{A.1}$$

while the real and imaginary parts can be derived from the polar form by

$$a = R\cos\phi \text{ and } b = R\sin\phi \tag{A.2a}$$

In electrical engineering the polar form is often written

$$R\angle\phi \tag{A.2b}$$

and described as "R angle ϕ".

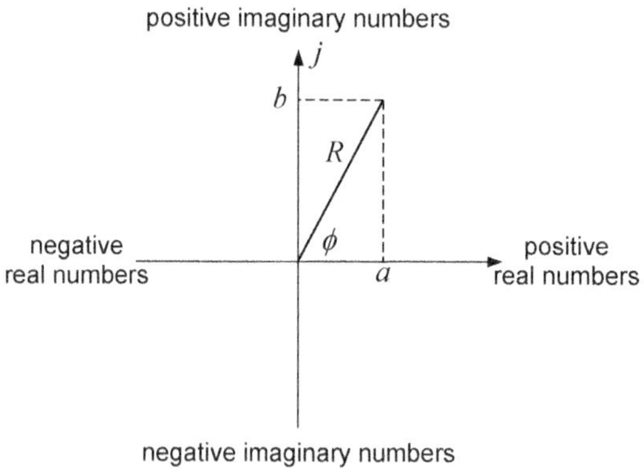

Fig. A.1 The Argand diagram for representing complex numbers

From (A.2a) we have

$$z = R(\cos\phi + j\sin\phi) \tag{A.3}$$

Interestingly, if we substitute the power series expansions for the cosine and sine functions in (A.3) an extremely important result emerges. From

$$\cos\phi = 1 - \frac{\phi^2}{2!} + \frac{\phi^4}{4!} - \ldots$$

$$\sin\phi = \phi - \frac{\phi^3}{3!} + \frac{\phi^5}{5!} - \dots$$

we have $$\cos\phi + j\sin\phi = 1 + j\phi - \frac{\phi^2}{2!} - j\frac{\phi^3}{3!} + \frac{\phi^4}{4!} + j\frac{\phi^5}{5!}\dots$$

The last expansion should be compared with

$$e^x = 1 + x + \frac{x^2}{2!} + \frac{x^3}{3!} + \frac{x^4}{4!} + \dots$$

when x is replaced by jx: $$e^{jx} = 1 + jx - \frac{x^2}{2!} - j\frac{x^3}{3!} + \frac{x^4}{4!} + \dots$$

Thus we have demonstrated that

$$\cos\phi + j\sin\phi = e^{j\phi} \tag{A.4}$$

This is referred to as *Euler's theorem* or formula, and is the basis of the most remarkable set of results involving complex numbers and sinusoidally time varying radiation. Using (A.4) we see that (A.3) gives us another representation of the complex number:

$$z = Re^{j\phi} \text{ or } z = R\exp j\phi \tag{A.5}$$

This exponential representation is convenient for multiplying and dividing complex numbers, using the properties of indices. Note for example that

$$z_1 z_2 = R_1 R_2 e^{j(\phi_1 + \phi_2)}$$

and $$\frac{z_1}{z_2} = \frac{R_1}{R_2} e^{j(\phi_1 - \phi_2)}$$

The polar form of the complex number in (A.2b) can be regarded as a short hand version of the exponential form, in which the e^j is understood[1]. Multiplication and division can thus also be expressed:

$$z_1 z_2 = R_1 R_2 \angle(\phi_1 + \phi_2)$$

$$\frac{z_1}{z_2} = \frac{R_1}{R_2} \angle(\phi_1 - \phi_2)$$

The number -1 appears on the Argand diagram of Fig. A.1 at a unit distance on the left hand real axis. In terms of the exponential descriptor of (A.5) that means R=1 and ϕ=π. Thus we have an important fundamental identity:

$$e^{j\pi} = -1 \tag{A.6a}$$

[1] In engineering this is often referred to as the *phasor* form.

Similarly we find that

$$e^{\pm j\frac{\pi}{2}} = \pm j \tag{A.6b}$$

and

$$e^{j2\pi} = 1 \tag{A.6c}$$

As seen the exponential (polar) form of the complex number is enormously powerful. It also allows the roots and powers of any number – real, complex or imaginary – to be found. For example

$$\sqrt{j} = j^{\frac{1}{2}} = \left[e^{j\frac{\pi}{2}}\right]^{\frac{1}{2}} = e^{j\frac{\pi}{4}} \equiv 1\angle 45^o = \cos 45^o + j\sin 45^o = 0.707 + j0.707$$

and

$$\frac{1}{j} = j^{-1} = \left[e^{j\frac{\pi}{2}}\right]^{-1} = e^{-j\frac{\pi}{2}} = -j$$

and

$$(1+j2)^2 = [\sqrt{5}e^{j63.4^o}]^2 = 5e^{j126.8^o} = 5\cos(126.8^o) + j5\sin(126.8^o)$$

$$= -3 + j4$$

This last result is demonstrated in the Argand diagram of Fig. A.2.

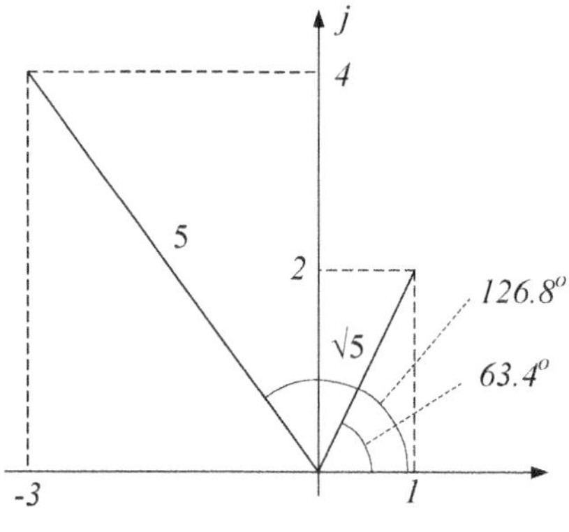

Fig. A.2. Summary of square calculation on the Argand diagram

The *complex conjugate* of a complex number is that with the sign of the imaginary component reversed. It is denoted with a superscript asterisk. Thus the conjugate of $z=a+jb$ is $z^*=a-jb$. As shown in Fig. A.3 this is the same as reversing the sign of the angle, so that if $z=Re^{j\phi}$ then its conjugate is $z^*=Re^{-j\phi}$. Conjugates play a very important role in electromagnetism and signal analysis. One interesting operation is the product of a number and its conjugate

$$zz^* = RRe^{j0} = R^2$$

or
$$zz^* = (a + jb)(a - jb) = a^2 + b^2 \equiv R^2$$

In other words, the product of a number and its conjugate is real and equal to the square of its magnitude.

Most signals in which we are interested with radar are sinusoidal or can be reduced to sinusoidal form. While sinusoids are not necessarily difficult to use, exponentials are much easier, particularly when calculus is involved. Note from (A.4) that

$$\cos\phi = \mathcal{Re}\{e^{j\phi}\}$$

so that a travelling sinusoid can be written

$$\cos(\omega t - \beta r) = \mathcal{Re}\{e^{j(\omega t - \beta r)}\}$$

Provided we remember to take the real part of the result of any operation, either explicitly or by implication, we can replace the sinusoidal form in any calculations by the exponential version; that makes analysis straightforward. There is one requirement for this: the system in which we are interested must behave as a linear system, which is the case for those considered in this book.

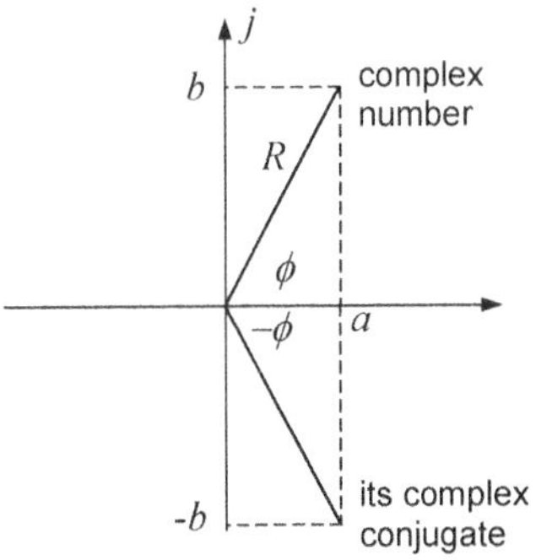

Fig. A.3. Definition of the complex conjugate

Sometimes a radio wave travels in a lossy medium, such that its amplitude decreases exponentially with distance travelled, in which case it is written

$$\begin{aligned} e^{-\alpha r}\cos(\omega t - \beta r) &= e^{-\alpha r}\mathcal{Re}\{\exp(j(\omega t - \beta r)\} = \mathcal{Re}\{\exp(-\alpha r + j(\omega t - \beta r)\}\} \\ &= \mathcal{Re}\{\exp(j\omega t - \gamma r) \end{aligned}$$

in which $\gamma = \alpha + j\beta$ is called the *propagation constant*, itself a complex number.

APPENDIX B

MATRICES

B.1 Matrices and Vectors, Matrix Multiplication

A *matrix* is an array of numbers arranged by rows (along the horizontal) and columns (down the vertical). Most frequently matrices arise in relation to sets of equations, or in linear transformations. For example, the simultaneous equations

$$2x - 7y = 10$$
$$5x + 2y = -15$$

can be expressed

$$\begin{bmatrix} 2 & -7 \\ 5 & 2 \end{bmatrix}\begin{bmatrix} x \\ y \end{bmatrix} = \begin{bmatrix} 10 \\ -15 \end{bmatrix}$$

or, symbolically

$$\mathbf{Mg} = \mathbf{c} \tag{B.1}$$

in which **M** is a 2x2 matrix and **g** and **c** are referred to as (2 element) *column vectors* – because they are columnar in nature. As a second example, the following transformation will rotate axes by an angle θ in the anti-clockwise direction

$$\begin{bmatrix} y_2 \\ x_2 \end{bmatrix} = \begin{bmatrix} \cos\theta & \sin\theta \\ -\sin\theta & \cos\theta \end{bmatrix}\begin{bmatrix} y_1 \\ x_1 \end{bmatrix}$$

which is the matrix form of the pair of equations

$$y_2 = \cos\theta\, y_1 + \sin\theta\, x_1$$
$$x_2 = -\sin\theta\, y_1 + \cos\theta\, x_1$$

This can be written as

$$\mathbf{g}_2 = \mathbf{M}\mathbf{g}_1$$

These examples also show how a column vector multiplies a matrix. In principle, a column vector is an *n*x1 matrix, where *n* is the number of rows (vertical elements). The result of the multiplication is obtained by multiplying the column entries of the vector, one by one, with the row entries of the matrix, one row at a time, and then adding the products. The result of each of those operations is a new vector element. This is illustrated in Fig. B1, along with a symbolic representation of the multiplication of two matrices, which follows the same pattern.

Note that Fig.B.1 introduces the *row vector*. The column vectors above have their elements arranged down a column, whereas a row vector has its elements arranged across a row. The difference is important because row vectors enter into multiplication in a different way, as illustrated. The product of a row vector and a column vector will also be different depending on the order in which they appear. If the row vector is on the left hand side the result is a simple scalar; if it is on the right hand side the result is a matrix.

$$\begin{bmatrix} 4 & -3 \end{bmatrix} \begin{bmatrix} 9 \\ 7 \end{bmatrix} = 15 \qquad \text{(B.2a)}$$

$$\begin{bmatrix} 9 \\ 7 \end{bmatrix} \begin{bmatrix} 4 & -3 \end{bmatrix} = \begin{bmatrix} 36 & -27 \\ 28 & -21 \end{bmatrix} \qquad \text{(B.2b)}$$

The order in which the matrices are multiplied is also important. **AB** for example will give a different result to **BA**, except in special circumstances. We say that **A** "pre-multiplies" **B** in **AB** whereas **B** "post-multiplies" **A**. Although the above examples were computed with 2 dimensional vectors and matrices the patterns the same for any orders so long as the order of the vector matches the relevant dimension of the matrix. For example, a 3x12 matrix (3 rows and 12 columns) can only be *post-multiplied* by a 12 element column vector and can be *pre-multiplied* by a 3 element row vector.

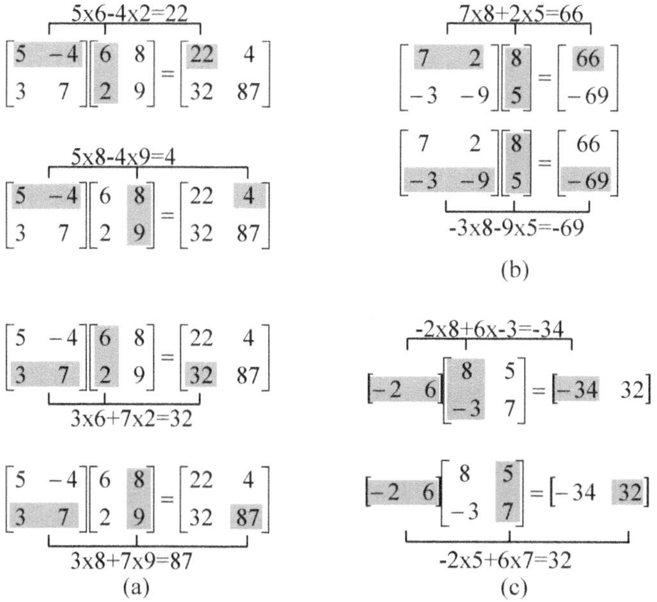

Fig. B.1. Illustrating the steps involved in matrix multiplication (a) two matrices, (b) a column vector post-multiplying a matrix and (c) a row vector pre-multiplying a matrix

B2. Indexing and Describing the Elements of a Matrix

In working with matrices it is important to be able to refer unambiguously to its elements. A double subscript notation is used in which the first refers to the row to which the element belongs and the second to its column. Thus

$$\mathbf{M} = \begin{bmatrix} m_{11} & m_{12} & m_{13} & .. \\ m_{21} & m_{22} & m_{23} & .. \\ m_{31} & m_{32} & m_{33} & .. \\ \vdots & \vdots & \vdots & \vdots \end{bmatrix}$$

The elements, referred to generically as m_{ij}, can be real or complex. The dots in this expression simply mean the matrix can be of any size, as determined by the problem being considered. If the matrix has as many rows as columns then it is called a *square matrix*.

Elements that lie on the same row and column, m_{ii}, are called diagonal elements and together define the *diagonal*, or principal diagonal of the matrix. All the other elements are referred to as *off-diagonal* elements.

B.3 The Kronecker Product

There is another matrix product sometimes used in radar imaging. Called the *Kronecker product*, it is best illustrated using algebraic entries. If A and B are the matrices

$$\mathbf{A} = \begin{bmatrix} a_{11} & a_{12} \\ a_{21} & a_{22} \end{bmatrix} \text{ and } \mathbf{B} = \begin{bmatrix} b_{11} & b_{12} & b_{13} \\ b_{21} & b_{22} & b_{23} \\ b_{31} & b_{32} & b_{33} \end{bmatrix}$$ then the Kronecker product is

$$A \otimes B = \begin{bmatrix} a_{11}\mathbf{B} & a_{12}\mathbf{B} \\ a_{21}\mathbf{B} & a_{22}\mathbf{B} \end{bmatrix} = \begin{bmatrix} a_{11}b_{11} & a_{11}b_{12} & a_{11}b_{13} & a_{12}b_{11} & a_{12}b_{12} & a_{12}b_{13} \\ a_{11}b_{21} & a_{11}b_{22} & a_{11}b_{23} & a_{12}b_{21} & a_{12}b_{22} & a_{12}b_{23} \\ a_{11}b_{31} & a_{11}b_{32} & a_{11}b_{33} & a_{12}b_{31} & a_{12}b_{32} & a_{12}b_{33} \\ a_{21}b_{11} & a_{21}b_{12} & a_{21}b_{13} & a_{22}b_{11} & a_{22}b_{12} & a_{22}b_{13} \\ a_{21}b_{21} & a_{21}b_{22} & a_{21}b_{23} & a_{22}b_{21} & a_{22}b_{22} & a_{22}b_{23} \\ a_{21}b_{31} & a_{21}b_{32} & a_{21}b_{33} & a_{22}b_{31} & a_{22}b_{32} & a_{22}b_{33} \end{bmatrix} \quad \text{(B.3)}$$

B.4 The Trace of a Matrix

The trace of a matrix is the sum of its diagonal terms, which for an *nxn* square matrix is expressed:

$$\text{trace}\, M \equiv \text{tr} M = \sum_{i=1}^{n} m_{ii} \quad \text{(B.4)}$$

B.5 The Identity Matrix

The identity matrix is a square matrix (i.e. with the same number of rows and columns) which is zero everywhere except down its diagonal, on which each element is unity. Multiplication of a vector by the identity matrix, which has the symbol **I**, leaves the vector unchanged. Thus

$$\begin{bmatrix} 1 & 0 & 0 \\ 0 & 1 & 0 \\ 0 & 0 & 1 \end{bmatrix} \begin{bmatrix} g_1 \\ g_2 \\ g_3 \end{bmatrix} = \begin{bmatrix} g_1 \\ g_2 \\ g_3 \end{bmatrix} \text{ and } \begin{bmatrix} g_1 & g_2 & g_3 \end{bmatrix} \begin{bmatrix} 1 & 0 & 0 \\ 0 & 1 & 0 \\ 0 & 0 & 1 \end{bmatrix} = \begin{bmatrix} g_1 & g_2 & g_3 \end{bmatrix}$$

or symbolically $\mathbf{Ig} = \mathbf{g}$ and $\mathbf{gI} = \mathbf{g}$ as appropriate. Similarly, multiplication of any matrix by the identity matrix leaves the matrix unchanged. Thus

$$\mathbf{MI} = \mathbf{M}$$

The identity matrix is the matrix equivalent of the real number "1".

B.6 The Transpose of a Matrix or a Vector

If the elements of a matrix are rotated about the diagonal the *transpose* of the matrix results. The transpose is represent by a superscript T (or sometimes t), so that if

$$\mathbf{M} = \begin{bmatrix} m_{11} & m_{12} & m_{13} \\ m_{21} & m_{22} & m_{23} \\ m_{31} & m_{32} & m_{33} \end{bmatrix}$$

then
$$\mathbf{M}^{\mathrm{T}} = \begin{bmatrix} m_{11} & m_{21} & m_{31} \\ m_{12} & m_{22} & m_{32} \\ m_{13} & m_{23} & m_{33} \end{bmatrix}$$

Vectors can also be transposed by rotating around their first element thus transforming a row vector into a column vector and vice versa. If

$$\mathbf{g} = \begin{bmatrix} g_1 \\ g_2 \\ g_3 \end{bmatrix}$$

then
$$\mathbf{g}^{\mathrm{T}} = \begin{bmatrix} g_1 & g_2 & g_3 \end{bmatrix}$$

Note that
$$\mathbf{g}^{\mathrm{T}}\mathbf{g} = g_1^2 + g_2^2 + g_3^2 = |\mathbf{g}|^2$$

In other words that operation gives the square of the magnitude of the vector. How can vectors have magnitude? To illustrate: the vector **g** could be the set of spectral measurements for a pixel in three dimensional multi-spectral space. Its magnitude is its overall brightness, or the length of the vector drawn from the origin to the point in the three dimensional space described by the vector elements as coordinates. Alternatively, **g** could be the target vector of (3.46), in which case the magnitude is the span of the vector.

If in (B.2) we put

$$\mathbf{g} = \begin{bmatrix} 4 \\ -3 \end{bmatrix} \text{ and } \mathbf{h} = \begin{bmatrix} 9 \\ 7 \end{bmatrix}$$

we see that

$$\mathbf{g}^{\mathrm{T}}\mathbf{h} = 15 = \text{scalar}$$

$$\mathbf{gh}^{\mathrm{T}} = \begin{bmatrix} 36 & 28 \\ -27 & -21 \end{bmatrix} = \text{matrix}$$

Sometimes the first of those expressions is called an *inner product*; less frequently, the second is called an *outer product*.

Sect. 2.16 shows that the vector transpose can be used to evaluate the dot product of two vectors: $\mathbf{A.B} = \mathbf{A}^{\mathrm{T}}\mathbf{B} = \mathbf{B}^{\mathrm{T}}\mathbf{A}$.

B.7 The Determinant

The determinant of the square matrix **M** is expressed

$$|\mathbf{M}| = \det \mathbf{M} = \begin{vmatrix} m_{11} & m_{12} & m_{13} & .. \\ m_{21} & m_{22} & m_{23} & .. \\ m_{31} & m_{32} & m_{33} & .. \\ : & : & : & \end{vmatrix}$$

It is a scalar quantity that, in principle, can be computed in the following manner. Unfortunately, in all but the simplest cases, this approach does not lead to an efficient method for determinant evaluation, and numerical methods must be used in practice.

First, we define the *cofactor* of a matrix element. The cofactor of the element m_{ij} is the determinant of the matrix formed by removing the ith row and j^{th} column from **M** and multiplying the result by $(-)^{i+j}$. Thus the cofactor of m_{21} is

$$\mathcal{M}_{21} = -\begin{vmatrix} m_{12} & m_{13} & m_{14} & .. \\ m_{32} & m_{33} & m_{34} & .. \\ m_{42} & m_{43} & m_{4} & .. \\ : & : & : & \end{vmatrix}$$

The classical method for evaluating the determinant is to express it in terms of the cofactors of its first row (or of any row or column). For a square matrix of size *nxn* this expansion is

$$|\mathbf{M}| = \sum_{j=1}^{n} m_{1j}\mathcal{M}_{1j}$$

The cofactors in this expression can be expanded in terms of their cofactors, and so on until the solution is found. The case of a 2x2 matrix is simple:

$$\det \mathbf{M} = \begin{vmatrix} m_{11} & m_{12} \\ m_{21} & m_{22} \end{vmatrix} = m_{11}m_{22} - m_{12}m_{21}$$

For matrices of larger dimensions this method for evaluating the determinant is grossly computationally inefficient and numerical methods are adopted.

If the determinant of a matrix is zero the matrix is called *singular*.

B.8 The Matrix Inverse

A matrix multiplied by its inverse gives the identity matrix. The inverse is represented by adding the superscript –1 to the matrix symbol. Thus

$$\mathbf{M}^{-1}\mathbf{M} = \mathbf{I} \tag{B.5}$$

Applying the inverse concept to (B.1) shows that the solution to the pair of simultaneous equations can be derived by pre-multiplying both sides by $\mathbf{M}^{-1}$

$$\mathbf{g} = \mathbf{M}^{-1}\mathbf{c}$$

provided the inverse matrix can be found. As with finding determinants, that is not a trivial task and generally approximations and numerical methods are used. However, also like determinants, there are theoretical expressions for the matrix inverse. It can be defined in terms of the *adjoint* (more recently called the *adjugate*) of the matrix, which is the transposed matrix of cofactors:

$$adj\mathbf{M} = \begin{bmatrix} \mathcal{M}_{11} & \mathcal{M}_{12} & \mathcal{M}_{13} & .. \\ \mathcal{M}_{21} & \mathcal{M}_{22} & \mathcal{M}_{23} & .. \\ \mathcal{M}_{31} & \mathcal{M}_{32} & \mathcal{M}_{33} & .. \\ \vdots & \vdots & \vdots & \end{bmatrix}^T = \begin{bmatrix} \mathcal{M}_{11} & \mathcal{M}_{21} & \mathcal{M}_{31} & .. \\ \mathcal{M}_{12} & \mathcal{M}_{22} & \mathcal{M}_{32} & .. \\ \mathcal{M}_{13} & \mathcal{M}_{23} & \mathcal{M}_{33} & .. \\ \vdots & \vdots & \vdots & \end{bmatrix}$$

with which the inverse of M is

$$\mathbf{M}^{-1} = \frac{adj\mathbf{M}}{|\mathbf{M}|} \tag{B.6}$$

From this we see that the matrix must be non-singular to have an inverse – i.e. its determinant must not be zero.

B.9 Special Matrices

A *symmetric* (square) matrix is equal to its own transpose: $\mathbf{M}=\mathbf{M}^T$, and thus $m_{ij}=m_{ji}$.

An *orthogonal* matrix is equal to the inverse of its transpose: $\mathbf{M}=(\mathbf{M}^T)^{-1}$. In other words, its transpose is its inverse.

A *conjugate* matrix, written here as $\overline{\mathbf{M}}$, has elements that are the complex conjugates of the original matrix, so that $\overline{m_{ij}} = m_{ij}^*$.

A *Hermitian* matrix, written here as $\mathbf{M}^*$, is equal to its own transposed conjugate matrix: i.e. $\mathbf{M} = \overline{\mathbf{M}}^{\mathrm{T}}$, sometimes also referred to as the conjugate transpose.

A *unitary* matrix is one in which the inverse is equal to the conjugate transpose.

B.10 The Eigenvalues and Eigenvectors of a Matrix

Equation (B.1) can be interpreted as the transformation of the column vector $\mathbf{g}$ by the matrix $\mathbf{M}$ to form a new column vector $\mathbf{c}$. We now ask ourselves whether there is any particular vector, say $\mathbf{g}_1$, for which multiplication by a scalar will produce the same transformation as multiplication by a matrix. In other words can we find a $\mathbf{g}_1$ such that

$$\lambda \mathbf{g}_1 = \mathbf{M}\mathbf{g}_1 \tag{B.7}$$

where λ is a constant, which is sometimes complex. We can introduce the identity matrix into this equation without changing its meaning:

$$\lambda \mathbf{I}\mathbf{g}_1 = \mathbf{M}\mathbf{g}_1$$

so that we can then re-arrange the equation to read

$$(\mathbf{M} - \lambda\mathbf{I})\mathbf{g}_1 = 0 \tag{B.8}$$

Equation (B.8) is actually a short hand version of the set of homogeneous simultaneous equations in the unknown components[1] of $\mathbf{g}_1$

$$\begin{aligned}
&(m_{11} - \lambda)g_{11} + m_{12}g_{21} + m_{13}g_{31} \dots = 0 \\
&m_{21}g_{11} + (m_{22} - \lambda)g_{21} + m_{23}g_{31} \dots = 0 \\
&\dots
\end{aligned}$$

For a set of homogeneous simultaneous equations to have a non-trivial solution the determinant of the coefficients of the unknowns must be zero, viz.

$$|\mathbf{M} - \lambda\mathbf{I}| = 0 \tag{B.9}$$

This is called the *characteristic equation* of the matrix M. It consists of a set of equations in the unknown λ. By solving (B.9) the values of λ can be found. They can be substituted into (B.8) to find the corresponding vectors $\mathbf{g}_1$. The λ are referred to as the *eigenvalues* (or sometimes proper values or latent roots) of the matrix $\mathbf{M}$ and the corresponding vectors $\mathbf{g}_1$ are called the *eigenvectors* (proper vectors or latent vectors) of $\mathbf{M}$.

---

[1] Note that we have indexed the components of the vector using a double subscript notation in which the first subscript refers to the component – i.e. the column for that component – and the second refers to the vector itself, in this case $\mathbf{g}_1$. Later we will have a $\mathbf{g}_2$, etc.

As a simple example consider the matrix

$$\mathbf{M} = \begin{bmatrix} 6 & 3 \\ 4 & 9 \end{bmatrix}$$

Substituting into (B.9) gives

$$\begin{vmatrix} 6-\lambda & 3 \\ 4 & 9-\lambda \end{vmatrix} = 0$$

i.e.

$$(6-\lambda)(9-\lambda)-12=0$$

or

$$\lambda^2-15\lambda+42=0 \qquad \text{(B.10)}$$

which has the roots 11.275 and 3.725. In (B.10) it is interesting to note that the coefficient of λ is the trace of **M** and the constant term is its determinant. Substituting the first eigenvalue into (B.8) gives

$$-5.275g_{11}+3g_{21}=0$$

so that

$$g_{11}=0.569\, g_{21} \qquad \text{(B.11a)}$$

Likewise substituting the second eigenvalue into (B.8) shows

$$4g_{12}+5.275g_{22}=0$$

so that

$$g_{12}=-1.319\, g_{22} \qquad \text{(B.11b)}$$

Note that the eigenvectors are not completely specified; only the ratio of the terms is known. This is consistent with the fact that a non-trivial solution to a set of homogeneous equations will not be unique.

The eigenvalues for this example are both (all) real. A matrix for which all the eigenvalues are real is called a *positive definite* matrix. If they could also be zero the matrix is called *positive semi-definite*. Most generally the eigenvalues are complex, in which case they will occur in conjugate pairs.

Even though we commenced this analysis based on matrices that transform vectors, the concept of the eigenvalues and eigenvectors of a matrix is more general and finds widespread use in science and engineering.

B.11 Diagonalisation of a Matrix

If we have computed all the eigenvalues of a matrix **M** and constructed the diagonal matrix

$$\mathbf{\Lambda} = \begin{bmatrix} \lambda_1 & 0 & 0 & .. \\ 0 & \lambda_2 & 0 & .. \\ 0 & 0 & \lambda_3 & .. \\ : & : & : & \end{bmatrix}$$

then (B.7) can be generalised to

$$\Lambda \mathbf{G} = \mathbf{M}\mathbf{G} \tag{B.12}$$

in which **G** is a matrix formed from the set of eigenvectors of **M**:

$$\mathbf{G} = [\mathbf{g}_1 \; \mathbf{g}_2 \; \mathbf{g}_3 \, ..]$$

Provided **G** is non-singular, which it will be if the eigenvalues of **M** are all distinct, then B.12 can be written

$$\Lambda = \mathbf{G}^{-1}\mathbf{M}\mathbf{G} \tag{B.13}$$

which is called the *diagonal form* of **M**. Alternatively

$$\mathbf{M} = \mathbf{G}\Lambda\mathbf{G}^{-1} \tag{B.14}$$

This last expression is very useful for computing certain functions of matrices. For example consider **M** raised to the power p:

$$\mathbf{M}^p = \mathbf{G}\Lambda\mathbf{G}^{-1}.\mathbf{G}\Lambda\mathbf{G}^{-1}.\mathbf{G}\Lambda\mathbf{G}^{-1}...\mathbf{G}\Lambda\mathbf{G}^{-1} = \mathbf{G}\Lambda^p\mathbf{G}^{-1}$$

The advantage of this approach is that the diagonal matrix **Λ** raised to the power p simply requires each of its elements to be raised to that power.

B.12 The Rank of a Matrix

The rank of a matrix is a number equal to the number of linearly independent rows or linearly independent columns it possesses; for a square matrix they are the same. The rows or columns are linearly independent if any one cannot be expressed as a linear combination of one more other rows or columns. If there are linearly dependent rows or columns then the determinant of the matrix will be zero. A test of the rank of a matrix therefore is to evaluate its determinant. If the determinant is non-zero then the rank is equal to the dimension of the matrix – i.e. a 3x3 matrix will have rank 3 if it has a non-zero determinant. If its determinant is zero then its rank will be smaller than 3 and equal to the size of the largest non-zero determinant within it.

The rank of a matrix is also equal to the number of its non-zero eigenvalues. As an illustration, the matrix

$$\begin{bmatrix} 3 & 2 & 5 \\ 4 & 1 & 3 \\ 8 & 2 & 6 \end{bmatrix}$$

has the eigenvalue set (sometimes referred to as the eigenvalue spectrum) of 45.6, –0.6, 0 so that its rank is 2. Note that the last row of the matrix is twice the second row; therefore they are not linearly independent.

APPENDIX C

SI SYMBOLS AND METRIC PREFIXES

Symbols for the fundamental quantities used in this book, in their standard International System (SI) forms, are shown in Table C.1.

Table C.1 SI Symbols

symbol	meaning	symbol	meaning	symbol	meaning
m	metre	Np	neper	A	ampere
s	second	rad	radian	V	volt
Hz	hertz	deg	degree	Ω	ohm
H	henry	W	watt	S	siemens
F	farad	J	joule	T	tesla
K	kelvin				

Prefixes are prescribed in the SI system to represent variations of base units by factors of 1000, as illustrated in Table C.2. Examples are given of how the fundamental unit of length is modified by the use of these prefixes. The same pattern is applied to any other SI unit, as shown for the fundamental unit of frequency.

Table C.2 Metric Prefixes

prefix	name	meaning	examples	
a	atto	$x10^{-15}$	am (attometre)	
f	femto	$x10^{-12}$	fm (femtometre)	
n	nano	$x10^{-9}$	nm (nanometre)	
μ	micro	$x10^{-6}$	μm (micrometre)	
m	milli	$x10^{-3}$	mm (millimetre)	
		x1	m (metre)	Hz (hertz)
k	kilo	$x10^{3}$	km (kilometre)	kHz (kilohertz)
M	mega	$x10^{6}$	Mm (megametre)	MHz (megahertz)
G	giga	$x10^{9}$	Gm (gigametre)	GHz (gigahertz)
T	tera	$x10^{12}$	Tm (terametre)	THz (terahertz)
P	peta	$x10^{15}$	Pm (petametre)	PHz (petahertz)

APPENDIX D

IMAGE FORMATION WITH SYNTHETIC APERTURE RADAR

D.1 Summary of the Process

This appendix summarises the key steps in the process of compressing recorded synthetic aperture radar data to form an image. It will be seen to involve two major stages, one to compress the signal in range and the other to compress it in azimuth. It is during the second stage that multi-look filtering is carried out to reduce speckle. Steps taken to reduce false target indications generated by spurious signals in the compression process are also described. We focus on forming an image of a point target since the processes involved apply equally to any other target or cover type.

Recall from the material in Chapt. 3 that a series of ranging chirps is transmitted and scattered from the target as the vehicle passes the target to the side. Scattered returns are received by the radar when the target comes into view of the antenna, and persist until the target is just lost to view as summarised in Fig. 3.9. The target also has to be within the swath of ground illuminated by the vertical beamwidth pattern of the antenna.

Consider the response to a single ranging chirp from a point target located about midway across the swath; the target could be single house or single large tree, or possibly a calibration device. Assume also there are no other targets in view, or that the surface on which the point target sits is specular so there is no backscatter from it. The received signal will be zero from the time equivalent of the near range position up to the point when the chirp scattered from the point target appears, after which it will be zero again out to the position corresponding the far edge of the swath[1].

The signal is now sampled and placed into one row of an array of memory set aside for later processing of the signals to form the image. As the platform travels forward it transmits successive ranging chirps and receives echoes from the point target, which are also sampled and loaded into the computer memory. There will be as many sampled range lines loaded into memory as there are transmitted pulses while the target is in view of the radar. Fig. D.1 shows symbolically what the computer memory might look like. It will hold zeros everywhere except where there is a sample of the returning chirp. In the figure we have extended the azimuth dimension of the memory well beyond the number of returns from the single point target. It represents the landscape corresponding to a significant portion of spacecraft travel. The number of memory cells in the azimuth direction that contain samples of the echoes from the point target is indicative of the actual azimuth beam pattern of the transmitting antenna.

There is an assumption in Fig. D.1: that is that the distance to the point target is the same for each range line. We know that not to be the case, as the target will be further from the radar when just acquired, and when last seen, than it will be at broadside. We will return to that actual situation later.

[1] The signal received is actually modulated onto a carrier at the operating frequency (wavelength) of the radar. It is shifted down to so-called base band for processing, which is the form assumed in the descriptions given in this Appendix.

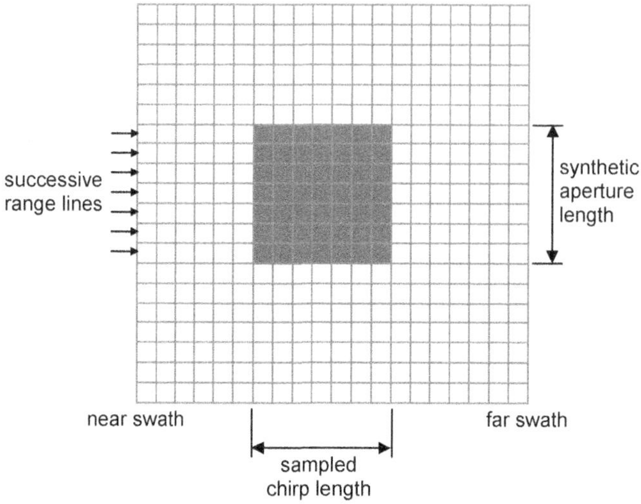

Fig. D.1. Energy (shaded cells) associated with the succession of backscattered chirps from a point target over the extent of the synthetic aperture

As seen in Fig. D.1 the backscattered energy from the point target is smeared out in range and azimuth. Our goal is to compress it in both directions so that it occupies just a single memory cell, and thus looks like the image of a point target. That two stage process is depicted in Fig. D.2. We now examine details of how the two compressions are carried out.

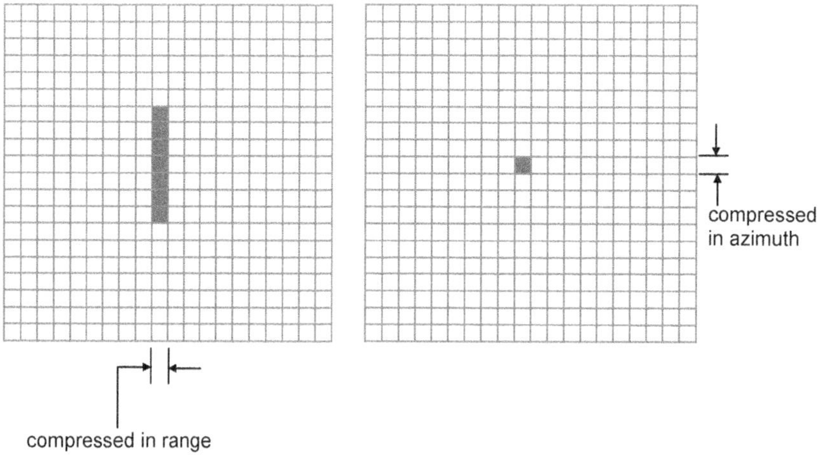

Fig. D.2. Showing the compression of the chirp energy in Fig. D.1 into a point as a two stage process involving the sequence of range compression followed by azimuth compression

D.2 Range Compression

Fig. 3.7 shows that range compression is performed by correlating the received chirps against a replica of what was transmitted. We now examine how that correlation operation is carried out. Although strictly not important for what is to follow, the cross-correlation of two signals $s(t)$ and $c(t)$ is given by

$$r(t) = \int s(t)c(t+\tau)d\tau \equiv s(t) * c(t)$$

where we have used the symbol * to represent the correlation process[2]. Very importantly, the *correlation theorem* says that correlation in the time domain is equivalent to multiplication of the Fourier transforms of the two signals, provided one transform is complex conjugated[3]. Thus if

$$r(t) = s(t) * c(t)$$

then

$$R(\omega) = S(\omega)C^*(\omega)$$

in which

$$R(\omega) = \mathcal{F}\{r(t)\}$$
$$S(\omega) = \mathcal{F}\{s(t)\}$$
$$C(\omega) = \mathcal{F}\{c(t)\}$$

where the symbol $\mathcal{F}\{..\}$ means the Fourier transform operation and the smaller * as a superscipt indicates complex conjugate. The result of the Fourier transform is complex in general.

The correlation theorem makes the correlation of the scattered signal $s(t)$ and the replica of what was transmitted $c(t)$ simple to perform. In fact, when the celebrated fast Fourier transform algorithm is used, the number of mathematical operations needed to perform the correlation in the so-called frequency domain is considerably smaller than if the correlation were done directly. Note that there needs to be an inverse Fourier transform operation as well to go from the result of the product $R(\omega)$ back to its time domain version $r(t)$. The Fourier transform of the chirp replica does not need to be performed each time it is used. It is sufficient to compute that operation once, take its conjugate, and store the result for use in range compression. Fig. D. 3 summarises those operations as they would be performed in the range compression step of a SAR correlator.

Although the correlation integral above has been shown as operating on continuous time functions, in practice we are dealing with sampled signals so that the Fourier operations indicated in Fig. D.3 are applied to the samples of the received chirps. Likewise the stored chirp conjugate Fourier transform is in the form of samples. In the figure we have shown a continuous version of what the sample sequences would look like, rather than the samples themselves. Finally, the compressed chirp in Fig. D.3 is shown as it emerges from the result of the inverse Fourier transform; it is then envelope detected so that only its overall shape remains as the result of the compression. The envelope can be seen in Fig. D.5.

---

[2] Note that correlation is commutative: $s(t) * c(t) = c(t) * s(t)$.

[3] When we take the Fourier transform of a *time domain* signal we say that we have created its *frequency domain* version.

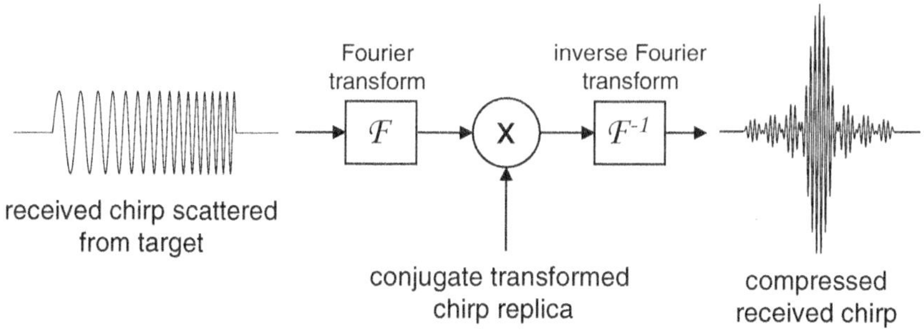

Fig. D.3. Range compression in the frequency domain

D.3 Compression in Azimuth

In principle, compressing the signal in the azimuth direction involves the same processes as in range compression. Recall from Sect. 3.6 that the motion of the platform induces a Doppler chirp on the transmitted waveform. If we can estimate what the Doppler induced chirp looks like then it can be used as a local replica in the correlation operation, just as for range compression.

There are two parameters of a chirp, apart from the centre frequency. They are its duration and rate. From Sect. 3.6 we can see that these are given by

$$T_a = L_a / v \tag{D.1a}$$

$$\beta = \frac{2v^2}{\lambda R_o} \tag{D.1b}$$

which depend on the platform velocity v, the length of the synthetic aperture L_a and the broadside distance to the target R_o; L_a in turn is also dependent on the slant range at broadside. Therefore, while we can, in principle, use correlation against a replica of the Doppler induced chirp to compress in azimuth, care needs to be taken with the two primary parameters of the chirp. If any of the factors that define those parameters change with time then the quality of the compression will be compromised. Moreover, R_o is different for each target position across the swath, so different estimates of β will be needed according to where the target, or pixel, is located. Fortunately, the important parameters – the Doppler centroid and rate β – can usually be estimated from the signal itself; the centroid is important if there is squint in the radar, as discussed in Chapt. 3. Usually, depending on the swath width, several values of those parameters are estimated, corresponding to sets of azimuth blocks across the swath.

D.4 Look Summing for Speckle Reduction

As discussed in Sect. 4.3.1 the coherent nature of the radiation used in radar means that the recorded data is heavily speckled. In some applications we work with the speckled imagery (using single look complex products for example) but most often speckle is reduced to improve the signal to noise ratio and make images easier to interpret visually. Speckle can be reduced by filtering the final image product, as was demonstrated in Fig. 4.17. However, it can also be reduced in the frequency domain during azimuth compression. Although that limits the flexibility of the final product (since the user is unable to revert to single look imagery) it is the approach most commonly used in SAR processing since it takes separate (and assumed independent) samples of the *same* resolution cell on the ground when forming the average. In contrast when speckle is reduced by averaging *adjacent* cells in the image itself one must assume that there is no significant variation in the average backscattering coefficient from cell to adjacent cell.

The process is very straightforward, but to appreciate the steps involved it is important to look at the Fourier transform of a chirp – i.e. its frequency spectrum. That is the output of the Fourier transform step in Fig. D.3 just before it is multiplied by the conjugate replica of itself to produce the compressed output.

The Fourier transform is a complex quantity, having both an amplitude and a phase term; we need only look at its amplitude here. Fig. D.4 shows what the amplitude of the chirp spectrum looks like. Essentially, it is a constant between a lower bound approximately equivalent to the lowest frequency in the chirp and an upper bound equivalent to the highest frequency in the chirp. The difference between those bounds is the *bandwidth* we saw in Chapt. 3.

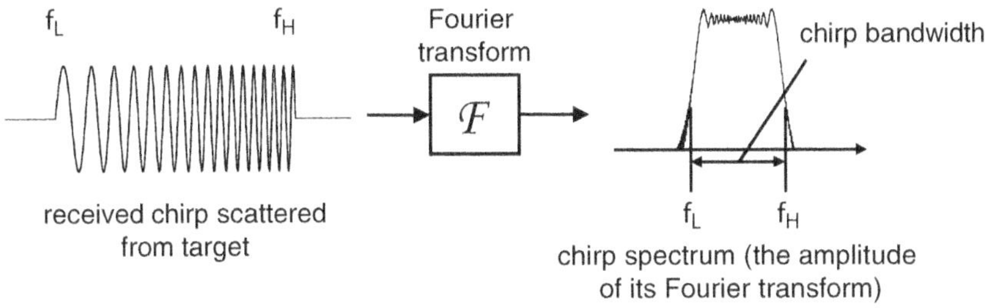

Fig. D.4. The Fourier transform of the chirp – called its frequency spectrum

Fig. D.5 shows the result of correlating the chirp against the replica of itself, except here the chirp is represented by its spectrum of Fig. D.4 rather than its time domain plot. The three examples given are for differing chirp bandwidths. As seen, the wider the chirp bandwidth the narrower the compressed pulse – that leads to better azimuth resolution, as discussed in Sect. 3.3. The compressed pulse is shown in two forms: as it comes from the correlator with the description in (3.4), and after envelope detection, which means just its overall envelope amplitude is displayed. The envelope detected form is used in practice. Although not readily discerned from the diagram, the effective width of the compressed pulse, and its envelope, is equal to the reciprocal of the chirp spectrum bandwidth. If the bandwidth is reduced by four then the compressed pulse width broadens by four.

Now consider some calculations involving an actual space borne SAR mission. For Seasat in 1978 the chirp bandwidth was 19MHz. From (3.5a) that shows the slant range resolution to have been 7.89m. At an incidence angle of 20° at the earth's surface that gave a ground range resolution of 23m. The azimuth length of its antenna was 10.7m, which would suggest from (3.8) a maximum azimuth resolution of 5.35m; in practice the azimuth resolution was a little poorer than that, but this figure suits our purposes.

The question is, why is there such a disparity in the ground range and azimuth resolutions? We have rectangular pixels with a 4:1 aspect ratio rather than square pixels, which would be much more useful. The answer lies in the means by which speckle was reduced in Seasat imagery. It was a "four look" system, in the way we will now demonstrate, that halves the speckle power if the azimuth resolution is degraded to about 22m (i.e. four times reduction in azimuth resolution).

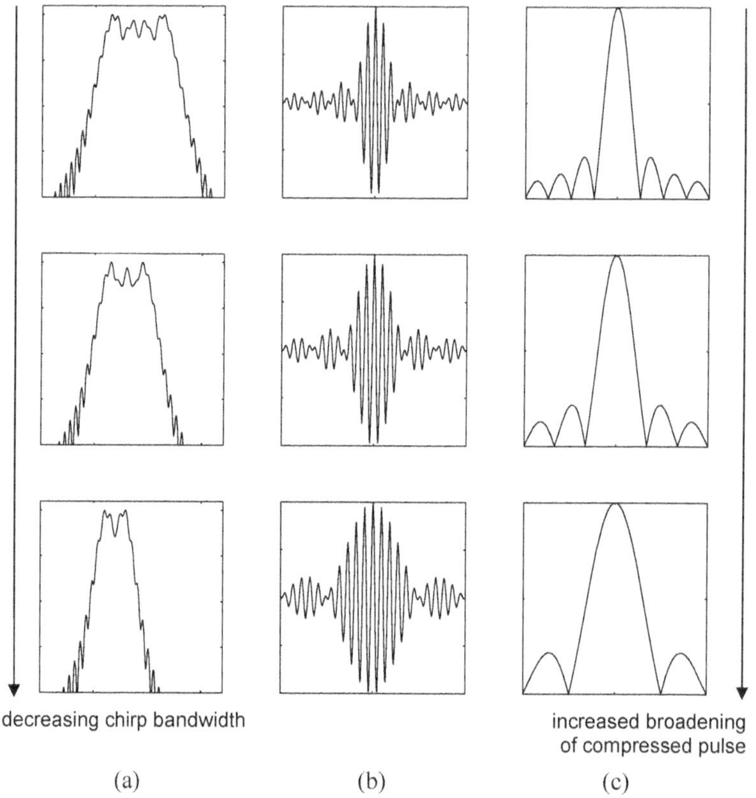

Fig. D.5. Illustrating the trade off between chirp bandwidth and the width of the compressed pulse after correlating with a replica of the chirp: (a) chirp spectrum (b) compressed pulse (c) envelope of the compressed pulse

Learning from Fig. D.5, if we use only one quarter of the available azimuth chirp spectrum then the compressed width after correlation will be four times that if the full chirp bandwidth were used. We could therefore cut up the chirp spectrum into four equal pieces and correlate each to generate a compressed pulse. In the case of Seasat that would give a pulse width equivalent to about 22m rather than the 5.35m that results if the full azimuth spectrum is used. Four of those individual compressed pulses are produced. Since each came from a different part of the original chirp spectrum they can be regarded as independent samples that can be averaged to reduce speckle in the radar image.

Fig. D.6 shows how four "look filters" are used to select the four independent segments of the chirp spectrum, each of which is separately correlated; the set is then summed (averaged) to reduce speckle. The subsequent image is then said to be four look averaged. In principle any number of looks could be used – specified at the time of the SAR system design – but most space craft systems operate with between about 3 and 6 looks.

As an alternative to look summing by segmenting the chirp spectrum as above, single look data could be produced with rectangular pixels. In the case of Seasat they would be 23m in range and 5.35m in azimuth. Four azimuth pixels could be averaged in the azimuth direction, as discussed earlier, to give 23x22m pixels.

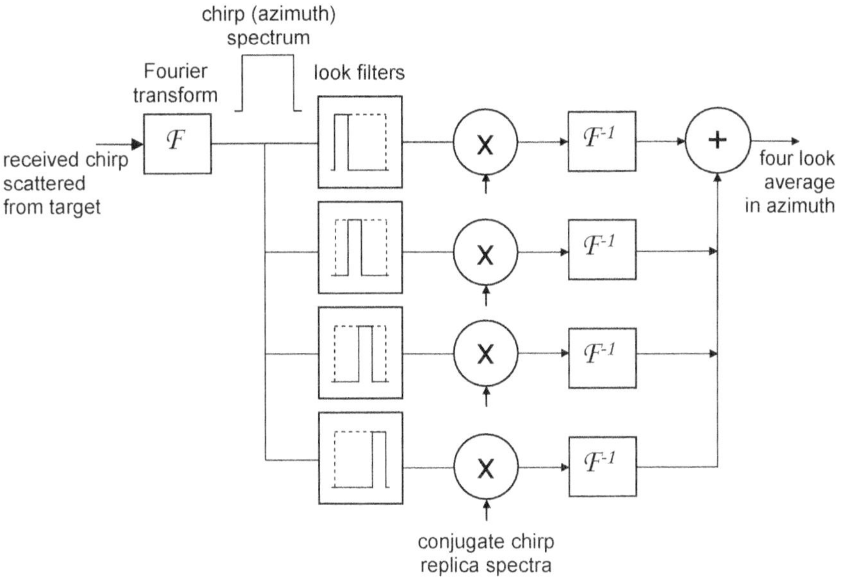

Fig. D.6. Segmenting the azimuth spectrum into four non-overlapping portions for look summing to reduce speckle; not shown in this diagram is a square law, envelope detector step at the outputs of the four paths just prior to the look summing (averaging)

D.5 Range Curvature

We now examine the assumption that the range to the point target doesn't change significantly over the duration of the period that the target is irradiated. From Fig. D.7 the slant range to the target is well approximated by

$$R(t) \approx R_o + \frac{x^2}{2R_o}$$

The largest value of x is half the length of the synthetic aperture L_a. Therefore the maximum slant range is

$$R(t) \approx R_o + \frac{L_a^2}{8R_o}$$

so that the largest difference between the actual range to the point target, and the value R_o assumed in the above treatment, is

$$\Delta R = \frac{L_a^2}{8R_o} \tag{D.2}$$

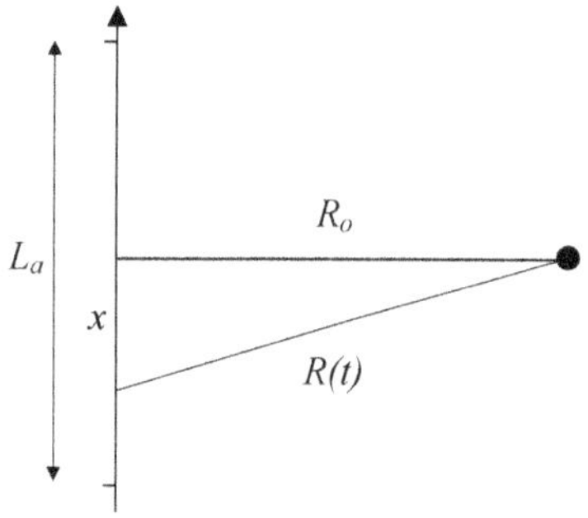

Fig. D.7. Geometry for determining the maximum change in slant range

Is this significant? That depends on whether, from the first line of the target energy shown in Fig. D.1 through to broadside, the change in range is greater than the equivalent of one of the memory cells in the range direction. Those memory cells in range are equivalent to the width of the compressed range chirp – i.e. they are equivalent in metres to the slant range resolution of the radar. Therefore the test of whether the maximum change in range to the point target over the synthetic aperture length is significant is to compare it to the slant range resolution. We thus define the ratio

$$M = \frac{\Delta R}{r_r} = \frac{L_a^2}{8R_o r_r} \tag{D.3}$$

where r_r is the slant range resolution. We can make this expression more meaningful by noting that the synthetic aperture length is the azimuth beamwidth of the antenna (λ/l_a) multiplied by the distance to the target at broadside R_o; thus $L_a = \lambda R_o / l_a$. Moreover, since $r_a = l_a/2$ we can substitute for l_a in this last expression so that (D.3) can be written

$$M = \frac{\Delta R}{r_r} = \frac{\lambda^2 R_o}{32 r_r r_a^2} \tag{D.4}$$

This is a useful measure since it allows us to compute the ratio of the change in range to slant range resolution in terms of the system resolutions and the broadside slant range. For Seasat, with r_a=6.25m (the actual value), r_r=7.89m, λ=0.235m and R_o=850km we find M=4.8. Thus, for Seasat data the range lines in Fig. D.1 will be 4.8 resolution cells further away from the radar when the target is first encountered, and when last seen, compared with broadside. The received chirp energy is therefore "curved" in the memory as depicted in Fig. D.8a. When compressed in range, which it still can be done because that is an operation carried out on range lines separately, the result is as shown in Fig. D.8b. Clearly, if we were now to attempt azimuth compression errors would occur, unless the curvature of the signal were corrected. Not surprisingly the effect is referred to as *range curvature*.

In contrast to Seasat, ERS-1 does not suffer significant range curvature. For ERS-1 r_r=6.6m, r_a=6.25m, R_o=853km, λ=0.054m, which gives M=0.3. It can be seen that the major difference between the two missions is the wavelength. In general, we can conclude that range migration is more likely to be a problem with long wavelength radars.

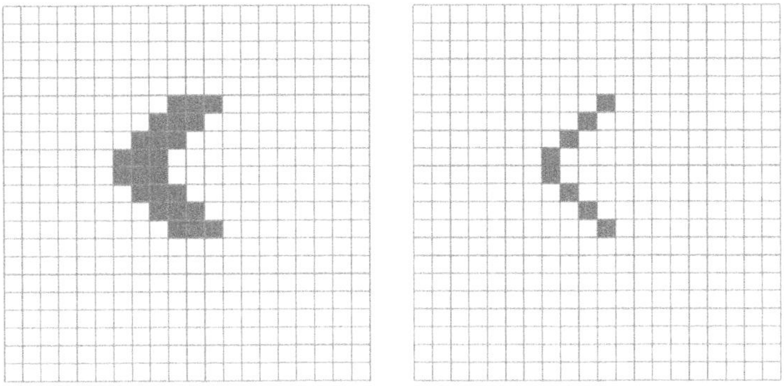

(a) (b)

Fig. D.8. Effect of range curvature (a) before and (b) after range compression

In addition to migration of the chirp energy in the range memory locations resulting from the variation in range to the point target, there can also be range variations caused by yaw of the spacecraft and other ephemeris variations, and by the rotation of the earth. Sometimes these additional mechanisms are said to lead to *range walk*.

In general the distribution of energy from the point target after range compression will not look parabolic as suggested in Fig. D.8a, but will have a more complex migration through the memory, much as depicted in Fig. D.9. The general effect is called *range migration* and the correction procedure is called *range migration correction*.

D.6 Side Lobe Suppression

Fig. D.5 shows that the compressed pulse has side lobes adjacent to the main pulse; it is possible therefore that the side lobes will be interpreted as false point targets adjacent to the one being imaged. For distributed scattering media, which is the more common situation in remote sensing, the presence of significant side lobes means that energy from a target within the designed resolution cell is distributed over that cell and its neighbours in the reconstructed image data. That is the classical point spread function effect experienced with any imaging device. The side lobes of the sinc function shown in Fig. D.5 are quite large; the first are just 13dB (twenty times) lower than the main lobe. That energy contributes to the neighbouring pixel response.

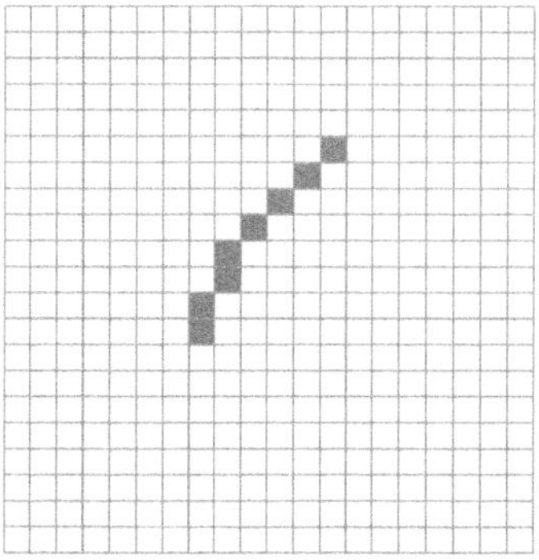

Fig. D.9. Range migration in general

Although not easily demonstrated here, it is well known in signal processing that the side lobes are related to the sharp turn on and turn off of the ranging chirp in Figs. D.3 and D.4. If the turn on and off can be smoothed the side lobes can be reduced. A common way to do that is to multiply the chirp by a weighting or window function, of which many candidates are available. A simple and often used weighting function is the Hann window or "raised cosine" defined by

$$w(t) = 0.5 - 0.5\cos(2\pi t/\tau) \quad -\tau/2 \leq t \leq \tau/2 \tag{D.5}$$

where τ is the duration of the chirp that is to be smoothed. Multiplying the chirp of (3.3) by (D.5) leads to the smoother version shown in Fig. D.10.

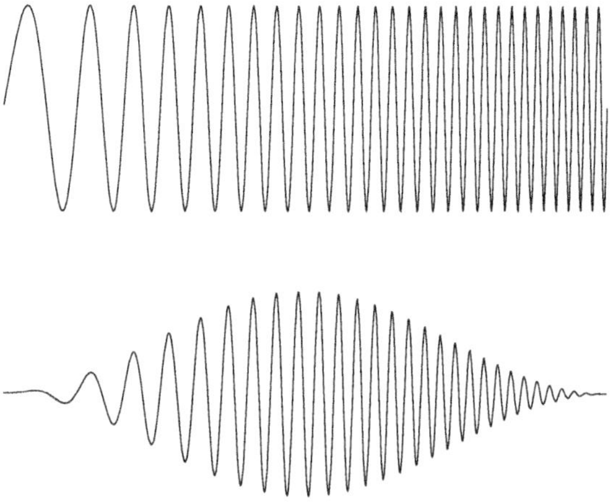

Fig. D.10. Original and weighted chirp signals

If smoothed chirps are used for the ranging pulses instead of abrupt ones the result of the compression step (for both the range and azimuth operations) will be as shown in Fig.

D.11. The side lobes are significantly reduced (to about 32dB below or just over 1000 times smaller than the main lobe), minimising the leakage of energy into adjacent resolution cells. This is at the expense of broadening the main lobe and thus degrading slightly the spatial resolution of the system.

To gauge the effect on the appearance of the final image product, Fig. D.12 shows a sequence of 9 pixels along a range line centred on a point target, both with and without chirp smoothing. They were created by averaging the compressed chirps of Fig. D.11 over windows equal in size approximately to the range resolution cell. As observed, without smoothing significant leakage of the point target energy into adjacent resolution cells occurs, whereas with smoothing it is only in the immediate vicinity of the point target that leakage occurs. Because of the ability to constrain the point spread function in this manner, SAR correlators employ weighted chirps, although not all will use the simple version of (D.5). Others are available that lead to greater side lobe suppression.

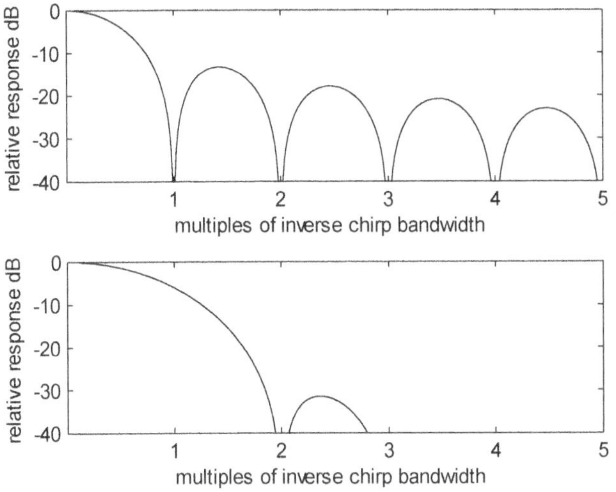

Fig. D.11. Compressed chirp detail (one side) without (top) and with (bottom) Hann weighting

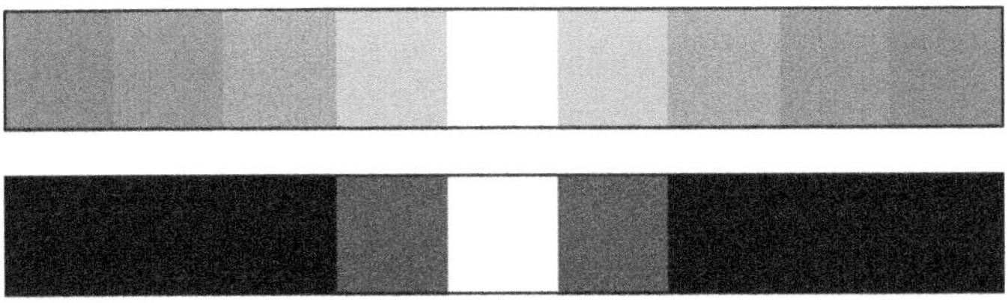

Fig. D.12. Adjacent pixel brightnesses in a point target response along a range line using unweighted chirps (top) and Hann weighted chirps (bottom)

APPENDIX E

BACKSCATTER AND FORWARD SCATTER ALIGNMENT COORDINATE SYSTEMS

When dealing with single polarisation radar the coordinate system used to describe the propagation of the forward and scattered waves is not especially significant. However, when multi-polarisation radar is of interest it is important to be clear about coordinate conventions for describing wave propagation, otherwise confusion can arise in respect of target descriptions. Unfortunately, there are two coordinate systems in common usage. They are related but have one significant difference, as will be seen in the following. In any given problem it is important to know and identify which system is being used.

The starting point is the convention adopted for describing the orientations of the two polarisation components of the electric field with respect to the direction of propagation. We will use the names horizontal and vertical polarisation for the two orthogonal components, but any pair at right angles to each other can be used. More precisely we should call our components the perpendicular and parallel polarised components, as discussed in Sect. 2.10, but we will stay with the horizontal and vertical descriptors because of common usage.

Fig. E.1 shows the respective orientations of the horizontal and vertical components of a field and the direction of propagation. As implied, the convention is that rotation from the orientation of the horizontal component to that of the vertical component should be in a clockwise sense when looking in the direction of propagation. This is known as the right hand rule since it emulates the forward motion of a screw driver when rotating it clockwise.

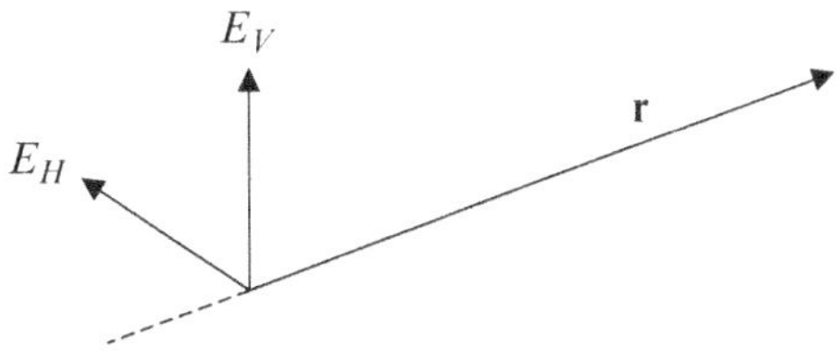

Fig. E.1 The coordinate convention for a polarised ray propagating in the **r** direction

Fig. E.2 shows such a ray incident on a target. Although it is not highly significant for this discussion, both components undergo a phase reversal, leading to a reversal of their polarisation directions. That is because of the negative reflection coefficients that describe scattering, as seen in (5.3). We ignore those changes here. The wave propagates beyond the target, obeying the same conventions of the right hand rule in Fig. E.1. This is a natural coordinate system description for the propagation of the wave under these circumstances and is referred to as *forward scatter alignment* (FSA) because of the forward scattering nature of the wave propagation, as drawn. It is also said to be a description of the propagation in *wave coordinates*, since the coordinate system is fully consistent with the wave propagation conventions of Fig. E.1.

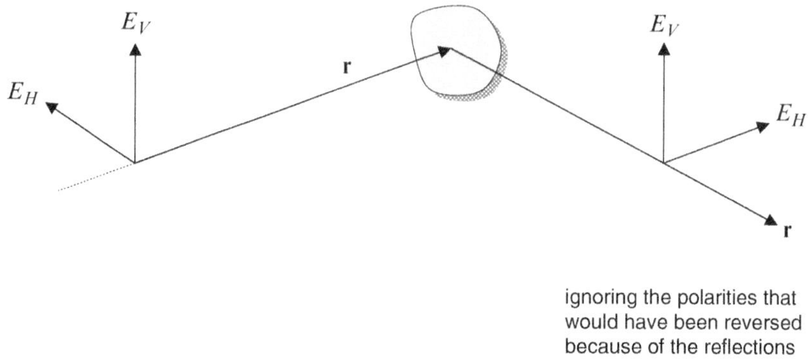

Fig. E.2 Forward scattering from a target, maintaining the same coordinate convention; this is called forward scatter alignment

In Fig. E.3 we show the reflected wave as backscattering – i.e. the scattered path is back to the radar antenna, which is the situation we have with monostatic radar. Since we just fold the outgoing path of Fig. E.2 over to align with the backscattered direction, as shown in Fig. E.3a the formal FSA convention is still acceptable as a descriptor of the back propagating wave. It is, however, more usual in (monostatic) radar theory to describe all propagation – both outwards and backscattered – in terms of the single vector **r** that points along the propagating pathway of the transmitted or outgoing ray. In the FSA case there are two of those vectors – one pointing in the transmitted direction and one in the backscattered direction.

The consequence of having just one directional vector for propagation (in the forward direction) is that the returning wave is travelling in the –**r** direction. To make that possible in terms of the right hand rule of Fig. E.1, the sense of one of the components of the backscattered wave has to be reversed. Conventionally we reverse the horizontal component as shown in Fig. E.3. This system is referred to as *backscatter alignment* (BSA) and, while it appears a little awkward, it is the most natural system when dealing with backscatter problems as Fig. 3.18 demonstrates. It also aligns with the fact that a single antenna is used for both transmission and reception and so we sometimes say the wave is described in terms of *antenna coordinates*.

Since the difference between the two systems lies in the opposite orientations of the horizontal received component of the field, the scattering matrix expressed in FSA can be derived from that for BSA by reversing the polarities of elements concerned with reception in the horizontal component – i.e. the entries on the first row. This is effected by the transformation

$$\mathbf{S}_{FSA} = \begin{bmatrix} -1 & 0 \\ 0 & 1 \end{bmatrix} \mathbf{S}_{BSA} \tag{E.1}$$

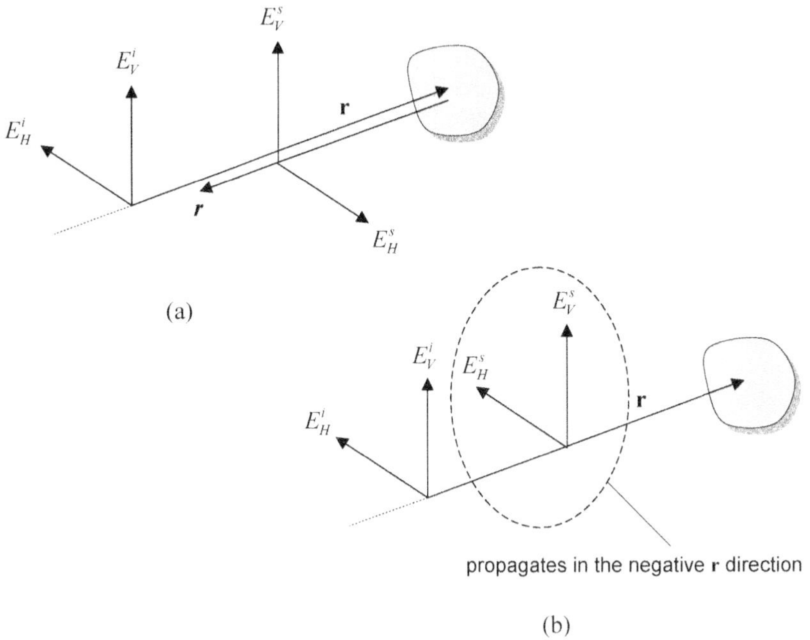

Fig. E.3 (a) Maintaining forward scattering alignment for backscattering and (b) using a backscatter alignment, so that there is a single propagation vector

INDEX

T